Modeling and Simulation-based

Life Cycle Engineering

Spon's Structural Engineering: Mechanics and Design series

Innovative structural engineering enhances the functionality, serviceability and life-cycle performance of our civil infrastructure systems. As a result, it contributes significantly to the improvement of efficiency, productivity and quality of life.

Whilst many books on structural engineering exist, they are widely variable in approach, quality and availability. With the *Structural Engineering: Mechanics and Design* series, Spon Press is building up an authoritative and comprehensive library of reference books presenting the state-of-the-art in structural engineering, for industry and academia.

Topics under consideration for the series include:
- Structural Mechanics
- Structural Dynamics
- Structural Stability
- Structural Reliability
- Structural Durability
- Structural Assessment
- Structural Renewal
- Numerical Methods
- Model-based Structural Simulation
- Composite Structures
- Intelligent Structures and Materials

Books published so far in the series:
1. Approximate Solution Methods in Engineering Mechanics
   A.P. Boresi and K.P. Chong 1-85166-572-2
2. Composite Structures for Civil and Architectural Engineering
   D.-H. Kim 0-419-19170-4
3. Earthquake Engineering
   Y.-X. Hu, S.-C. Liu and W. Dong 0-419-20590-X
4. Structural Stability in Engineering Practice
   Edited by L. Kollar 0-419-23790-9
5. Innovative Shear Design
   H. Stamenkovic 0-415-25836-7

Potential authors are invited to contact the Series Editors, directly:

Series Editors

Professor Ken P. Chong
National Science Foundation,
4201 Wilson Blvd, Room 545,
Arlington, VA 22230,
USA
e-mail: kchong@nsf.gov

Professor John E. Harding
Department of Civil Engineering,
University of Surrey,
Guildford, GU2 5XH,
UK
e-mail: j.harding@surrey.ac.uk

Modeling and Simulation-based Life Cycle Engineering

Edited by

Ken P. Chong
Sunil Saigal
Stefan Thynell
Harold S. Morgan

CRC Press
Taylor & Francis Group
Boca Raton London New York

CRC Press is an imprint of the
Taylor & Francis Group, an **informa** business

A SPON PRESS BOOK

CRC Press
Taylor & Francis Group
6000 Broken Sound Parkway NW, Suite 300
Boca Raton, FL 33487-2742

First issued in paperback 2019

© 2002 Ken P. Chong, Sunil Saigal, Stefan Thynell & Harold S. Morgan
CRC Press is an imprint of Taylor & Francis Group, an Informa business

No claim to original U.S. Government works

ISBN-13: 978-0-415-26644-4 (hbk)
ISBN-13: 978-0-367-39636-7 (pbk)

British Library Cataloguing in Publication Data
A catalogue record for this book is available from the British Library

Library of Congress Cataloging-in-Publication Data

Modeling and simulation-based life cycle engineering / Ken P. Chong...[et al.].
    p. cm.
  Includes bibliographic references and index.
  ISBN 0-415-26644-0 (alk. paper)
    1. Manufacturing processes--Mathematical models. 2. Product life cycle. I. Chong K.
  P. (Ken Pin), 1942-

TS183 .M59 2002
658.5'038--dc21

                                                                        2001049841

**Visit the Taylor & Francis Web site at
http://www.taylorandfrancis.com**

**and the CRC Press Web site at
http://www.crcpress.com**

Contents

Preface

This book is a collection of Life Cycle Engineering [LCE] projects from the three-year collaborative research initiative between the National Science Foundation [NSF] and Sandia National Laboratories [Sandia]. Sandia has the responsibility for engineering systems that have profound impact on national security and defense, and helps to assure operability of other national systems, particularly under conditions of stress. This responsibility spans the "life cycle" of a variety of engineered systems, where "life cycle" for the system includes defining its requirements, establishing the concepts to meet the requirements, proposing designs, verifying that the design satisfies the requirements, manufacturing the system, operating and maintaining the system, and finally dismantling and disposing of the system.

With the advent of teraflop, massively parallel computers, Sandia is moving toward an engineering process in which decisions will increasingly be based on computational simulations with decreasing experimental validation. These simulations are of a magnitude unprecedented in computational size, scope of technical issues, spatial and temporal resolution, complexity in terms of coupled multiphysics phenomena, and comprehensiveness in terms of parameter-space that is being explored.

The NSF mission is to advance the fundamental science and engineering base of the United States, including a commitment to the further development of engineering processes using computer modeling and simulation. The two organizations have entered into a collaborative program to fund research projects that are focused on advancing the fundamental knowledge base needed to support advanced computer simulations.

The LCE grantees' meeting held at Albuquerque, NM in June 2001 provided an opportunity for researchers to present progress in research work and also provide a forum for the exchange of ideas. Sandia and NSF have been funding proposals that address modeling and simulation advances in several focus areas including Solid Mechanics, Thermal Transport and Engineering Design (which includes the sub-areas of Design Theory and Modeling and Simulation Uncertainty). Detailed information of the scope of each of the focus areas is given in the SCOPE AND INTRODUCTARY REMARKS. Researchers supported by NSF and Sandia, program managers from NSF and Sandia as well as contacts at Sandia attended the meeting. This life-cycle engineering research effort is both world class and on track to meet national goals.

The Editors

Ken P. Chong; National Science Foundation*
Sunil Saigal; Carnegie Mellon University
Stefan Thynell, National Science Foundation
Harold S. Morgan; Sandia National Labs

*also Guest Researcher, National Institute of Standards &Technology

Acknowledgments

The NSF/Sandia LCE initiative was started by **Tim Tong** of NSF [currently Dean of Engineering, George Washington University] and **Russell Skocypec** of Sandia. Other Program Directors and Managers that have been involved in this initiative are:

From NSF:

Stefan Thynell
Ken P. Chong
George Hazelrigg
Ash Emery
Sunil Saigal
Cliff Astill

From Sandia:

Harold S. Morgan
Wahid Hermina
Tony E. P. Chen
Charles Hartwig
Thomas Bickel
Paul Hommert

The assistance provided by the Publisher, including Richard Whitby, Alice Hudson and Jessica Haggerty as well as the grantees' conferences support provided by Tammy Eldred of Sandia, and presentations by grantees are gratefully acknowledged.

The views of this book are those of the grantees and editors, not necessary those of the NSF or Sandia.

Scope and Introductory Remarks

Ken P. Chong[1], Stefan Thynell[1], Harold S. Morgan[2] and George Hazelrigg[1]

INTRODUCTION

With the advent of teraflop, massively parallel computers, research is moving toward an engineering process in which decisions will increasingly be supported by computational simulations with decreasing experimental validation. Emerging, high fidelity simulations are of a magnitude unprecedented in computational size, scope of technical issues, spatial and temporal resolution, complexity in terms of coupled multiphysics phenomena, and comprehensiveness in terms of parameter-space that is being explored. However, the use of advanced simulations on teraflop computers, is impeded by an inadequate knowledge base. To expand this knowledge base, advances are required in the fundamental sciences and engineering that form the foundation of all computational analyses. Advances are needed in broad classes of technical development: the fidelity of the simulation models, experimental discovery necessary for the determination of the models and their validations, uncertainty quantification of the resulting computations, and computational techniques for the solution of the simulation models on high performance computing platforms. Sandia and NSF jointly have funded proposals that address these modeling and simulation advances in several focus areas: Solid Mechanics, Thermal Transport and Engineering Design (which includes the sub-areas of Design Theory and Modeling and Simulation Uncertainty). Detailed information of the scope of each of the focus areas is given below.

---

[1] National Science Foundation
[2] Sandia National Labs.

SOLID MECHANICS

This focus area seeks to improve and expand fundamental computational and material mechanics knowledge in the areas of nonlinear, large deformation, deterioration of materials, quasistatics and transient dynamics. The shift from a test-based to a simulation-based design environment requires accurate, robust and efficient computer codes which model large ranges of loadings, deformation amplitudes and rates, length- (including nano-, micro-, meso- and macro-scales) and time-scale mechanics, and damping of mechanical interfaces and joints. It seeks to develop a basic engineering understanding of numerical solution methods including finite elements, boundary elements, and gridless Lagrangian methods for challenging simulation problems such as in impact and penetration, thermomechanical aspects of material processing and manufacturing, crack initiation, propagation and arrest, design optimization and uncertainty analysis, including accurate constitutive description of materials. The solution algorithms must be robust, reliable, efficient, and scalable on parallel computing platforms. Carefully designed experimental investigations to validate and otherwise support the above technology areas are also needed.

THERMAL TRANSPORT

Thermal transport plays a central role in many engineering applications such as thermal control of engineering systems, manufacturing and materials processing, power conversion and storage, biological, and micro/quantum scale thermal/fluid processes. Because the analysis of the related thermal processes often requires excessive computing times, the thermal analysis and design of systems involving these processes are difficult. With the advent of teraflop computers, it is now possible to exploit high performance computing methodologies to address these issues, assuming that the models are accurate and correctly implemented. When analyzing existing systems, the models can be modified by comparison with data and the errors minimized, leading to an improved understanding of the process. However, when designing systems, if the conditions under which the model was validated differ from those associated with the design, errors can result. This focus area is particularly interested in proposals which emphasize the development of analytical and computational methods that represent critical thermal transport phenomena and processes with appropriate resolution, dimensionality, coupling with other physical processes, and diversity of length and time scales. Topics suitable for consideration include, but are not limited to, convective heat transfer coupled with moving boundaries and possibly with participating media radiative heat transfer, interfacial heat transfer, phase change systems, interaction of heat transfer and material processing (such as crystallization, levitation, machining). The development and application of advanced experimental methods to better characterize critical thermal transport phenomena are also appropriate.

ENGINEERING DESIGN

There are new and emerging challenges that engineering designers face. Increasing and global competition demand that designs push limits of materials and processes, leaving less room for conservatism, and customers always want more for less. A fast-moving marketplace rapidly diminishes the value of new technologies, so that shortening the time from concept to market is increasingly important. On top of these demands, society also demands that new systems offer higher levels of safety and reliability, and lower environmental impact. These challenges have pushed conventional design approaches to their limits.

On the other hand computational capabilities are emerging that truly were inconceivable only a few years ago. Coupled with emerging models, such as finite element techniques, that represent engineered systems, these capabilities offer significant advantages. Computational design support tools enable the examination and comparison of wide ranges of design alternatives rapidly and inexpensively. The hope is that these tools will enable increased competitiveness in all the aspects noted above.

Still, despite the enormous power of computational models, they are far from perfect. All models are only abstractions of the realities that they are intended to represent. As such, the model-predicted performance of a system and the actual system performance will deviate at some level. When we use models to facilitate the understanding of nature, such deviations can be controlled and minimized. But in the case of design-support models, such control is not possible. The significant difference is that scientific models of nature are developed to fit extant data, whereas engineering design models are intended to predict future performance of systems. Studies of the inaccuracies in our ability to predict the behavior of engineered systems produce alarming results. Errors are considerable, and they cannot be controlled or minimized beyond modest limits. Thus, it is important to model the errors inherent in engineering design models and to develop a framework that will accommodate and use probabilistic results.

Recently, a framework has emerged that provides the capability to make use of probabilistic information in the context of engineering design. It is a decision theoretic framework. Under this framework, the role of decision making in engineering design is explicitly recognized. As such, three subactivities of the design process are recognized: (1) the generation of a set of design alternatives from which the preferred design will be chosen, (2) the estimation of expectations, that is, the performance expected from each design choice, and (3) the determination of human values relevant to the design and the use of these values to effect the selection of a preferred design. Much of an axiomatic base for decision-based design has been laid in the fields of mathematics and economics. For example, the von Neumann-Morgenstern axioms define a mathematics of value valid under conditions of risk and uncertainty that appear to apply to engineering design. And probability theory appears to provide a framework for analysis of uncertainty and risk.

Research in Engineering Design has been supported in two topical areas:

Design Theory

Proposals in this focus area of Design Theory were judged in terms of their ability/potential to provide sweeping theories that would cover and regularize wide ranges of engineering design. Under this activity, research has been supported to create and/or implement an axiomatic basis for engineering design. One acceptable approach would be to build upon the concept of decision-based design and the axioms that define von Neumann-Morgenstern utility. However, other rigorous approaches have also been entertained. There is a need for a theory of value applicable to design decision-making under conditions of uncertainty and risk. There is a need for a theory for the estimation of system performance given imperfect ability to perform system modeling, imperfect estimation of model data, and imperfect knowledge of the future environment within which engineered systems must operate. And there is a need for a theory of option creation or creativity. The first such theory may already exist primarily within the fields of economics and decision theory. The second is likely find a basis in the mathematics of probability theory and forecasting. The third is more speculative, and progress in this area is likely to be difficult.

Modeling and Simulation Uncertainty

Proposals in this focus area of Modeling and Simulation Uncertainty were judged in terms of their ability/potential to provide general approaches to, and new theories for, uncertainty estimation in modeling and simulation-based design. The goal of the uncertainty estimation methodology is to understand the impact of uncertainties on modeling and numerical simulation activities and thereby increase the confidence in decision-based design methods.

Present design methodology for engineered systems is based on incremental changes and improvements of previously successful designs. In addition, present design practice relies heavily on extensive testing of components, subsystems and prototype systems. With rapidly increasing computational capability, modeling and simulation based design is taking on increased responsibility for the success of new engineering systems. This is a fundamental paradigm shift; one whose risks and uncertainties must be assessed during the design process. This research activity addresses fundamental issues relating to the inclusion of quantitative estimation of uncertainty in mathematical modeling and computational simulation and the ascription of uncertainty to model and data elements. It complements the above mentioned research in design theory. All sources of uncertainty and error may be considered. Furthermore, techniques are sought that incorporate uncertainty quantification in the development of constitutive models for stochastic or uncertain subsystems. A variety of methods are sought to estimate the global impact of uncertainty/error sources on confidence in a design.

SCOPE

This book is a collection of Life Cycle Engineering [LCE] projects from the three-year collaborative research initiative between the National Science Foundation and Sandia National Laboratories. It addresses the LCE modeling and simulation advances in the focus areas of Solid Mechanics, Thermal Transport and Engineering Design (which includes the sub-areas of Design Theory and Modeling and Simulation Uncertainty) as described above.

Research Needs

Sunil Saigal[1], Ken P. Chong[2] and Harold S. Morgan[3]

The research needs session was chaired by the three authors listed above. The observations below are based on the contributions made by several authors in this book during the workshop as well as in written communications afterwards. The following is a summary of the discussions.

1. JOINT EFFORTS TOWARDS GRAND CHALLENGES

It was proposed that joint university/government/industry efforts are needed towards developing and validating computational models of complex mechanical phenomena, such as multi-scale problems, fracture, laminar-turbulent transition, etc.

Many of the engineering challenges that remain to be addressed are complex problems. Industry has found work-arounds to our imperfect understanding, but these work-arounds add cost and risk and limit design flexibility. Government laboratories are usually better funded to achieve a better understanding of the physics of these problems, but have still been limited to the use of empirical or semi-empirical work-arounds, which involve similar costs. Universities are well-placed to work on the fundamental mechanics of these challenges, but are rarely in a position to understand what portions of the complex problems are the most important and would allow practical design improvements. Thus, universities all too often work on scientifically interesting problems with less-than-optimal payoffs, and government and industry continue to be saddled with limited empirical work-arounds to complex problems.

The tremendous advances in computing power that have occurred over the last several decades have enabled the development of complex computational algorithms that could permit much improved solutions to these challenging problems. However, even with this greatly increased power, practical codes will still require many simplifying assumptions, especially for design. The predictions can be improved by creating codes that simulate the actual physical mechanisms. It needs to be determined which mechanisms need be simulated and to what level of accuracy? Improving these simulations will require close cooperation among computational, theoretical, and experimental researchers and engineers.

Efficient progress towards these improved simulations will require cooperation between industry, government, and universities. Industry can supply an understanding of the critical design issues and the critical limitations of existing prediction methods. Government agencies can also supply an understanding of design issues, usually with a better feel for how improved

[1] Carnegie Mellon University
[2] National Science Foundation
[3] Sandia National Labs.

simulations might change the design process. Universities can focus their research on problems whose solution is more likely to impact real systems. A continuing technical dialogue and interaction among these three groups is needed to make efficient progress, and to make the continuing case for funding research in these challenging problems.

2. MULTISCALE MODELING

The notion of multiscale material modeling has attracted much attention over the past decade. Several approaches exist: (i) the use of atomistic or discrete dislocation calculations to extract parameters used in macroscopic, phenomenological material models. (ii) quasicontinuum methods to span length scales from atoms to the continuum. (iii) Purely numerical approaches, such as meshless methods, to resolve fine scale fields. A fourth approach that is now gaining in importance is the notion of embedding fine scale theories in coarse scale ones. The fine scale theories of interest could include: (i) continuum formulations that resolve mechanics phenomena at length scales in the sub-micron to nanometer range such as the recently proposed strain gradient plasticity theories, and (ii) microforce theories. These theories are often applied to boundary value problems at length scales in the sub-micron to nanometer range. However there is a growing interest in applying them to phenomena such as microvoiding, microcracking, microscopic shear band formation, problems involving inclusions and internal surfaces, grain evolution and texture evolution in macroscopic bodies. Some of these fine scale theories analytically attain the proper response in the limit of macroscopic deformations. However, caution must be exercised in actually solving initial and boundary value problems on macroscopic dimensions (say, meters) with these theories. With any numerical method, the range of discretization will span from nanometers to centimeters or meters. This would be highly expensive, inefficient and possibly non-robust.

One alternative is to develop mathematical techniques for embedding the fine scale theories in coarser-scaled formulations. For instance strain gradient theories at the sub-micron scale might be embedded in the classical macroscopic formulation. The coarse scale formulation is thereby modified to yield a multiscale one. The variational framework is particularly well-suited to such manipulations. They can also be applied to solving boundary value problems with microforce theories while maintaining the tight coupling between the micro and macro scales.

Nanotechnology is emerging as a key area of technology innovation for the 21st century, with the hope of achieving unparalleled improvement on the standard of living of mankind. This technology hinges upon efficient and reliable nanoscale modeling and simulation methods which effectively link continuum and atomistic models to nanoscale material behaviors. It is critical to develop a novel and practical methodology combining the best features of continuum theory and atomistic simulations. As a widely used experiment for probing nanoscale mechanical properties, nanoindentation can be chosen as a focal point of study for the development and validation of the proposed methodology.

The combined continuum/atomistic methods will have high academic and industrial impact on the current pursuit of nanotechnology because of its unique methodology for predicting nanoscale mechanical properties. To continue the trend and advances indicated by the Moore's Law, the next generation of semiconductor technology will be built at the nanometer scale. The capability of predicting mechanical properties of the materials at this scale offers critical information for the designer of devices to select materials and processing techniques. The framework of continuum-atomistic linkage will have far-reaching significance in nanoscale engineering analysis.

3. FAILURE OF STRUCTURES

Failure of structures is often associated with fracture or fatigue of one or more of the elements. In the field of micro- and nano-structures, failure has numerous other manifestations. For example, micro-electro-mechanical systems (MEMS) exhibit large surface area to volume ratios. When two surfaces are brought together, the natural attraction between them, perhaps due to condensed water vapor or even van der Waal's forces, can be sufficient to overcome the relatively small restoring forces inherent in the structures. When this happens, the surfaces stick together and the MEMS has failed. A thorough understanding of this cohesion in MEMS is not available, depending upon a variety of fields, including tribology, materials science, surface chemistry and engineering.

A number of other failure modes have been observed in the mechanical elements of MEMS. Wear debris or other foreign particles entering the small gaps between components can cause the moving parts to seize. Thermally-induced or stress-induced warping of beams or plates can be quite large causing these elements to be well out of alignment with other elements of the system.

Failure of a system can also be identified as its inability to perform within the range for which it was designed. Many MEMS involve resonant structures. Uncertainties in fabrication parameters can lead to uncertain material and geometric properties, which in turn lead to uncertain response.

This variability suggests that the range of system response needs to be considered in assessing the success rate or yield of a particular process.

4. UNCERTAINTY ANALYSIS AND DESIGN CONSIDERATIONS

The research projects in the current Sandia/NSF Life Cycle Engineering program have dealt separately with advanced computational mechanics methods and uncertainty propagation methods. Most the advanced computational mechanics research has been done under deterministic assumptions, whereas the uncertainty analysis research has used simple mechanics models. The next step is to integrate the advances in these two directions. This presents new challenges in computational implementation, scale-up and efficiency. Advances in both directions have led to methods that are computationally intensive. Simple wrapping' of uncertainty analysis methods around deterministic mechanics analysis may not be efficient. The implementation will need to include parallel computing and other new computational science paradigms.

Propagation of uncertainty information through advanced computational mechanics models requires the development of new methods. Monte Carlo and response surface methods may be either too expensive or too approximate, since high-fidelity mechanics computation itself is likely to be computationally intensive. Sensitivity-based analytical approximations have been constructed through stochastic finite element and stochastic boundary element methods. Similar uncertainty propagation methods need to be developed in concert with high-fidelity mechanics methods, such as stochastic mesh-free methods, stochastic cohesive element methods, etc.

One of the challenges in model-based uncertainty simulation is the quantification of modeling error. Some work has been done to quantify the discretization error in traditional finite element analysis. Similar work is needed in the context of high-fidelity computational mechanics models, for example, mesh-free methods, methods using cohesive elements etc. Within uncertainty analysis, attempts have been made to treat computational approximations, biases, and errors as random variables; these approaches have not been entirely satisfactory and may require extensive (and expensive) calibration effort. Therefore, new efforts are needed for comprehensive characterization of model uncertainty, especially in the context of integration of advanced computational mechanics and uncertainty analysis methods. These efforts should also lead to rational approaches for the validation and acceptance/rejection of integrated models for large systems.

The use of advanced methods also presents new data requirements. Consider a simple example --- material property data for stress analysis of a beam. An isotropic strength of materials approach only requires one Young's modulus. An anisotropic linear finite element analysis requires moduli in several directions. A nonlinear analysis requires more comprehensive load-deformation data. An uncertainty analysis requires statistical data on the material properties. Thus the data requirements are different for different levels of analysis. Therefore, the integration of advanced methods will be affected by data limitations and information uncertainty. Comprehensive uncertainty quantification needs to include and develop methods to handle various types of information uncertainty.

Most of the research in the LCE effort has concentrated on developing advanced *analysis* techniques, both deterministic and non-deterministic. The next step is to develop design and life cycle management approaches that incorporate these techniques. There is a considerable gap between advances in analysis and design approaches for engineering systems. Engineering design is still mostly done using traditional deterministic analysis with simple system models calibrated with experience and design safety margins, without incorporating many of the advances in computational mechanics or uncertainty propagation. As model-based simulation gains ground for large and complex systems, the design approaches for such systems will need to incorporate these advances. Only then will the significant progress made in the LCE effort be fruitful. Therefore, new design approaches need to be developed using integrated high-fidelity computational mechanics and uncertainty analysis. Further, these approaches need to be implemented and demonstrated for *realistic systems*, in order to facilitate acceptance by the engineering community.

5. AGING AND LONG-TERM BEHAVIOR

For a large number of engineering components accumulation of damage occurs under cyclic loads far below loading conditions that would lead to instantaneous catastrophic failure. The development of modeling and simulation tools capturing the evolution of these incrementally deteriorating events is obviously essential if the entire life cycle of engineering components is to be predicted.

In the investigation of highly critical components, it is indispensable to account for the nucleation, and subsequent propagation of crack like defects. Essentially all currently available methodologies to analyze such problems are based on empirical laws, e.g. the Basquin or the Coffin-Manson Laws for crack initiation, or the Paris Law for crack propagation. While these equations are widely used, it is important to recall that they provide data correlation schemes rather than predictive capabilities. As an example, current fatigue crack growth methodologies encounter limits of applicability if small-scale yielding assumptions are not valid any longer such as at interfaces, or in layered materials, or if short cracks are of concern. No generally accepted methods exist in which crack initiation and subsequent growth - e.g. from sharp corners and intersections of interfaces with free surfaces - can be analyzed in a unified framework. To overcome these limitations, several attempts were made in the past to establish micromechanical models for fatigue failure. However, past models commonly are based on strongly simplifying assumptions, and did not take advantage either of recent advances in constitutive description of materials, or of the current state of computational tools in solid mechanics.

The main challenge in the advancement of fatigue failure is the development of the description of material separation under cyclic loading, and the determination of the associated material parameters. In an ideal world one would like to be able to predict the response of a specimen to cyclic loads from simulations on the atomistic level. For real engineering components under cyclic loading this seems to remain an elusive goal for some time to come. It seems more realistic to describe the micromechanisms of material separation processes at the crack tip by the use of appropriate constitutive models embedded in a continuum framework. Crack initiation, growth and arrest should then be computed without external interference of the analyst and depend solely on the definition of constitutive framework. To accomplish such an approach, additional and novel experimental input is needed to determine the cyclic material deterioration locally at the crack tip, and developments in computational mechanics are required to enable computations spanning the time scale of the individual damage event to the final time to failure of a component.

While the development of predictive models for aging and long-term behavior of materials and components remains, by itself, a high research priority, this development can be greatly enhanced by leveraging new technologies in information and computation, and by incorporating models into a system-level design and decision-making framework. More details are given in the following.

Synthesis of computational models with laboratory and field data

Advances in computational methods have resulted in impressive high-resolution models of materials. To make these models more useful in life-cycle engineering, however, it is necessary to provide high-resolution information on the parameters of the models. Such parameters describe spatial material properties including the distribution of initial flaws, geometrical properties, load processes arising from system operation and natural and man-made hazards, and degradation processes within the material or system. A research challenge is in developing appropriate experimental programs in the laboratory and field that would provide the type of data needed to quantify and calibrate these parameters. Since it is not possible to develop sensors to exhaustively measure the full set of high-resolution parameters inherent in the computational models, an inferential process may be needed. Such inferential processes would use a reduced set of parameters to effectively link laboratory and field experiments with high-resolution computational models.

Interface with information technology

With the rise of large-scale applications of information technology to engineering problems, such as that envisioned in the NSF-funded NEES network, there is a research opportunity in effectively using such information to develop more accurate predictive models for long-term behavior. Statistical tools would have to be supplemented with data mining, fusion and other modern information extraction and processing techniques to be effective in this research effort.

Design under uncertainty within a decision-making framework

The uncertainties inherent in the aging and long-term behavior of the materials and components of a system make it impossible to provide precise predictions of life-cycle costs. Within an axiom-based design framework, the uncertainties could, in principle, be incorporated into an optimal design that minimizes life-cycle costs, including those associated with construction, maintenance, and field-data collection and processing. There are, however, open research problems that must be addressed to bring this design framework for life-cycle engineering to fruition. These research problems are associated with developing the utility functions appropriate for life-cycle engineering, searching efficiently within a high-dimensional design space, and incorporating predictive models with uncertain parameters into the decision-making processes at the construction and maintenance life-cycle phases.

6. POLYMERS AND ELASTOMERS

In the chapter *Life–Cycle and Durability Predictions of Elastomeric Components*, it was shown that the mechanical properties of a natural vulcanized rubber are altered (degraded) due to microstructural changes, namely the scission and re-cross linking of macromolecular network junctions, that occur at sufficiently high temperatures. In most of the existing experimental studies of the consequences of these microstructural changes, the focus has been on uniaxial extensions. These studies form a solid foundation for the development of a model that can be used in numerical life–cycle and durability predictions for elastomeric components. Research is needed in the following topics in order to continue the development of the model. First, it is necessary to determine the consequence of microstructural changes at high temperatures on the response under biaxial and shear deformations. Second, there is a paucity of experimental results on heat transfer in elastomers, and how it is affected by large deformation and microstructural changes. Third, there have been few studies that document the development of 'hot spots' in typical elastomeric components, the degradation of properties in these 'hot spots' and their influence on component performance.

Most of the existing studies on the degradation of elastomers do not consider classical viscoelastic phenomena such as stress relaxation and creep. Yet, elastomers exhibit these properties. Indeed, the conversion of mechanical energy to heat due to viscoelastic effects contributes significantly to temperature rise. Polymeric solids are also composed of cross-linked networks of macromolecules, but are stiffer than elastomers. Their viscoelastic properties need to be considered in their applications. The events leading to degradation in elastomers can also be expected to occur in polymers and to alter their viscoelastic properties. Thus, another important research need is a study of the consequences of the scission and re-cross linking process on the viscoelastic response of elastomers and polymers.

An important related subject involves the solidification or curing of a polymeric liquid to form a polymeric solid. This occurs in fabrication processes such as the casting of elastomeric components, forming of epoxy based composite materials and encapsulation of microelectronic devices. During the curing process, macromolecules cross link to form networks as the liquid evolves into a solid. This process occurs continuously in time and depends on temperature. There are volume changes as the microstructure and constituents evolve. Stresses develop in the networks that depend on their viscoelasticity and the degree of cure. The spatial and temporal variation of the changes in volume, degree of cure, heat transfer and microstructure determine the final properties of the cured product as well as its residual stresses and dimensional stability. These, in turn, determine its life-cycle and durability properties. Extensive research is needed in the development of a model that can account for these factors.

In conclusion, elastomeric and polymeric components undergo microstructural changes that influence their life-cycle and durability properties. There is need for research into the development of models that can be used in numerical simulations to predict these properties.

7. MAPPING DISCRETE MODELS TO SYSTEM RESPONSE

With the exception of idealized, canonical cases, a model for a system is nearly always in discrete form. Intrinsic to each discrete model is a set of assumptions and approximations that are used in the model parameters, constitutive relations and computational procedure. Since the formulation of this set of assumptions and approximations is dependent on the modeller's point of view and on the available computational resources, there are, conceptually, an infinite number of discrete models corresponding to each system. Furthermore, there is no "exact" discrete model due to the intrinsic approximations; it is even difficult to determine a "best" model since more than one criterion for ranking the accuracy or suitability of the models may be of relevance.

Given that a modeller has, at his or her disposal, a large collection of models or potential approaches to modelling, the general question is: Which model or subset of models should be used? To formulate this question into a research problem, the context must be well defined. The following issues are of interest: the impact of modelling errors in the context of system failure or inefficient design, the nature, availability and cost of laboratory or field data which may be needed to calibrate the models, the definition of an appropriate utility function that could be used to develop a well-defined ranking in relation to life-cycle or other relevant costs, and the propagation of parameter and model uncertainties that lead a modeller to rely on more than a single model to assess system performance. Central to this class of research problems is the relationship, or mapping between the collection of discrete models and the characteristics of the system response.

Part I

Solid and Structural Dynamics

Computational Cohesive Zone Modeling Of Polymer Interfacial Failure

P. Rahul-Kumar[1], S. Muralidhar[1], A. Jagota[2], S.J. Bennison[2], S. Saigal[1]

1. INTRODUCTION

Linear elastic fracture mechanics (LEFM), based on singular stress fields and energy release rates, has proven to be quite successful in describing fracture in brittle materials (Lawn 1993). A cohesive zone description of near-tip process zones emerged early in the development of the field (Dugdale, 1960, Barenblatt, 1962), but remained relatively unused in the treatment of brittle fracture. LEFM has also been used successfully under well-defined conditions for polymers (Williams, 1984). However, it becomes limited in its treatment for polymers that exhibit significant inelasticity, large deformations, and time-dependence. Analytical challenges make the cohesive zone approach relatively difficult to use, although it is fundamentally capable of handling these complications (Dugdale, 1960, Knauss, 1973, Schapery, 1975). Implementation of the cohesive zone approach in a computational context provides a means of analyzing previously intractable problems of materials failure in a way that connects local process zones with the macroscopic deformation. For these reasons, computational cohesive zone modeling has attracted attention for brittle & elastic-plastic materials (Xu & Needleman, 1994, Camacho & Ortiz, 1996, Tvergaard & Hutchinson, 1992, Espinosa *et al.*, 1998, and Needleman and Rosakis, 1999). Failure in polymers, which is often accompanied by time-dependent and large deformation, presents unique challenges for computational analysis. In this chapter we present an implementation of cohesive zones as cohesive finite elements for tackling issues in modeling polymer interfacial fracture.

The cohesive zone approach overcomes some of the limitations of the stress intensity factor and J-integral methods that form the basis of LEFM. It can provide a description of the different stages of material failure starting from initiation to macroscopic crack propagation, while accounting for energy balance. By specifying a local cohesive law, one is able to partition the macroscopic

---

[1] Department of Civil and Env. Eng., Carnegie Mellon University, Pittsburgh, PA 15213
[2] CR&D, E.I. DuPont de Nemours Inc., Wilmington, DE 19880-0356

fracture resistance into its component parts, providing the sort of information important for materials design and understanding adhesion.

The examples presented in this chapter demonstrate the versatility of the computational implementation to describe crack propagation in hyperelastic/viscoelastic materials, crack "pop-in", dynamic crack propagation, and 3D geometries. Specifically, we present analyses of peel testing of elastomers (Rahul-Kumar *et al.*, 1999), adhesion measurement using a compressive shear test (Jagota *et al.* 1999, Rahulkumar *et al.*, 2000), and analysis of a through-cracked tension test (S. Muraldhar *et al.* 2000).

2. COHESIVE ZONE MODEL AND NUMERICAL IMPLEMENTATION

The cohesive zone model provides local tractions that oppose separation and relative sliding of crack faces. The model can represent various crack tip phenomena (Rahul-Kumar, 1999). In comparison to LEFM where, usually, fracture is tied to a single number (*e.g.*, K_c or G_c), the cohesive zone description of the process introduces an additional parameter. The work of separating the crack faces (per unit area) corresponds to G_c; the peak stress (or characteristic opening) experienced during separation is an additional descriptor of the failure process.

Specific choices of the cohesive law and parameters have to be dictated by the particular material and failure mechanism. Here we restrict ourselves to a finite element implementation of a phenomenological, rate-independent, cohesive zone model (Xu & Needleman, 1994). Figure 1 shows two bulk finite elements that share a common face. The cohesive element defines tractions acting across this face in terms of opening and sliding displacements: $\Delta_n, \Delta_{t1}, \Delta_{t2}$. Tractions typically vanish when this displacement vector is (0,0,0). The tractions are derived from a potential, $\phi(\Delta_n, \Delta_{t1}, \Delta_{t2})$ as:

$$T_n = -\partial\phi/\partial\Delta_n, T_{t1} = -\partial\phi/\partial\Delta_{t1}, T_{t2} = -\partial\phi/\partial\Delta_{t2} \qquad (1)$$

A reduced version of the Xu-Needleman (1994) potential is given as:

$$\phi(\Delta_n, \Delta_{t1}, \Delta_{t2}) = \Gamma_0\left[1 - \left(1 + \frac{\Delta_n}{\delta_{cr}}\right)\exp\left(-\frac{\Delta_n}{\delta_{cr}}\right)\right]\exp\left(-\frac{\Delta_{t1}^2 + \Delta_{t2}^2}{\delta_{cr}^2}\right) \qquad (2)$$

where δ_{cr} is the characteristic critical opening displacement and Γ_0 is the fracture energy of the interface. In order to penalize interpenetration of cohesive surfaces under compression the potential in equation (1) is augmented by the term $[H(\Delta_n) - 1]\kappa\Delta_n^3$, where $H(\Delta_n)$ is the Heaviside step function, and κ is a penalty parameter. This modification does not affect the work of interfacial separation as it operates only under compression. When modeling interfacial separation in three dimensions, the crack front is identified as the locus of points along the interface where the combination of openings (Δ_n, Δ_{t1}, Δ_{t2}) satisfies the relationship:

$$\left[1 - \left(1 + \frac{\Delta_n}{\delta_{cr}}\right)\exp\left(-\frac{\Delta_n}{\delta_{cr}}\right)\right]\exp\left(-\frac{\Delta_{t1}^2 + \Delta_{t2}^2}{\delta_{cr}^2}\right) = (1 - 2/e) \tag{3}$$

where, e is the Euler number exp(1). This identifies a locus of openings corresponding to identical work of separation. Under pure normal separation ($\Delta_{t1} = \Delta_{t2} = 0$) the work evaluates to $(1-2/e)\Gamma_0$, for $\Delta_n = \delta_{cr}$.

The numerical implementation of a cohesive zone model for interface fracture within an implicit finite element framework is accomplished using cohesive elements. Element nodal displacements define the kinematics of deformation. Cohesive element nodes are shared with neighboring bulk finite elements that model the continuum (Figure 1). Therefore, a damage process described by a specific type of cohesive element is coupled to the deformation occurring in the bulk material.

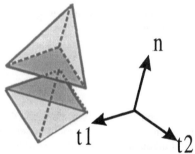

Figure 1. Schematic drawing of the local frame of reference used to define cohesive element tractions and stiffness

If cohesive tractions are given by $\{T\} = \{T_{t1}\ T_{t2}\ T_n\}^T$, finite element nodal forces can be computed as:

$$\{F\} = \int_{S_0} [Q]^T [A]^T \{T\} dS_0 \tag{4}$$

where, S_0 is the initial cohesive surface defined as the mid-plane between the two faces of the cohesive element. Matrix [Q] rotates the force vector from coordinates local to the cohesive surface to global coordinates, and [A] is a matrix of element interpolation functions that relates opening displacements $\{U\}^T = \{\Delta_n, \Delta_{t1}, \Delta_{t2}\}^T$ to current nodal coordinates $\{x\}_{np}$ as

$$\{U\} = [A]\{x\}_{np} \tag{5}$$

It has been assumed in (4) that the magnitude of cohesive tractions is per unit undeformed area. This ensures that the work of separation is independent of the in-plane stretch of the element, which is appropriate for modeling fracture in solids. An approximation to the cohesive element stiffness,

$$[J]_{ij} = \left[\frac{\partial F_i}{\partial x_j} \right] \tag{6}$$

is formed by neglecting the variation in [Q] and is

$$[J] = \int_{S_0} [Q]^T [A]^T [C] [A] [Q] \, dS_0 \tag{7}$$

where [C] is the cohesive material Jacobian that relates incremental tractions to the incremental displacement jumps

$$\{T\} = [C]\{dU\} \quad [C_{ij}] = \left[\frac{\partial T_i}{\partial u_j} \right] \tag{8a,b}$$

The nodal force vector and the cohesive element stiffness matrix are used for Newton-Raphson iterations in an implicit solution scheme. Further details can be found elsewhere (Rahulkumar *et al.*, 1999, and Rahul-Kumar *et al.*, 2000.). Implementation of this formulation has been separated into two functional units. The kernel specifies cohesive element tractions and the cohesive element Jacobian as a function of current opening displacement. The shell consists of numerical quadrature, transformations, and the like for calculating the stiffness matrix. A family of cohesive elements has been implemented for use with static and implicit-dynamic procedures in the commercial finite element package ABAQUS™, 1997.

3. APPLICATIONS

The implementation described has been used to study interfacial fracture in three experiments for adhesion of viscoelastic elastomers: the T-peel, compressive shear strength (CSS), and through crack tension (TCT) tests.

3.1 T-Peel test modeling

The peel test is employed routinely to measure the strength of elastomeric adhesives and joints (Kinloch 1987, Gent, 1996). Interpretation of peel tests is usually based on the following two assumptions: 1) the intrinsic interface resistance is *a rate-independent* quantity, and 2) bulk dissipation in the peel arms is the major contributor to the work of separation and its rate sensitivity (Gent, 1996). T-peel experiments have been analyzed using cohesive elements to represent the intrinsic adhesion and a viscoelastic constitutive description for the bulk. The analysis allows a decoupling of bulk dissipation from the intrinsic and near-crack process zone contributions to fracture resistance.

Figure 2 shows how peel energy varies with normalized peel rate when the bulk is modeled by a standard viscoelastic linear solid, for which the viscoelastic bulk and shear relaxation modulus are given by, $K(t) = K, G(t) = G_\infty + (G_0 - G_\infty) \exp(-t/\tau_0)$, where G_∞ is the rubbery modulus and G_0 is the instantaneous modulus. The normalized velocity, $v*$, is defined as: $v* = v\tau/h$, where v is the actual velocity, τ is the characteristic relaxation time and h is the thickness of the peel arm. In the limits of low and high peeling rates, $v* \to 0, v* \to \infty$, the entire peel arm behaves elastically (Regions I and IV). Peel energy then equals the intrinsic cohesive energy. For intermediate rates, dissipation in the bulk contributes to the overall peel energy.

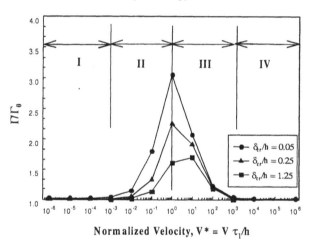

Figure 2. Predicted ratio of global fracture energy to intrinsic fracture energy and the effect of cohesive zone parameter δ_{cr} on the predicted fracture energy as a function of normalized crack velocity, $V*$ (peeling rate).

In region II, the fracture energy at a given velocity is sensitive to the value of δ_{cr}; smaller δ_{cr} results in greater dissipation at lower velocities. The predicted fracture energies fall off rapidly in Region III and are less sensitive to δ_{cr}. The direct problem solved here, one in which cohesive zone properties are specified and total dissipation is computed, shows that total dissipation depends on *both* cohesive zone parameters. This implies that the important inverse problem, that of partitioning measured global energy dissipation into bulk and near-tip components, cannot be resolved uniquely without the use of a cohesive zone model.

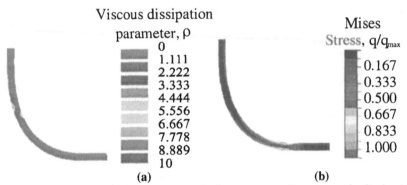

Figure 3. Deformed shapes at a velocity corresponding to peak dissipation showing (a) contours of parameter ρ, representing dissipation, and (b) von Mises stress q.

Figure 3 shows the computed peel shapes at a velocity corresponding to peak dissipation. Figure 3(a) shows contours of a parameter, ρ, which is a measure of viscous dissipation. It is defined as:, $\rho = [(\varepsilon - \varepsilon_0)/(\varepsilon_\infty - \varepsilon_0)]$, where $\varepsilon = \sqrt{2/3 \mathbf{e} : \mathbf{e}}$, and $\mathbf{e}$ is the total deviatoric strain tensor. The quantities, ε_0 and ε_∞, are obtained by dividing the von Mises stress, q, at the material point by the three times the unrelaxed and relaxed shear moduli respectively of the material. The Mises stress is defined as, $q = \sqrt{3/2 \mathbf{s} : \mathbf{s}}$, where $\mathbf{s}$ is the deviatoric stress tensor. At very slow and large peel rates the parameter ρ approaches 1 and 0, respectively. When loading is reversed, for a viscoelastic material, the parameter can exceed these bounds and become large. In the peel simulations, a contour plot of this parameter is a signature of the intensity and location of viscous dissipation. Figure 3(b) shows contours of the von Mises stress, q, under the same conditions.

In the low and high peel rate limits, not shown here, the peel arm deforms as an elastic material; the parameter ρ remains in the limits (0,1). Figure 3 shows that at the velocity corresponding to peak peel energy, the region of maximum dissipation is completely distinct from the region of highest stress; dissipation is localized some distance behind the crack tip. This is consistent with the prediction (Rahulkumar *et al.*, 2000), based on de Gennes' analysis of viscoelastic crack propagation (de Gennes, 1996), that dissipation is localized roughly at a distance vτ from the crack tip. These results can guide the design of adhesives. Since the relaxation time is related to the glass transition temperature of the material and the test temperature, an adhesive may be designed with an optimum glass transition temperature for specified use temperature and loading rates.

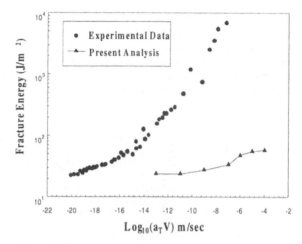

Figure 4. Experimental fracture energies on peeling of butadiene elastomeric sheets (Gent, 1996) and predictions based on fixed fracture energy and rate-dependent bulk viscoelastic loss.

The cohesive element computational was used to model experiments on peeling of polybutadiene elastomers (Chang, 1980, and Gent, 1996). The peel energy measured in these experiments qualitatively appears to be a combination of rate-independent intrinsic work of fracture and viscoelastic bulk losses. Experimentally observed increase in peel energy with peel velocity has been attributed to bulk viscoelastic losses. Here the cohesive element technique is used to test this hypothesis. Details on the characterization of the bulk viscoelastic constitutive model for the polybutadiene elastomer along with cohesive element model details are given in Rahulkumar *et al.*, 1999. As shown in Figure 4, the predicted increase in macroscopic interfacial fracture energy (adhesion) with peel rate is several orders of magnitude lower than observed experimentally. These results suggest that the interpretation of Gent's measurements of peel energy as a combination of rate-independent losses at the crack tip cohesive zone and a rate-dependent loss due to bulk viscoelasticity cannot be supported. Some additional failure mechanism appears to be operating at the crack tip that has not yet been clearly identified.

3.2 3D analysis of crack propagation in CSS test

Adhesion between an elastomer and a rigid substrate may be probed using a compressive shear strength (CSS) test shown schematically in Figure 5(a). The CSS test consists of loading a 3-ply laminate, substrate/polymer/substrate, in combined compression and shear and monitoring the force-displacement characteristics. The laminate used in the current study comprised of glass and plasticized-polyvinyl butyral (Butacite®), which is used in the manufacture of laminated safety glazing for automotive windshields, and architectural glazing. Figure 5(b) is an *in-situ* micrograph showing polymer deformation and associated interfacial debonding during loading. Observations made during a loading a sequence reveal that an interfacial crack nucleates readily and propagates stably until reaching a critical size at which unstable debonding ensues. Modeling of such crack growth behavior is difficult and requires methods such as those presented by Jagota *et al.* (2000) and in the current contribution.

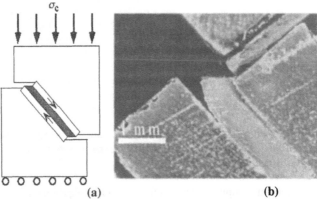

Figure 5. (a) Schematic drawing of a compressive shear test for glass/elastomer adhesion. (b). *in situ* micrograph showing polymer deformation and stable interface debonding prior to unstable failure of the interface.

A 3D analysis of the CSS test has been performed by employing 3D cohesive elements and a hyperelastic material model for the polymer. The 3D finite element discretization employed in the study is shown in Figure 6. The dimensions of the 3D CSS test specimen used in the 3D numerical study are: polymer thickness, h = 0.76 mm, length = 25 mm, and width = 4.56 mm (6h). Based on symmetry about the x-y plane, one half of width (3h) is modeled in the z direction. The polymer is modeled using 8-node hybrid brick elements, and glass is modeled using regular displacement based 8-node brick elements. A layer of 3D 8-node cohesive elements is placed along the interface between the polymer and glass. The size of cohesive elements in the fine discretized region in Figure 16 at the free edge, z=3h, are of dimension 0.047h along the length and 0.094h along the width. The 3D discretization consists of 48,900 degrees of freedom. The boundary conditions for the nodes on the bottom surface of the glass elements are specified as, $T_x = 0$, $u_y = 0$, $T_z = 0$. The nodes on the symmetry face, z=0, have, $u_z=0$. The loading consists of specifying displacements, $u_x = 0$, $u_y = \gamma h/\sqrt{2}$, and $u_z = 0$, for the nodes on the top surface of bulk polymer elements, where, γ is the equivalent shear strain in the polymer. It has been modeled as a neo-Hookean hyperelastic material with material constant, $C_{10} = (1/2)G = 1.66 \times 10^5$ Pa. Glass has been modeled as a rigid substrate. The cohesive zone parameters are: $\Gamma_0 = 4.95$ N/m, and

$\delta_{cr} = 10.0$ μm. An implicit-dynamic analysis has been performed to simulate stable and unstable crack growth along the interface.

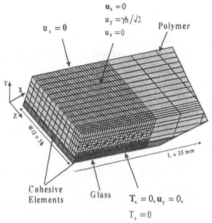

Figure 6. Finite element discretization of one half of the CSS test specimen.

The evolution of an interfacial crack with initial size of $a_0 = 0.25h$ is shown in Figure 7 for various levels of applied shear strain.

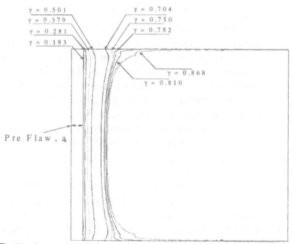

Figure 7. Evolution of failure and crack front along the interface in CSS test specimen for polymer modeled as Neo-Hookean hyperelastic material.

The crack front is defined by employing equation (3). It is observed that failure initiates ahead of the pre-flaw near the specimen edges at an applied shear strain of, $\gamma \sim 0.183$. At an applied shear strain, $\gamma \sim 0.281$, the crack tunnels through the width of the specimen. The crack front at this instant is relatively straight. For progressively increasing shear strain, the crack front moves along the interface by developing a finite curvature at the free edge. Some distance from the free edge (~ 2h) the crack front becomes straight, implying plane strain conditions. Moving from the interior towards the free edge, crack length increases, attains a maximum value, and decreases near the free edge. At an applied shear strain of $\gamma=0.704$ instabilities occur in the 3D simulation due to a local "pop-through" of the crack front at the free edge. At an applied strain of $\gamma=0.782$ the crack front at the free edge undergoes a complete pop-through and the crack front curvature changes sign. At this instant the crack is longer at the free edge when compared to interior of the specimen. Similar crack front profiles have been observed experimentally for crack growth along Glass/Epoxy interface (Swadener & Liechti 1998). Finally, unstable crack growth and failure of the interface occurs at a strain of $\gamma = 0.868$.

3.3 Through cracked tension test – short crack response

Mechanical behavior of cracked glass bridged by an elastomeric ligament is central to properties of cracked glass-polymer laminates and other composites. A through cracked tension (TCT) test is representative of such a situation wherein a glass/polymer/glass laminate with single aligned cracks in each glass plate is subjected to tensile loading. The objective of studying this test is to develop a bridging force-displacement relationship in terms of bulk polymer constitutive properties and interfacial toughness. This is of utility for at least two reasons: (a) to extract interfacial properties from an experiment that is directly relevant to laminate performance, and (b) to build models for cracked laminates based on the behavior of individual bridging ligaments.

A schematic drawing of a tension test for a through cracked specimen is shown in Figure 8(a). The composite specimen consists of two glass sheets of thickness h_g each, held together by a polymer layer of thickness $2h$. The width, b, of the glass sheets and the length, L, of the polymer layer are large compared to the thickness $2h$. Glass plates are laminated with a plasticized PVB interlayer. The tensile specimens are scored and pre-cracked prior to the experiments. One end of the specimen is held in fixed grips, and displacement is applied to the other end at a fixed velocity. The force applied at the lower end, P, is measured as a function of the applied displacement, 2δ. More details of the experimental procedure may be found in (S. Muralidhar et al., 2000).

The test involves stable delamination at the interface between the polymer and glass. As the interfacial crack length increases, the force required to maintain steady state crack propagation reaches a constant value. An analytical result has been established that allows the extraction of adhesion in terms of the steady value of the measured force and the hyperelastic polymer constitutive parameters (S. Muralidhar et al., 2000). The hyperelastic representation of the polymer implies that the extracted adhesion value combines interfacial toughness and bulk viscoelastic losses. The steady state response for long cracks is independent of the cohesive zone size. However, in viscoelastic materials, the magnitude of dissipation is determined by the size of the cohesive zone. The simulation of the TCT test using cohesive elements, therefore, has two uses: (i) to extract an adhesion value that combines all the dissipation in the bulk, and (ii) to estimate a cohesive zone size based on the force-displacement response for short cracks. The latter would be an *equivalent* size if all the dissipation were to occur within the cohesive zone. The effect of the cohesive zone size on the force-displacement response is difficult to incorporate in analytical models.

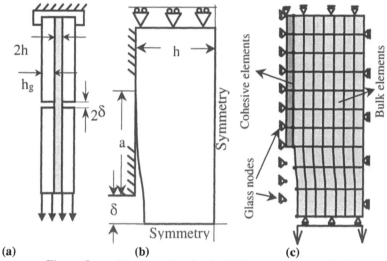

Figure 8. (a) Schematic sketch of a TCT test specimen. (b) Quarter model of the TCT specimen. (c) Cohesive element model of TCT test specimen.

Utilizing symmetry, only a quarter of the configuration has been modeled, as shown in Figure 8 (b). In the corresponding finite element model shown in Figure 8(c), the polymer layer has been discretized using 4-node quadrilateral (plane-strain) finite elements. Glass, treated as rigid, is represented by a set of fixed nodes. The interface between the two has been joined by cohesive elements. The cohesive elements provide tractions to the continuum elements that depend on the displacement jumps across the cohesive surface. The polymer is modeled as a reduced polynomial hyperelastic material, with constants $C_{10} = 1.23$ MPa and $C_{20} = 36$ KPa (S. Muralidhar et al., 2000). The objective is to match the force-displacement response observed in an experiment with proper choices for the values of adhesion and cohesive zone size.

Figure 9 shows a comparison of the experiment and simulation. The parameters used in the simulation are $\Gamma_0 = 284$ J/m^2 and $\delta_{cr} = 150$ μm. Jagota et al. (1999) have reported an adhesion value of 100 J/m^2 for the same interface after accounting for the bulk viscoelastic losses, using the compressive shear strength test. The value extracted here indicates significant viscoelastic losses in the bulk. The estimated cohesive zone size needs to be interpreted correspondingly.

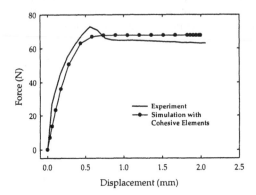

Figure 9. Comparison of simulated and measured force-displacement response of a TCT test specimen.

4. CONCLUSIONS

A family of cohesive finite elements for the simulation of crack nucleation and growth has been developed. Cohesive elements potentially allow one to study crack nucleation and growth in interfacial failures that are accompanied by bulk inelasticity and large strains. The examples considered here demonstrate that it is possible to apply cohesive element methods in situations that are not usually analyzed under the framework of fracture mechanics, e.g., where large deformations and material nonlinearity dominate. Application of cohesive elements has been illustrated by three problems in interfacial fracture of polymers. The analysis of peel testing shows how the interaction between intrinsic fracture energy and macroscopic energy loss may be understood. The example of the compressive shear test shows that cohesive elements can be used to analyze problems involving crack growth along bi-material interfaces with large strain deformation and dynamic crack pop-through behavior can be successfully captured. In the final example of the TCT test, cohesive elements are used to simulate the force-displacement response with a finite cohesive zone. This effect is otherwise very difficult, in general, to incorporate using analytical tools. From this response, an overall adhesion value that combines bulk dissipation is estimated.

REFERENCES

ABAQUS, Version 5.7, (1997). Theory and User Manuals I, II & Hibbitt, Karlsson, & Sorensen, Inc., 1080 Main Street, Pawtucket, R.I., 02860 – 4847, USA.

Barenblatt, G.I., (1962). The Mathematical Theory of Equilibrium Cracks in Brittle Fracture. *Advances in Applied Mechanics* **7**, 55-129, Academic Press.

Camacho, G.T., Ortiz, M., (1996). Computational Modeling of Impact Damage in Brittle Materials. *International Journal of Solids and Structures* **33**(20-22), 2899 – 2939.

Chang, Rong-Jong (1980). Effect of Interfacial Bonding of the Strength of Adhesion Between Elastomer Layers. *Ph.D Dissertation*, University of Akron., U.S.A.

de Gennes, P.G., (1996). Soft Adhesives. *Langmuir* **12**, 4497 – 4500.

de Gennes, P.G. (1997), *Soft Interfaces*, The 1994 Dirac Memorial Lecture, Cambridge University Press

Dugdale, D.S., (1960). Yielding of Steel Sheets Containing Slits. *Journal of the Mechanics and Physics of Solids,* **8**, 100.

Gent, A.N., (1996). Adhesion and Strength of Viscoelastic Solids. Is There a Relationship Between Adhesion and Bulk Properties ? *Langmuir* **12**, 4492-4496.

Hui, Chun-Yuen, Xu, Da-Ben, Kramer, E.J., (1992). A Fracture Model for Weak Interface in a Viscoelastic Material (Small Scale Yielding Analysis). *Journal of Applied Physics* **72**(8), 3294 – 3304.

Jagota A., Bennison, S.J., Smith, C.A. (2000) Analysis of a Compressive Shear Test for Adhesion Between Elastomeric Polymers and Rigid Substrates, *Int. J. of Fracture,* **10** 105-130.

Kinloch, A.J. (1987). Adhesion and adhesives: science and technology. Chapman & Hall, New York.

Lawn, Brian., Fracture of Brittle Solids. Cambridge Solid State Science Series, Cambridge University Press, 1993.

Knauss, W.G., (1973). On the Steady Propagation of a Crack in a Viscoelastic Sheet: Experiments and Analysis. *Deformation and Fracture of High Polymers*, Eds. H.H. Kausch and R. Jaffee, Plenum Press, New York, 501 – 541.

Needleman, A., Rosakis, A.J. (1999) The effect of bond strength and loading rate on the conditions governing the attainment of intersonic crack growth along interfaces. *Journal of the Mechanics and Physics of Solids*, **47**(12), 2411-2449.

Rahul-Kumar, P. (1999) Computational Fracture Mechanics Using Cohesive Element Formulations, *Ph.D. Dissertation*, Carnegie Mellon University, Pittsburgh, U.S.A

Rahul-Kumar, P. Jagota, A., Bennison, S.J., Saigal, S., S. Muralidhar. (1999) Polymer Interfacial Fracture Simulations Using Cohesive Elements. *Acta-Materialia.* **47** (15/16), 4161-4169.

Rahulkumar, P. Jagota, A., Bennison, S.J., Saigal, S. (2000) Cohesive Element Modeling of Viscoelastic Fracture: Application to Peel Testing of Polymers. *Int. Journal. of Solids and Structures.* **37**, 1873-1897.

Rahul-Kumar, P. Jagota, A., Bennison, S.J., Saigal, S. (2000) Interfacial Failures in a Compressive Shear Strength Test of Glass/Polymer Laminates. *Int. Journal of Solids and Structures.* **37**, 7281-7305.

Seshadri, Muralidhar, (2001) Mechanics of Glass-Polymer Laminates using Multi Length Scale Cohesive Zone Models, *Ph.D. Dissertation*, Carnegie Mellon University, Pittsburgh, USA.

S. Muralidhar, A. Jagota, S. J. Bennison and S. Saigal. (2000) Mechanical Behaviour in Tension of Cracked Glass Bridged by an Elastomeric Ligament, *Acta Materialia*, **48**, 4577-4588.

Schapery, R.A., (1975). A Theory of Crack Initiation and Growth in Viscoelastic Media. *Int. J of Fracture.*, **11**(1), 141 – 159.

Swadener, J.G., Liechti, K.M. (1998) Asymmetric Shielding in Mixed-Mode Fracture of a Glass/Epoxy Interface. *J. Applied Mechanics.*, **65**, 25-29

Williams J.G., (1984) *Fracture Mechanics of Polymers*, Ellis-Horwood Ltd. Chichester (UK).

Tvergaard, V., Hutchinson, J.W., J. of Mech. Phys. Solids, 1992, **40**(6), 1377 – 1397.

Xu, X.P., Needleman, A., (1994). Numerical Simulations of Fast Crack Growth in Brittle Solids. *Journal of the Mechanics and Physics of Solids* **42**(9), 1397–1434.

A Strain-Gradient
Virtual-Internal-Bond Model

A.Vainchtein[1], P. A. Klein[2], H. Gao[3] and Y. Huang[4]

1 INTRODUCTION

Nonlinear continuum models with nonconvex elastic energies result in equilibrium equations that lose ellipticity at some critical level of deformation. Beyond this critical strain, discontinuous-strain solutions emerge, and due to the absence of an internal length scale, the equilibria computed using finite element methods strongly depend on the selected mesh size. In particular, this problem presents itself in nonlinear models of fracture. One such example is the Virtual Internal Bond (VIB) model (Gao and Klein, 1998; Klein and Gao, 1998, 2000; Zhang et al., 2001), where the constitutive law is found by averaging over a random network of cohesive bonds with nonconvex Lennard-Jones-type potentials. The model successfully predicts critical stress level and direction of the deformation zone for nucleation of fracture that appears as a localized zone of high strain. However, without an internal length scale it cannot predict the size of the localized deformation zone.

One way to introduce a length scale into the VIB model is to employ a strain-gradient nonlinear elasticity theory. Phenomenological models incorporating higher-order gradient term in the constitutive law have been used to study localization phenomena in materials (Coleman; 1983; Aifantis, 1984; Coleman and Hodgdon, 1985; Fleck and Hutchinson, 1998; Shi et al., 2000). Recently, a derivation of such theories from periodic lattice microstructures has been presented by Triantafyllidis and Bardenhagen (1993) and Bardenhagen and Triantafyllidis (1994). This derivation introduces a natural length scale into the continuum model: a characteristic lattice spacing ε.

The derivation of the strain-gradient theory in the one-dimensional case given by Triantafyllidis and Bardenhagen (1993) is reviewed and discussed in this paper. It is shown that the positive sign of the strain-gradient coefficient is crucial for the existence of strain localization zones that replace strain discontinuities resulting from the local approximation. On the other hand, the positive sign of strain-gradient term requires a rather special and not always realistic choice of the long-range interaction potentials. It also leads to a wrong qualitative behavior of the dispersion relation.

[1] Department of Mathematics, University of Pittsburgh, Pittsburgh, PA 15260

[2] Sandia National Laboratories, Livermore, CA 94551

[3] Division of Mechanics and Computation, Stanford University, Palo Alto, CA 94305

[4] Department of Mechanical and Industrial Engineering, University of Illinois, Urbana, IL 61801

In the second part of the paper, the derivation of the strain-gradient theory in higher dimensions given by Bardenhagen and Triantafyllidis (1994) for a periodic lattice is generalized to a random network of bonds. The possible choices of the strain-gradient coefficient that ensure the existence of the localized deformation zones in equilibria are briefly discussed.

Finally, an alternative approach is proposed to introduce a length scale into the VIB model from the theory of mechanism-based strain gradient plasticity (Gao et al., 1999b; Huang et al., 2000a,b; Qiu et al., 2001b).

2 ONE-DIMENSIONAL MODEL

Triantafyllidis and Bardenhagen (1993) consider a one-dimensional chain of atoms connected by nonlinear springs. Each atom is connected to its nearest neighbour and to a number of next-to-nearest neighbours, via springs of length $p\varepsilon$, $p=1,2,...,q$. The chain has a finite length $L=N\varepsilon$, where $N+1$ is the number of atoms, or nodes. The energy densities of the springs are $w_p(e_p)$, where $p=1,2,...,q$. The strain in a spring of length $p\varepsilon$ attached to node i is given by

$$e_p^{i+} = \frac{u_{i+p} - u_i}{p\varepsilon} \quad \text{or} \quad e_p^{i-} = \frac{u_i - u_{i-p}}{p\varepsilon}. \tag{1}$$

Here e_p^{i+} and e_p^{i-} are the strains in the springs to the right and to the left of node i, respectively, and u_j denotes the displacement of the jth node. The force in a spring of length $p\varepsilon$ attached to node i is given by $f_p(e_p) \equiv w_p'(e_p)$.

The kinematic definitions (1) are valid for nodes that are sufficiently far from the ends of the chain, i.e., $0 \le i - q$ and $i + q \le N$. Otherwise, the following kinematic relations hold:

$$e_p^{i+} = \frac{2u_N - u_{2N-i} - u_i}{p\varepsilon} \quad \text{for} \quad i + p > N$$

$$e_p^{i-} = \frac{u_i + u_{p-i} - 2u_0}{p\varepsilon} \quad \text{for} \quad i - p < 0. \tag{2}$$

These relations ensure the existence of a trivial equilibrium solution $\bar{u}_i$ of equal relative displacements: $\bar{u}_{i+1} - \bar{u}_i = const$.

The equation of equilibrium for an interior node i is then

$$\sum_{p=1}^{q} \left[f_p(e_p^{i+}) - f_p(e_p^{i-}) \right] = 0. \tag{3}$$

It is assumed that there is a sufficiently smooth continuous function $u(x)$ coinciding with all equilibrium displacements u_j at nodal points $x_j \equiv j\varepsilon$, and $\varepsilon/L = 1/N \ll 1$. One can then expand the strains $e_p^{i\pm}$ in Taylor series about x_i:

$$e_p^{i+} = \frac{u(x_{i+p}) - u(x_i)}{p\varepsilon} = u_x + \frac{1}{2}p\varepsilon u_{xx} + \frac{1}{6}(p\varepsilon)^2 u_{xxx} + \frac{1}{24}(p\varepsilon)^3 u_{xxxx} + \ldots$$

$$e_p^{i+} = \frac{u(x_{i+p}) - u(x_i)}{p\varepsilon} = u_x + \frac{1}{2}p\varepsilon u_{xx} + \frac{1}{6}(p\varepsilon)^2 u_{xxx} + \frac{1}{24}(p\varepsilon)^3 u_{xxxx} + \ldots, \quad (4)$$

where all derivatives of $u(x)$ are evaluated at $x=x_i$. After substituting (4) into (3) and subsequently expanding the result in terms of ascending powers of ε, one obtains the following equation:

$$\sum_{p=1}^{q} p \left\{ f_p'(u_x)u_{xx} + \frac{(p\varepsilon)^2}{6} \left[f_p'''(u_x)\frac{u_{xx}^3}{4} + f_p''(u_x)u_{xx}u_{xxx} + f_p'(u_x)\frac{u_{xxxx}}{2} \right] \right\} = 0,$$

$$(5)$$

where the term $O(\varepsilon^4)$ has been neglected. The macroscopic functions $W(u_x)$ and $h(u_x)$ are defined by the following relations to the discrete energy density $w_p(e_p)$:

$$W(u_x) \equiv \sum_{p=1}^{q} pw_p(u_x), \qquad h(u_x) \equiv \sum_{p=1}^{q} -\left(p^3/12\right)w_p''(u_x). \quad (6)$$

It is clear from (6) that $W(u_x)$ thus defined can be viewed as the continuum energy density. By combining (5) and (6), we obtain the following equilibrium equation for the continuum model:

$$\left[W'(u_x) - \varepsilon^2 u_{xxx} h(u_x) - \frac{\varepsilon^2}{2} u_{xx}^2 h'(u_x) \right]_x = 0. \quad (7)$$

It can be shown that (7) is the Euler-Lagrange equation for the following energy functional:

$$E = \int_0^L \left[W(u_x) + \frac{\varepsilon^2}{2} h(u_x)u_{xx}^2 \right] dx. \quad (8)$$

Minimizing this functional subject to the displacement boundary conditions

$$u(0) = 0, \qquad u(L) = d, \quad (9)$$

we obtain the natural boundary conditions

$$\varepsilon^2 u_{xx}(0) = 0, \qquad \varepsilon^2 u_{xx}(L) = 0. \quad (10)$$

The ε^2-terms in (7) and (10) are called the *couple stresses*.

It is known that one can define more than one continuum energy density for the same discrete model (Kunin, 1982; Triantafyllidis and Bardenhagen, 1993). In particular, if one derives the energy density directly from the discrete system via

$$\overline{W}(u_x) \equiv \frac{1}{2\varepsilon} \sum_{p=1}^{q} \left[p\varepsilon w_p\left(e_p^{i+}\right) + p\varepsilon w_p\left(e_p^{i-}\right) \right], \quad \text{one arrives at a different energy}$$

functional

$$\overline{E} = \int_0^L \left[W(u_x) - \frac{3\varepsilon^2}{2} h(u_x)u_{xx}^2 + \frac{\varepsilon^2}{6} \sum_{p=1}^{q} p^3 w'(u_x)u_{xxx} \right] dx. \quad (11)$$

However, it can be shown that the two functional (8) and (11) differ by a null Lagrangian, i.e., a functional whose Euler-Lagrange equation is identically zero. Due to its simpler form, the energy functional (8) is usually preferred.

3 DISPERSION RELATION

A dispersion relation is derived in this section for the atomic chain under consideration. The density of the system is defined by $\rho \equiv M/L = (N+1)m/(N\varepsilon)$. Here M and m are the total mass of the system and the mass of each atom, respectively (assuming massless springs). Then the linear momentum balance can be written as

$$\frac{N\varepsilon}{N+1}\rho\ddot{u}_j = \sum_{p=1}^{2}\left[f_p\left(e_p^{i+}\right) - f_p\left(e_p^{i-}\right)\right], \tag{12}$$

where $e_p^{i\pm}$ are defined in (1), $\ddot{u}_j$ denotes the second time derivative of the displacement u_j, and for simplicity the number of neighbours $q=2$. The boundary conditions are

$$u_0(t) = 0, \quad u_N(t) = d. \tag{13}$$

The equilibrium equation (3) has a uniform strain (trivial) solution

$$\bar{u}_j(t) = j\varepsilon d/L \tag{14}$$

for each value of d in (13). The linearization of (12) about (14) gives

$$\frac{N\varepsilon}{N+1}\rho\ddot{v}_j = w''\left(\frac{d}{L}\right)\frac{v_{j+p} - 2v_j + v_{j-p}}{p\varepsilon}, \tag{15}$$

where $v_j \equiv u_j - \bar{u}_j$. The boundary conditions for (15) are $v_0(t) = v_N(t) = 0$.

If one seeks solutions to (15) in the form of plane waves $v_j(t) = exp(i\omega t)sin(kj\varepsilon)$, the following discrete dispersion relation is obtained:

$$\omega^2 = \frac{4(N+1)}{\rho N\varepsilon^2}\left(\beta_1 sin^2\frac{k\varepsilon}{2} + \frac{\beta_2}{2}sin^2 k\varepsilon\right), \tag{16}$$

where $\beta_p = f_p'(d/L)$ $(p=1,2)$ are the moduli of the nearest and near-to-nearest interaction potentials. For $k\varepsilon <<1$, the Taylor series of (16) gives

$$\omega^2 = \frac{N+1}{\rho N}\left[k^2(\beta_1 + 2\beta_2) - \frac{k^4}{12}(\beta_1 + 8\beta_2)\varepsilon^2 + o(\varepsilon^2)\right]. \tag{17}$$

On the other hand, the continuum model derived in the previous section results in the following linearized dynamic equation:

$$\begin{aligned}\rho v_{tt} &= W''(d/L)v_{xx} - \varepsilon^2 h(d/L)v_{xxxx} \\ v(0,t) &= v(L,t) = v_{xx}(0,t) = v_{xx}(L,t) = 0\end{aligned}. \tag{18}$$

If one seeks a plane-wave solution of (18), the dispersion relation is obtained

$$\omega^2 = \frac{1}{\rho}\left[k^2(\beta_1 + 2\beta_2) - \frac{k^4}{12}(\beta_1 + 8\beta_2)\varepsilon^2\right] \tag{19}$$

for the continuum model, with the aid of (6). The relation (19) agrees with the expanded dispersion relation (17) for the discrete model up to a factor of $(N+1)/N$ which tends to one at the continuum limit $N \to \infty$.

The relation (19) implies that the trivial solution $u(x) = dx/L$ is stable as long as

$$\beta_1 + 2\beta_2 - \frac{k^2}{12}(\beta_1 + 8\beta_2)\varepsilon^2 > 0, \tag{20}$$

where $\beta_p = w_p''(d/L)$. If one considers only nearest-neighbour interactions [$\beta_1 = w_1''(d/L), \beta_2 = 0$], the condition (20) with $\beta_1 > 0$ is reduced to

$$1 - \frac{k^2}{12}\varepsilon^2 > 0 \tag{21}$$

which implies that the high-k modes ($k > \sqrt{12}\varepsilon$) will result in instability of the trivial state even though the modulus β_1 is positive. This instability is of pure nonphysical nature since the continuum approximation, from which the stability condition (2) is derived, requires that $k\varepsilon \ll 1$ in the discrete model, and therefore the mode numbers satisfying (21) should not be admissible. So the instability of the trivial state is a result of extrapolating the continuum model derived for low modes onto the higher modes for which the model is, strictly speaking, not applicable. If the next-to-nearest neighbour interactions are included with $\beta_2 > 0$, the trivial state in the strain-gradient model looses its stability at even lower k; the trivial state is unstable at k higher than

$$k_{cr} = \frac{\sqrt{12}}{\varepsilon}\sqrt{\frac{\beta_1 + 2\beta_2}{\beta_1 + 8\beta_2}}. \tag{22}$$

To fix this problem and stabilize the trivial state for $\beta_1 > 0$ in the strain-gradient model, Triantafyllidis and Bardenhagen (1993) propose to include the near-to-nearest-neighbour interactions into the model and choose their pair potential $w_2'(e)$ in such a way as to ensure that

$$\beta_1 + 8\beta_2 < 0. \tag{23}$$

In conjunction with (6), the condition (23) is equivalent to $h(d/L) > 0$. This requires that β_2 be negative. While this is plausible for the next-to-nearest-neighbour interactions in case of large interatomic distances (so-called "improper crystals"), it is by no means always the case. In addition, this choice results in the well-known "wrong sign problem" in the dispersion relation. At small wave numbers k the squared-frequency ω^2 increases with k for both the discrete model and strain gradient model. At large k, however, ω^2 in the discrete model reaches a maximum and then starts to decrease, while the strain-gradient model predicts that ω^2 will always increase. Hence the strain-gradient theory with moduli satisfying (23) restores the stability of the trivial state (as long as the effective modulus

$\beta_1+2\beta_2$ is positive) at the expense of a wrong qualitative behavior at large wave numbers (where the model derived by Taylor expansion is actually not applicable).

There are alternative remedies for the problem of loss of stability of the trivial state. For example, Kunin (1982) suggests using other (nonpolynomial) approximations of the dispersion relation. In particular, in case of the nearest-neighbour interactions with elastic modulus β one can use functions

$$\Phi(k) = 2\beta l^{-2}\left[1 - exp\left(-\frac{l^2 k^2}{2}\right)\right] \tag{24}$$

and

$$\Phi(k) = 6\beta l^{-2}\left[1 - \frac{sin(lk)}{lk}\right] \tag{25}$$

to approximate the dispersion relation over a wide range of wave numbers. Here l is a scale parameter depending on the curvature of the dispersion curve at small k. Unlike polynomial expansions, these approximations give a qualitatively correct behavior at large wave numbers and predict stability of the trivial state at positive β with respect to all modes, even though the dispersion relations are still significantly different from that of the discrete model. Moreover, such approximations result in nonlocal models which are hard to analyze.

Another approach is to adopt a quasicontinuum approximation of the dynamics of the discrete chain with only nearest-neighbour interactions by using (4) (Rosenau, 1986). This results in

$$\rho u_{tt} = L_D f'[u_x(x,t)], \tag{26}$$

where $L_D \equiv L_A D$, $L_A \equiv 1+\frac{1}{12}\varepsilon^2 D^2 + \ldots$ and $D \equiv \frac{\partial}{\partial x}$. The operator L_A is then inverted, with

$$L_A^{-1} \equiv 1 - \frac{1}{12}\varepsilon^2 D^2 + \frac{1}{240}\varepsilon^4 D^4 + \ldots, \tag{27}$$

and (26) becomes (up to the second order of ε):

$$\rho u_{tt} = f'[u_x(x,t)] + \frac{1}{12}\rho\varepsilon^2 u_{xxtt}. \tag{28}$$

The second term expresses the leading effects of the discreteness in the continuum model. One can also derive (28) by using the Lagrangian density

$$L = \frac{1}{2}\rho u_t^2 - f(u_x) + \frac{1}{24}\rho\varepsilon u_{xt}^2. \tag{29}$$

The third term in (29) gives rise to the second term in (28) and contains the discreteness effects. It represents an interesting mixture of kinetic and potential energy. This term is sometimes called microkinetic energy (Theil and Levitas, 2000) because it takes into account microscopic fluctuations which can be interpreted as heat. The dispersion relation for (28) is

$$\omega^2 = \frac{k^2}{1+\frac{1}{12}\varepsilon^2 k^2}. \tag{30}$$

This relation is compared to the ones resulting from the discrete model and its polynomial approximation. While both approximations start deviating from the discrete model at the same value of k, the behavior of (30) is clearly more physical (ω^2 tends to a constant and therefore is bounded as k approaches infinity) than the polynomial approximation with next-to-nearest-neighbour interactions satisfying (23). Unlike the polynomial approximation with positive moduli, this model does not result in the unphysical loss of stability of the trivial solution since ω^2 is always positive. The beauty of this model is that it results in a linear additional term and, unlike the strain-gradient models, it does not require extra boundary conditions. However, the discreteness effects in this model disappear in the statics case, which brings the problems that arise in statics in some nonlinear models.

4 STRAIN-GRADIENT TERM AS A REGULARIZATION

It is recalled that the idea of using the higher-order-gradient continuum theory is partially motivated by the problems with only the nearest-neighbour interactions in which the equilibrium equation loses its ellipticity. In these problems the stress-strain curve $\sigma(u_x) = W'(u_x)$ (in the one-dimensional case considered here) is non-monotonic. For example, in nonlinear models of fracture, the stress increases from zero to some maximum in the interval of strains $[0, e_m]$ and then decreases to zero as strain tends to infinity. At the critical point $u_x = e_m$ one has $W''(e_m) = 0$ and the equilibrium equation $W'(u_x) = constant$ changes type. This results in equilibria with discontinuity strain. It can be shown that in the absence of the strain-gradient term in the energy functional (8), the energy-minimizing equilibria at the loading exceeding the critical strain e_m will have discontinuous displacement fields (hence δ-functions in strain) (Truskinovsky, 1996). This causes the undesired mesh-size dependence of the finite-element computations mentioned in the introduction. The location of displacement discontinuities can be seen as cracks in a bar. The energy of the states with displacement discontinuities, or the *fracture energy*, is zero in this case, and any number of cracks may form.

The introduction of strain-gradient term as in (8) with $h(u_x) > 0$ [or equivalently, condition (23) satisfied at each d] ensures that the equilibrium equation (7) with nonzero ε is elliptic despite the fact that $W(u_x)$ is still nonconvex. This is because the ellipticity is determined by the sign of the highest-order term, which is now always positive. The ellipticity of the equilibrium equation, in turn, results in replacing the δ-functions in strain field by the strain localization zones whose size scales with the interatomic distance ε. A positive $h(u_x)$ also ensures the stability of the trivial state in the interval $[0, e_m]$, as seen in the previous section. In fact, it can be shown that the trivial state will be stable in a small interval of strains beyond the critical deformation (Triantafyllidis and Bardenhagen, 1993).

To demonstrate these ideas, the following potential suggested by Triantafyllidis and Bardenhagen (1993) is selected:

$$w_1(e) = -7 + e + 6\ln(1+e) + \frac{7}{1+e}, \qquad w_2(e) = -\frac{1}{2}\left(-1 + e + \frac{1}{1+e}\right). \qquad (31)$$

The corresponding force $f_2(e) = w_2'(e)$ for the next-to-nearest-neighbour interactions decreases monotonically with e. The resulting negative $w_2''(e)$ at each e restores the ellipticity of the equilibrium equation. It can be shown that

$$W(e) = 6\left[\ln(1+e) - \frac{e}{1+e}\right], \qquad h(e) = \frac{1}{2}\frac{e}{(1+e)^3}. \qquad (32)$$

The maximum stress (cohesive strength) occurs at $e_m = 1$ where $W''(e) = 6(1-e)/(1+e)^3$ vanishes and the equilibrium equation (7) loses ellipticity when $\varepsilon = 0$. The choice of w_2 in (31) ensures that $h(u_x)$ is positive and ellipticity of (7) is restored when ε is nonzero.

The trivial solution $u(x) = dx/L$ obviously satisfies (7), (9) and (10) for any value of the end displacement d. This solution becomes unstable in the softening part of the stress-strain curve ($u_x > e_m$). It can be shown that there is also a countable infinite number of nontrivial solution branches with strain localization bifurcating from the trivial one. Along a nontrivial branch, the size of the strain localization zone is on the order of ε. In addition to introducing a length scale, this model predicts a nonzero fracture energy which decreases and approaches to zero with ε. It is recalled that, when $\varepsilon = 0$, the fracture energy vanishes. For a nonvanishing ε, both fracture energy and the size of strain-localization zone in the finite-element computation become mesh-independent. Only one-crack solutions are stable at $d/L > e_m$ in absence of imperfections, in agreement with the behavior of the discrete chain (Truskinovsky, 1996) and cohesive-type models with monotonically decreasing cohesive energy (Del Piero and Truskinovsky, 1998).

It is also observed that $h(u_x)$ in (32) decreases rapidly as the strain grows and thus the contribution of strain-gradient terms for localized strain solutions is very small.

In summary, selecting the negative-moduli springs for the next-to-nearest neighbours to ensure $h(u_x) > 0$ restores ellipticity and thus regularizes the model. There are many phenomenological models (e.g., Triantafyllidis and Aifantis, 1986; Leroy and Molinari, 1993) in which the strain gradient term is postulated to be positive-definite for the same reason. On the other hand, this same choice results in a nongeneric dispersion relation, as seen in the previous section.

5   GENERALIZATION OF THE VIRTUAL-INTERNAL-BOND MODEL TO INCORPORATE THE STRAIN-GRADIENT TERMS

The higher-order gradient theory is introduced in higher dimensions in this section by generalizing the Virtual-Internal-Bond (VIB) model (Gao and Klein, 1998; Klein and Gao, 1998, 2000; Zhang et al., 2001) to incorporate the higher-order-gradient terms. While a strain-gradient theory has been derived for periodic lattices

(Bardenhagen and Triantafyllidis, 1994), the following derivation is more general since it involves a less ordered network of bonds.

A network of randomly distributed bonds is considered. A bond of initial length l_0 lies in the direction $\boldsymbol{\xi}$ (not necessarily a unit vector). The current length of the bond is l, and the stretch is $\lambda = l/l_0$. The bond energy is denoted by $U(l)$, and $w(\lambda) = U(l_0\lambda)$. One end of the bond has the coordinates X in the undeformed configuration, while the other end is $X + \varepsilon x$. Here ε is a characteristic length scale, for example, the smallest bond size. Hence the initial bond length is

$$l_0 = \varepsilon \|\boldsymbol{\xi}\|. \tag{33}$$

The deformation function and deformation gradient are denoted by $\chi(X)$ and $F(X) \equiv \partial \chi / \partial X$, respectively. The ends of the deformed bond have coordinates $x = \chi(X)$ and $\chi(X + \varepsilon \boldsymbol{\xi})$. The stretch of the bond is given by

$$\lambda(\varepsilon) = \frac{l}{l_0} = \frac{\|\chi(X + \varepsilon \boldsymbol{\xi}) - \chi(X)\|}{\varepsilon \|\boldsymbol{\xi}\|}. \tag{34}$$

For $\varepsilon/L << 1$ where L is a characteristic size of the material specimen, one can expand $\lambda(\varepsilon)$ in Taylor series about X, with ε being a small parameter:

$$\lambda(\varepsilon) = \frac{1}{\|\boldsymbol{\xi}\|} \left\{ \begin{array}{l} \|F\boldsymbol{\xi}\| + \varepsilon \dfrac{F\boldsymbol{\xi} \cdot \nabla F \boldsymbol{\xi}\boldsymbol{\xi}}{2\|F\boldsymbol{\xi}\|} \\[2mm] + \varepsilon^2 \left[\dfrac{\nabla F \boldsymbol{\xi}\boldsymbol{\xi} \cdot \nabla F \boldsymbol{\xi}\boldsymbol{\xi}}{8\|F\boldsymbol{\xi}\|} - \dfrac{(F\boldsymbol{\xi} \cdot \nabla F \boldsymbol{\xi}\boldsymbol{\xi})^2}{8\|F\boldsymbol{\xi}\|^3} + \dfrac{F\boldsymbol{\xi} \cdot \nabla\nabla F \boldsymbol{\xi}\boldsymbol{\xi}\boldsymbol{\xi}}{6\|F\boldsymbol{\xi}\|} \right] + o(\varepsilon^2) \end{array} \right\}. \tag{35}$$

The bond potential $w[\lambda(\varepsilon)]$ can also be expanded in powers of ε:

$$w[\lambda(\varepsilon)] = w\left(\frac{\|F\boldsymbol{\xi}\|}{\|\boldsymbol{\xi}\|} \right) + \varepsilon w'\left(\frac{\|F\boldsymbol{\xi}\|}{\|\boldsymbol{\xi}\|} \right) \frac{F\boldsymbol{\xi} \cdot \nabla F \boldsymbol{\xi}\boldsymbol{\xi}}{2\|F\boldsymbol{\xi}\|\|\boldsymbol{\xi}\|}$$

$$+ \varepsilon^2 \left\{ \begin{array}{l} w''\left(\dfrac{\|F\boldsymbol{\xi}\|}{\|\boldsymbol{\xi}\|} \right) \dfrac{(F\boldsymbol{\xi} \cdot \nabla F \boldsymbol{\xi}\boldsymbol{\xi})^2}{8\|F\boldsymbol{\xi}\|^2\|\boldsymbol{\xi}\|^2} - w'\left(\dfrac{\|F\boldsymbol{\xi}\|}{\|\boldsymbol{\xi}\|} \right) \dfrac{(F\boldsymbol{\xi} \cdot \nabla F \boldsymbol{\xi}\boldsymbol{\xi})^2}{8\|F\boldsymbol{\xi}\|^3\|\boldsymbol{\xi}\|} \\[3mm] + w'\left(\dfrac{\|F\boldsymbol{\xi}\|}{\|\boldsymbol{\xi}\|} \right) \dfrac{\nabla F \boldsymbol{\xi}\boldsymbol{\xi} \cdot \nabla F \boldsymbol{\xi}\boldsymbol{\xi}}{8\|F\boldsymbol{\xi}\|\|\boldsymbol{\xi}\|} + w'\left(\dfrac{\|F\boldsymbol{\xi}\|}{\|\boldsymbol{\xi}\|} \right) \dfrac{F\boldsymbol{\xi} \cdot \nabla\nabla F \boldsymbol{\xi}\boldsymbol{\xi}\boldsymbol{\xi}}{6\|F\boldsymbol{\xi}\|\|\boldsymbol{\xi}\|} \end{array} \right\}$$

$$+ o(\varepsilon^2) \tag{36}$$

The continuum energy density can be defined following (6) as the average of the bond potential $w[\lambda(\varepsilon)]$ using (36) and neglecting the terms of third and higher orders in ε:

$$\overline{W}(F, \nabla F, \nabla\nabla F) \equiv \langle w[\lambda(\varepsilon)] \rangle, \tag{37}$$

where the average is defined as

$$\langle ... \rangle = \iiint (...) D(l_0, \theta, \phi) \sin\theta \, dl_0 \, d\theta \, d\phi, \tag{38}$$

$D(l_0, \theta, \phi)$ is the bond density function and θ and ϕ are the angles in the spherical coordinates. For example, the bond density function for the nearest-neighbour and next-to-nearest-neighbour interactions has the form

$$D(l_0, \theta, \phi) = [\delta(l_0 - \varepsilon) + \delta(l_0 - 2\varepsilon)]\hat{D}(\theta, \phi). \qquad (39)$$

In this case the bonds have initial lengths ε and 2ε. It should be emphasized, however, that the theory in general permits considering network of bonds with different initial lengths.

As a result of the averaging procedure, one obtains the following expression for the continuum strain energy density per unit referential volume:

$$\overline{W}(F, \nabla F, \nabla\nabla F) = W(F) - \frac{\varepsilon^2}{2} \left[\begin{array}{c} 3B_{iklmnr}(F)(\nabla F)_{ikl}(\nabla F)_{mnr} \\ + C_{ik\,ln}(F)(\nabla\nabla F)_{ik\,ln} \end{array} \right], \qquad (40)$$

where $W(F) = \langle w(\overline{\lambda}) \rangle$ is the nearest-neighbour energy density, $\overline{\lambda} = \dfrac{l}{l_0} = \dfrac{\|F\xi\|}{\|\xi\|}$, and the tensors $B(F)$ and $C(F)$ are given in component form by

$$B_{iklmnr}(F) = -\left\langle \left\{ \frac{1}{\overline{\lambda}^2 \|\xi\|^4} \left[w''(\overline{\lambda}) - \frac{w'(\overline{\lambda})}{\overline{\lambda}} \right] F_{ij}\xi_j F_{mp}\xi_p + \frac{w'(\overline{\lambda})\delta_{im}}{\overline{\lambda}\|\xi\|^2} \right\} \frac{\xi_k \xi_l \xi_n \xi_r}{12} \right\rangle \qquad (41)$$

and

$$C_{ik\,ln}(F) = -\left\langle \frac{w'(\overline{\lambda})}{3\overline{\lambda}\|\xi\|^2} F_{ij}\xi_j \xi_k \xi_l \xi_n \right\rangle. \qquad (42)$$

A point symmetry has been assumed when obtaining (40), i.e., for each bond in the direction ξ there is another in the direction $-\xi$, with the same potential such that the ε-terms cancel out after averaging.

The energy density (40) now depends not only on the deformation gradient but also on its first and second gradients. Equations (40)-(42) reduce to the expressions derived by Bardenhagen and Taiantafyllidis (1994) in the special case of a crystal lattice.

Similar to the one-dimensional case, it can be shown that there is another way to define energy density, namely,

$$\overline{W}(F, \nabla F) = W(F) + \frac{\varepsilon^2}{2} B_{iklmnr}(F)(\nabla F)_{ikl}(\nabla F)_{mnr}; \qquad (43)$$

the energy functionals with (43) and (40) differ by a null Lagrangian. This can be verified (Bardenhagen and Triantafyllidis, 1994) using the fact that

$$C_{\alpha jkl}, \qquad B_{\alpha jk\beta mn} = \frac{1}{4}\frac{\partial C_{\alpha jkl}}{\partial F_{\beta n}}, \qquad \frac{\partial B_{\alpha jk\beta mn}}{\partial F_{\gamma p}} \qquad (44)$$

are symmetric with respect to any interchange of Latin or Greek indices. Due to its simpler form and no dependence on the third deformation gradient, (43) is preferred here. However, it is noted that both energy densities give the same equilibrium equation (Bardenhagen and Triantafyllidis, 1994):

$$\frac{\partial}{\partial X_s}\left[\frac{\partial W}{\partial F_{rs}} - \frac{\varepsilon^2}{2}\left(\frac{\partial B_{ijklmn}}{\partial F_{rs}}\right)u_{i,jk}u_{l,mn} + 2B_{ijkrsn}u_{i,jkn}\right] = 0, \qquad r=1,2,3, \qquad (45)$$

where u_i are the components of the displacement field $\boldsymbol{u}$.

It is assumed that the equilibrium equations are strongly elliptic at $\boldsymbol{F}=\boldsymbol{I}$ when $\varepsilon=0$, i.e.,

$$\frac{\partial^2 W(\boldsymbol{I})}{\partial F_{ij}\partial F_{kl}}m_i n_j m_k n_l > 0 \qquad (46)$$

for any unit vectors $\boldsymbol{m}$ and $\boldsymbol{n}$, but lose ellipticity at some level of strain $\boldsymbol{F_c} \neq \boldsymbol{I}$. This means (Bardenhagen and Triantafyllidis, 1994) that there exist unit vectors $\boldsymbol{m^c}$ and $\boldsymbol{n^c}$ such that

$$\frac{\partial^2 W(\boldsymbol{I})}{\partial F_{ij}\partial F_{kl}}m_i^c n_j^c m_k^c n_l^c = 0. \qquad (47)$$

In order to restore the ellipticity of equilibrium equations using the strain-gradient theory ($\varepsilon \neq 0$), one has to check if there is a neighbourhood of the critical deformation $\boldsymbol{F_c}$ for which the strong ellipticity condition holds for $\hat{W}(\boldsymbol{F},\nabla\boldsymbol{F})$, there is $\delta > 0$ such that

$$B_{iklmnr}(\boldsymbol{F_c} + \boldsymbol{H})m_i m_m n_k n_n n_l n_r > 0 \qquad (48)$$

for any unit vectors $\boldsymbol{m}$ and $\boldsymbol{n}$ and for any second-order tensors $\boldsymbol{H}$ such that $\|\boldsymbol{H}\| \equiv (H_{ij}H_{ij})^{1/2} < \delta$. If the condition (48) is satisfied for some positive δ, it can be shown (Triantafyllidis and Aifantis, 1986) that a localized deformation equilibrium solution can be constructed, as long as the deformation gradient $\boldsymbol{F}(\boldsymbol{X})$ remains within the δ-neighbourhood at all points $\boldsymbol{X}$.

The ellipticity condition (48) is clearly not satisfied if the short and long-range interaction potentials are the same for all bonds. This is analogous to the one-dimensional case considered in the previous section, where such a choice results in violating the condition (23) and thus loss of ellipticity of the higher-order equilibrium equations. The ellipticity domain of $\boldsymbol{B}$ in two dimensions for a perfect crystal model with a finite-range potential that is the same for all bonds has been studied by Bardenhagen and Triantafyllidis (1994) and no neighbourhood of $\boldsymbol{F_c}$ for which $\hat{W}(\boldsymbol{F},\nabla\boldsymbol{F})$ is strongly elliptic has been found.

Hence in order to restore the ellipticity, one has to pick different short and long-range interaction potentials, e.g., with bond density (39) one may choose

$$w(\lambda) = w_1(\lambda), l_0 = \varepsilon, \qquad w(\lambda) = w_2(\lambda), l_0 = 2\varepsilon. \qquad (49)$$

While the first potential w_1 comes from the physics of the problem at hand, the second potential w_2 must be chosen in such a way as to satisfy (48). Such a choice is clearly not easy to make since the ellipticity condition (48) has to be satisfied for all deformations in a certain neighbourhood and $\boldsymbol{B}$ given by (41) depends on the deformation tensor and bond coordinates not only through the bond potentials, unlike the one-dimensional case.

Since the choice of w_2 is motivated only by the desire to satisfy the ellipticity condition (48) and is not physically relevant for most crystals, one may as well consider a purely phenomenological strain-gradient coefficient $B(F)$, as has been already done in the literature. For example, if the material is isotropic, one may choose (Leroy and Molinari, 1993)

$$B_{iklmnr}(F) = c(I, II)\delta_{im}\delta_{kn}\delta_{lr} , \qquad (50)$$

where I and II are the first and second invariants of the right Cauchy-Green tensor $F^T F$. Any positive function $c(I, II)$ in (50) will satisfy the ellipticity condition (48) for any H. Due to the lack of information about the constitutive law $c(I, II)$ for real materials, it has been chosen to be a constant (e.g., Triantafyllidis and Aifantis, 1986; Leroy and Molinari, 1993). This choice, however, is not good for modelling localized deformations occurring in fracture. In this case the strain gradient ∇F rapidly grows as the maximum deformation gradient in a localized zone increases without bound. With $c(I, II) = const$, this growth is penalized through the strain-gradient term in the energy functional, no matter how small is the constant strain-gradient coefficient. This penalization increases with increasing strain gradient. Thus, a uniform strain solution eventually becomes energetically more favourable as we increase the external loading. To avoid this non-physical situation, one must choose a function $c(I, II)$ decreasing as I and II grow due to the increased F. The function should decrease faster than the term $\nabla F_{ikl}\nabla F_{mnr}$ in (43) with B_{iklmnr} given by (50), thus making the energy penalization smaller as the deformation gradient grows.

6 INTRODUCTION OF AN INTERNAL MATERIAL LENGTH VIA THE THEORY OF MECHANISM-BASED STRAIN GRADIENT PLASTICITY

The internal material length introduced in the previous sections for the elastic strain-gradient VIB model is essentially the bond length, which is on the order of atomic spacing ($10^{-10} m$) for crystalline materials and is many orders of magnitude smaller than any representative lengths in engineering problems. An alternative way to introduce an internal material length into the VIB model is via the theory of mechanism-based strain gradient (MSG) plasticity. The internal material length in MSG plasticity is on the order of $10^{-6} m$, which is four orders of magnitude larger than the atomic spacing and is much closer to the characteristic length for fracture nucleation in engineering materials.

Microscale: $\tilde{\varepsilon}, \tilde{\sigma}$

Mesoscale: $(\varepsilon, \sigma) \quad (\eta, \tau)$

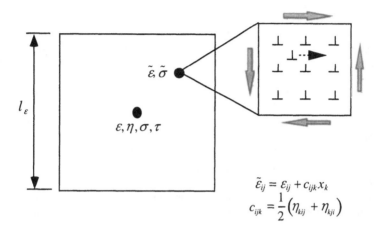

$$\tilde{\varepsilon}_{ij} = \varepsilon_{ij} + c_{ijk} x_k$$

$$c_{ijk} = \frac{1}{2}\left(\eta_{kij} + \eta_{kji}\right)$$

Figure 1 A schematic diagram of the multiscale framework to link the theory of mechanism-based strain gradient (MSG) plasticity to the Taylor dislocation model.

The mesoscale continuum theory of MSG plasticity is established from the microscale Taylor dislocation model (Taylor, 1938) via a multiscale, hierarchical framework shown in Fig. 1 (see Gao et al., 1999b; Huang et al., 2000a,b; Qiu et al., 2001b for infinitesimal deformation; Hwang et al, 2001 for finite deformation). The density of geometrically necessary dislocations is related to the gradient of plastic strain (e.g., Ashby, 1970), while the density of statistically stored dislocations is determined via the uniaxial stress-strain curve of the solid (Nix and Gao, 1998). In conjunction with the Taylor dislocation model, the flow stress in MSG plasticity is given by

$$\sigma = \sqrt{\left[\sigma_{ref} f\left(\varepsilon^p\right)\right]^2 + 18\alpha^2 \mu^2 b \eta^p} = \sigma_{ref} \sqrt{\left[f\left(\varepsilon^p\right)\right]^2 + l^{MSG}\eta^p} \,, \tag{51}$$

where $\sigma = \sigma_{ref} f\left(\varepsilon^p\right)$ is the uniaxial stress-strain curve, σ_{ref} is a reference stress (e.g., yield stress) in uniaxial tension, ε^p and η^p are the effective plastic strain and plastic strain gradient, respectively, μ the shear modulus, b the Burgers vector (~ atomic spacing), and α the empirical coefficient between *0.1* and *0.5*. The internal material length in MSG plasticity has been identified from (51) as

$$l^{MSG} = 18\alpha^2 \left(\frac{\mu}{\sigma_{ref}}\right)^2 b. \tag{52}$$

This internal material length in MSG plasticity represents a natural combination of the effects of elasticity (shear modulus μ), plasticity (reference stress or yield stress σ_{ref}) and atomic nature of solids (Burgers vector b), and is on the order of microns for typical metallic materials ($b\sim0.2nm$, $\mu\sim100\sigma_{ref}$). On the micron and submicron scales, the MSG plasticity theory has shown excellent agreements with the micro-indentation hardness experiments of various metallic materials (Huang et al., 2000b; Qiu et al., 2001a; Saha et al., 2001), the micro-bend and micro-twist tests (Gao et al., 1999a), and particle-reinforced composite materials (Xue et al., 2001). Once the characteristic length of deformation becomes much larger than the internal material length in (52), the strain gradient effect disappears and the MSG plasticity theory naturally degenerates to the classical plasticity theory.

Incorporating plasticity into the hyperelastic response of the VIB model begins with a multiplicative decomposition of the deformation gradient introduced by Lee (1969) and advanced by Pierce et al. (1983). The total deformation gradient is separated into successive mappings as given by

$$F=F^eF^p, \tag{53}$$

where F^e is the elastic deformation gradient which governs the elastic stretching of the lattice, and F^p is the isochoric mapping due to plastic deformation related to dislocation activities. In the elastic-plastic strain-gradient VIB model, the elastic deformation gradient F^e replaces the total deformation gradient F in the elastic VIB model (Gao and Klein, 1998; Klein and Gao, 1998, 2000; Zhang et al., 2001) and therefore links the bond potential to the elastic strain energy density stored in the lattice. On the other hand, the plastic deformation gradient F^p and its spatial gradient ∇F^p have been related to the effective plastic strain ε^p and effective plastic strain gradient η^p, respectively, and therefore are linked to the flow stress in (51) and the stress tensor (e.g., second Piola-Kirchhoff stress) in the MSG plasticity theory (Hwang et al., 2001). However, unlike the elastic strain-gradient VIB model which gives an explicit relation between stresses and strains in terms of the bond potential as in the previous sections, the elastic-plastic strain-gradient VIB model leads to a fully implicit constitutive model which has to be solved numerically.

ACKNOWLEDGMENTS

H.G. and Y.H. gratefully acknowledge insightful discussions with Dr. E.P. Chen from Sandia National Laboratories and the support from *Life Cycle Engineering Program*, NSF (grants #CMS-9979717 and #9983779, respectively).

REFERENCES

Aifantis, E.C., 1984, On the microstructural origin of certain inelastic models. *Journal of Engineering Materials and Technology*, **106**, pp. 326-330.

Ashby, M.F., 1970, The deformation of plastically non-homogeneous alloys. *Philosophical Magazine*, **21**, pp. 399-424.

Bardenhagen, S. and Triantafyllidis, N., 1994, Derivation of higher order gradient continuum theories in 2,3-D non-linear elasticity periodic lattice models. *Journal of the Mechanics and Physics of Solids*, **42**, pp. 111-139.

Coleman, B.D., 1983, Necking and drawing in polymeric fibers under tension. *Archives of Rational Mechanics Analysis*, **83**, pp. 115-137.

Coleman, B.D. and Hodgdon, M.L., 1985, On shear bands in ductile materials. *Archives of Rational Mechanics Analysis*, **90**, pp. 219-247.

Del Piero, G. and Truskinovsky, L., 1998, A one-dimensional model for localized and distributed failure, *Journal de Physique IV*, **8**, pp. 95-102.

Fleck, N.A. and Hutchinson, J.W., 1998, A discussion of strain gradient plasticity theories, and application to shear bands. In *Material Instabilities in Solids*, edited by De Borst, R. and Van Der Giessen, E. (Chichester: Wiley), pp. 507–519.

Gao, H. and Klein, P., 1998, Numerical simulation of crack growth in an isotropic solid with randomized internal cohesive bonds. *Journal of the Mechanics and Physics of Solids*, **46**, pp. 187-218.

Gao, H., Huang, Y. and Nix, W.D., 1999a, Modeling plasticity at the micrometer scale. *Naturwissenschaften*, **86**, pp. 507-515.

Gao, H., Huang, Y., Nix, W.D. and Hutchinson, J.W., 1999b, Mechanism-based strain gradient plasticity. I. – Theory. *Journal of the Mechanics and Physics of Solids*, **47**, pp. 1239-1263.

Huang, Y., Gao, H., Nix, W.D. and Hutchinson, J.W., 2000a, Mechanism-based strain gradient plasticity. II. – Analysis. *Journal of the Mechanics and Physics of Solids*, **48**, pp. 99-128.

Huang, Y., Xue, Z., Gao, H., Nix, W.D. and Xia, Z.C., 2000b, A study of micro-indentation hardness tests by mechanism-based strain gradient plasticity. *Journal of Materials Research*, **15**, pp. 1786-1796.

Hwang, K.C., Jiang, H., Huang, Y. and Gao, H., 2001, Finite deformation analysis of mechanism-based strain gradient plasticity: torsion and crack tip field. *International Journal of Plasticity* (in press).

Klein, P. and Gao, H., 1998, Crack nucleation and growth as strain localization in a virtual-bond continuum. *Engineering fracture mechanics*, **61**, pp. 21-48.

Klein, P. and Gao, H., 2000, Study of crack dynamics using the Virtual Internal Bond method. In *Multiscale Deformation and Fracture in Materials and Structures, the James R. Rice 60th Anniversary Volume*, edited by Chuang, T.-J. and Rudnicki, J.W. (Dordrecht, Kluwer Academic Publishers), pp. 275-309.

Kunin, I.A., 1982, *Elastic Media with Microstructure I (One-dimensional models)*, (New York: Springer-Verlag).

Lee, E.H., 1969, Elastic-plastic deformations at finite strain. *Journal of Applied Mechanics*, **36**, pp. 1-6.

Leroy, Y.M. and Molinari, A., 1993, Spatial patterns and size effects in shear zones: a hyperelastic model with higher gradients. *Journal of the Mechanics and Physics of Solids*, **41**, pp. 631-664.

Nix, W.D. and Gao, H., 1998, Indentation size effects in crystalline materials: a law for strain gradient plasticity. *Journal of the Mechanics and Physics of Solids*, **46**, pp. 411-425.

Pierce, D., Asaro, R.J. and Needleman, A., 1983, Material rate dependence and localized deformation in crystalline solids. *Acta Metallurgica et Materialia*, **31**, pp. 1951-1976.

Qiu, X., Huang, Y., Nix, W.D., Hwang, K.C. and Gao, H., 2001a, Effect of intrinsic lattice resistance in strain gradient plasticity (submitted for publication).

Qiu, X., Huang, Y., Wei, Y., Gao, H. and Hwang, K.C., 2001b, The flow theory of mechanism-based strain gradient plasticity (submitted for publication).

Rosenau, P., 1986, Dynamics of nonlinear mass-spring chains near the continuum limit. *Physics Letters A*, **118**.

Saha, R., Xue, Z., Huang, Y. and Nix, W.D., 2001, Indentation of a soft metal film on a hard substrate: strain gradient hardening effects. *Journal of the Mechanics and Physics of Solids* (in press).

Shi, M., Huang, Y. and Hwang, K.C., 2000, Plastic flow localization in mechanism-based strain gradient plasticity. *International Journal of Mechanical Sciences*, **42**, pp. 2115-2131.

Taylor, G.I., 1938, Plastic strain in metals. *Journal of Institute of Metals*, **62**, pp. 307-324.

Theil, F. and Levitas, V.I., 2000, A study of a Hamiltonian model for martensitic phase transformations including microkinetic energy. *Mathematics and Mechanics of Solids*, **5**, pp. 337-368.

Triantafyllidis, N. and Aifantis, E.C., 1986, A gradient approach to localization of deformation. I. hyperelastic materials. *Journal of Elasticity*, **16**, pp. 225-237.

Triantafyllidis, N. and Bardenhagen, S., 1993, On higher order gradient continuum theories in 1-D nonlinear elasticity. Derivation from and comparison to the corresponding discrete models. *Journal of Elasticity*, **33**, pp. 259-293.

Truskinovsky, L., 1996, Fracture as a phase transition. In *Contemporary Research in the Mechanics and Mathematics of Materials*, edited by Batra, R.C. and Beatty, M.F. (Barcelona: CINME), pp. 322-332.

Xue, Z., Huang, Y. and Li, M., 2001, Particle size effect in metallic materials: a study by the theory of mechanism-based strain gradient plasticity (submitted for publication).

Zhang, P., Klein, P.A., Huang, Y., Gao, H. and Wu, P.D., 2001, Numerical simulation of cohesive fracture by the virtual-internal-bond model. *Computer Modeling in Engineering and Sciences* (in press).

Simulation-Based Design Environment by Meshfree-Particle Methods

Wing Kam Liu[1]

1   INTRODUCTION

A multiscale meshfree particle method is developed, which includes recent advances SPH and other meshfree research efforts. Key features will include linear consistency, stability, and both local and global conservation properties. In addition, through the incorporation of standard moving least squares (MLS) enhancement and wavelet techniques, the method will have the flexibility of resolving multiple scales in the solution of complex, multiple physics processes. We present the application of this approach to the following areas: 1) simulations on propagation of dynamic fracture and shear band; 2) impact and penetration; 3) fluid dynamics and 4) nano-mechanics.

Meshfree methods are a new class of methods that offer the potential to treat difficult multi-scale problems, including problems of material failure, fracture, and fragmentation. Since failure and fracture processes are associated with damage evolution, it is important to choose the correct computational multi-physics material models. Generally, damage evolution and failure start from one or more local regions in a material, wherein a crack growth begins due to localized damage, thereby leading to damage concentrations that finally result in material fracture. Modeling the failure process can be idealized by studying the discrete energy transfer from the larger scales to the smaller scales. Meshfree methods can deal with the multi-scale phenomena inherent in failure of materials more naturally than finite elements, thus providing an added impetus to simulating these problems with realistic physical models. We expect that each scale can reproduce a particular physical phenomenon such as nucleation, high amplitude strain and damage concentration, or movement of dislocations. Therefore, major goals are to advance multi-scale meshfree methods for the modeling of material failure, and to link the micro-scale notion of dislocations, the meso-scale plastic strains and strain gradient, and the macro-scale continuum mechanics within a multi-scale framework. To the authors' knowledge, current simulation methods do not have multi-scale capabilities.

Methodologies can be developed for treating problems involving widely varying scales, such as shear band formation and localized deformations, which are important in the prediction of failure. To meet the demands of modern software,

[1] Northwestern University, Department of Mechanical Engineering
2145 Sheridan Road, Evanston, IL 60201, USA
Email: w-liu@northwestern.edu, webpage: http://www.tam.nwu.edu/wkl/liu.html

error estimators, hp adaptivity, multiresolution analysis, sampling approximations, and edge detection, among others should be developed. The proposed approach is fundamentally different from the traditional finite element methods. It could play a significant role in the next generation of discretization methods for fracture and fragmentation in proof-test simulations. The main reasons that promote this speculation, in our opinion, are twofold: the methods' ability to abandon the mesh generation and their flexibility to embed multi-physics of the specific problem into the interpolation function basis.

2 THE MULTISCALE MESHFREE PARTICLE METHOD

Despite its success in the analysis of geometric and material nonlinear behavior in structures and solids, the widely used finite element methods exhibit a number of shortcomings in handling problems involving large deformation, high gradients, or moving discontinuities such as cracks, fracture, fragmentation and multi-scale analysis. Recently, a new generation of numerical methods called "meshfree" methods has emerged and is now profoundly influencing almost every branch of engineering and the physical science. One of the most distinguished features about meshfree methods is that no explicit mesh is needed in the formulation.

As one part of this family, the Reproducing Kernel Particle Methods originally evolved from wavelet theory and SPH method. It has been applied successfully to a broad range of problems. In addition to SPH and wavelet theory, meshfree methods modifies the kernel function by introducing a correction function in order to enhance its accuracy near or on the boundary of the problem domain. Due to this correction function suggested by Liu, et al. (1995), the consistency condition is satisfied. Liu, et al. (1995) demonstrated the application of meshfree methods to structural dynamics, and the method was used successfully for large deformation simulations (Chen, et al. 1996, Liu, et al. 1996, Jun, et al. 1998, Liu, et al. 1998), computational fluid dynamics (Liu, et al. 1995). Through the flexibility of choice of window function and dilation parameter, meshfree methods can be used extensively in error estimation, image processing, edge detection and hp-adaptivity algorithms.

3 SIMULATION ON PROPOGATION OF DYNAMIC FRACTURE AND SHEAR BAND

In 1996, Zhou, Rosakis et al. (1996a) performed a set of experiments in which a high-speed cylindrical projectile impacted a pre-notched rectangular plate. Their major finding was that as the impact velocity increased, the mode of failure went from brittle (crack propagation) to ductile (dynamic shear band propagation), which contradicts the traditional school of thought. Others have conducted similar experiments, notably Mason et al (1994), Kalthoff and Winkler (1987), and Ravi-Chandar (1995).

The numerical simulations of the Zhou experiment have had mixed results. Some researchers, such as Zhou, Rosakis et al. (1996b) have been able to predict the dynamic shear band propagation, while others, notably Klein (1999) and Belytschko et al (1996) have been able to capture the correct crack initiation angle. The numerical results of Zhou and others have suffered from mesh dependence. Although both studies predict shear band propagation, the shear band is constrained to propagate in a straight line along the element boundaries. In contrast, experimental results show that the shear band actually propagates in a curved direction, and that the curvature is a function of the impact speed. Thus the ability to capture both the failure mode/transition and the dynamic shear band propagation remained an open problem.

These shortcomings motivated researchers at Northwestern to further analyze the Zhou and Rosakis experiment. Using RKPM, a meshfree numerical method, Li and Liu (2001) were able to predict for the first time:

- The ductile-to-brittle failure mode transition observed that is observed experimentally
- A curved shear band whose character changes as a function of the projectile impact speed (see Fig. 1)
- A periodic temperature profile matching that obtained experimentally (Rosakis, 2000)
- An intense, high strain rate region in front of the shear band tip, which we believe to cause a stress collapse that drives the formation and propagation of a dynamic shear band (see Fig. 3)
- A phenomenon called "temperature reflection" that can only be seen in 3D calculations. In this phenomenon, the temperature at the end of the plate has already risen before the shear band tip has reached that region, which is similar to the spalling phenomenon of a target/projectile problem.
- Multiple-scale decomposition and analysis of the shear band. In analyzing the shear band at different scales, one can first observe the macroscopic prediction of the curved shear band. Then by magnifying the shear band, one can gradually resolve the detail of the shear band to the point where the periodic temperature profile of the shear band can clearly be seen. (see Fig. 2)

Preliminary results related to dynamic shear band propagation can be found at: http://www.tam.nwu.edu/wkl/liu.html.

In most computer simulations of strain localization, shear band formation is the outcome of a bifurcated solution sought in numerical computations due to material instability. However, how to simulate the propagation of such material instability is still an open problem, and most earlier numerical simulations fail in simulating the propagation of shear band formation. The key technical ingredient in such simulations, we believe, is how to simulate the collapsing state of shear band formation. And it is found that the stress collapse in the newly formed localization zone (shear band tip) promotes or triggers the shear band's further propagating. To simulate stress collapse state inside the shear band, a so-called shear band damage model is introduced, which conforms to physicists' belief that instead of being a bifurcated mathematical solution a shear band is a physical entity, within which there is significant weakening, or changes in material properties which may even be identified as phase transformation. In order to simulate the collapsing state of shear band formation, a thermal-viscous Newtonian fluid damage model was adopted by Li and Liu (2001), similar to that proposed by Zhou et al (1996a). The choice of such a multi-physics model is necessitated by the many mechanisms which drive the formation and propagation of a shear band.

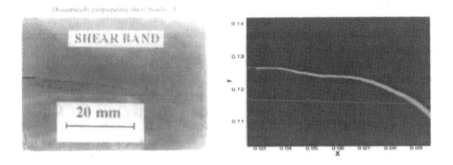

Figure 1 Comparison of experimental result with computation

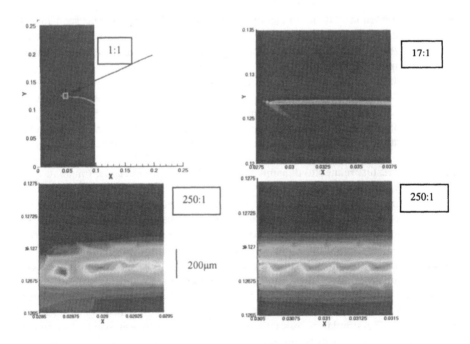

Figure 2 Multiple scale analysis of shear band temperature field

In addition, there are also a lot of new physics revealed from this simulation. Shown in Fig 3 is the distribution of the strain rate and a detail look at the structure. From the numerical results, we found that a self-similar strain rate field moves with the propagation of the shear band tip, this is quite similar to the self-similar singular stress field moving in front of a crack tip. We propose that the concentrated strain rate field in front of the shear band tip might be the driving force for the formation of shear band under high speed impact. Therefore, a close link between the strain rate field concentration and stress collapse must be established in the proposed modeling.

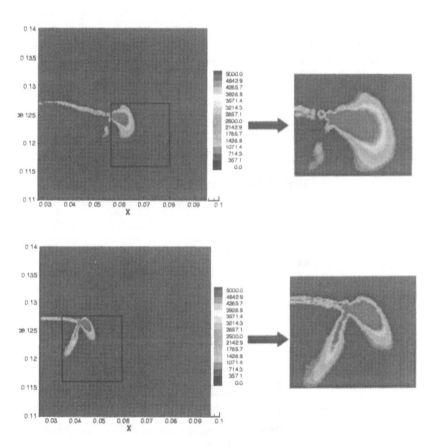

Figure 3 Strain rate contour and detail structure at 36 μs (top) and 48 μs (bottom) (impact speed v=37m/s). Note the appearance of a stress collapse zone at the shear band front.

4 IMPACT AND PENETRATION PROBLEMS

A dynamic, explicit, large deformation version of meshfree particle method has been also developed for three dimensions. To test it in a real world engineering application, it has been applied to a high-speed penetration simulation. The computation has been compared with experimental data obtained by the Army Water Station. The problem statement is given in Fig 4. Two cases have been studied: a single projectile case and a multiple projectile case. In both cases the projectile has an initial velocity more than Mach 5. The target is made of rebar reinforced concrete. A multi-scale damage constitutive law is used in the simulation. Figures 5a and b show damage contours for the single and multiple penetrator cases, respectively. Figure 6 shows the result of the computed penetrator depth compared with the experimental data. Note in 5a and b that the particles that have been displaced by the penetrator have been squashed, representing the degree of damage in the material element lumped at these particles. The mass of the material is thus preserved, and no artificial erosion algorithm is necessary. The complete MPFEM 3D large deformation formulation will be provided in the subsequent paper (Hao et.al.2001).

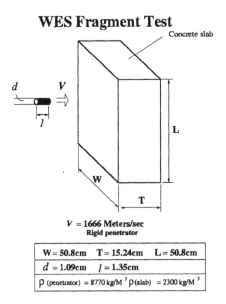

WES Fragment Test

$V = 1666$ **Meters/sec**
Rigid penetrator

W = 50.8cm	T = 15.24cm	L = 50.8cm
d = 1.09cm	l = 1.35cm	
ρ (penetrator) = 8770 kg/M³	ρ(slab) = 2300 kg/M³	

Figure 4 Problem statement for penetration simulation

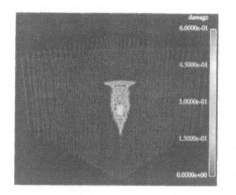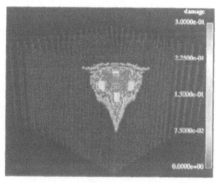

Figures 5a and 5b Damage contours for single and multiple penetrator case

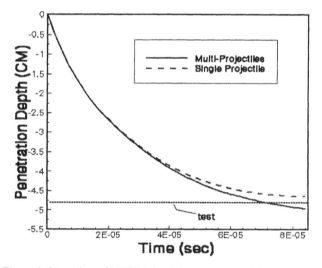

Figure 6 Comparison of MPFEM simulation to experimental result

5   FLUID DYNAMICS

A parallel computational implementation of the Reproducing Kernel Particle Method (RKPM) is used for 3-D implicit CFD analysis. Even though meshfree methods involve more computation time due to the shape functions, parallelization scheme enables us to solve large size problems in a reasonable amount of time. A uniform flow past a cylinder is simulated as an example problem. A cylinder with a diameter of 1.5cm is placed at 4.5cm upstream. The dimension of the computational domain is *21.5cm×14cm×4cm* as shown in Fig.7.

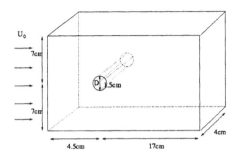

Figure 7

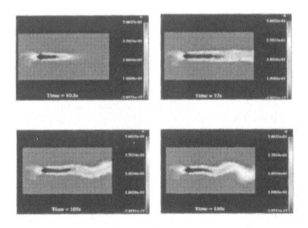

Figure 8

The domain is discretized using 2236 nodes and 11628 quadrature points. The discretization is finer around the cylinder than it is further away. The initital condition is the uniform flow speed far from the cylinder U_0 with a Reynolds number of 100. The inflow boundary condition is uniform flow of speed U_0, while the outflow is a zero-stress boundary. Strouhal number, is calculated to be approximately 0.174 with a vortex-shedding frequency of 0.5-0.6, for Reynolds number of 100. Velocity in the x-direction along the cross-sectional plane of z=2cm at different time steps can be seen in Fig. 8. The vectors for the velocity field are shown in Fig.9. The recirculation can be seen in the boxed area in the plots. The drag coefficients are plotted against different Reynolds Numbers that range from 10 to 1000 and are compared with the values obtained from experiments, FEM and RKPM without enrichment in Fig. 9. In the algorithm of RKPM without enrichment, the essential boundary is treated using the "corrected collocation" method. As shown on the plot, using the enrichment method to implement the essential boundary yields results closer to the experimental values than the one without, but both methods are more accurate than FEM. A model of the same geometry problem that has 15447 nodes and 87646 quadrature points was built for comparison purposes, shown in Fig.10. Our results indicate that even the coarse discretization is sufficiently fine to obtain accurate results using RKPM, while the FEM solution is not fully resolved even with the fine mesh. The parallel performance on the Cray T3E for this problem is shown in Fig.11. Although the calculation performance can not reach the ideal case of linear speedup, we are still be able to see the speedup when running the fixed-size problem using up to 32 processors.

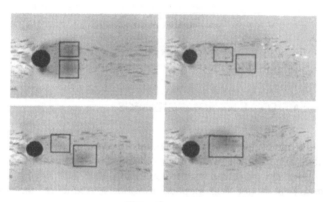

Figure 9

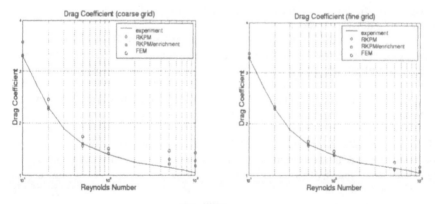

Figure 10

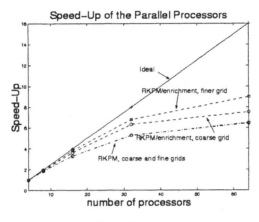

Figure 11.

6 COMPUTATIONAL NANOMECHANICS

Materials at the nanoscale have demonstrated impressive physical and chemical properties, thus suggesting a wide range of areas for application. For instance, carbon nanotubes are projected to be remarkably stiff and strong and conduct electricity and heat better than copper at room temperature, suggesting their eventual role in light weight, high-strength materials such as cabling and discontinuous fiber reinforced composites, and as both devices and nanowires in molecular electronics. They are under intensive study as an efficient storage medium for systems ranging from alkali ions for nanoscale power sources to hydrogen for fuel cells. Future possible application of nanotubes includes as accurate drug delivery systems.

The existence of fullerenes inside nanotubes has been observed as a product of different processes of synthesizing nanotubes. For instance, Smith et al (1998) used pulsed laser evaporation (PLV) of a catalyst-containing graphite target, Sloan et al (2000) and also Zhang et al (1999) who used arc vaporization of carbon with a mixed Ni/Y catalyst . Thus fullerenes inside of nanotubes is an experimental reality from work in the last few years, and there has been no reported modeling of these structures in the literature to our knowledge. One may speculate that fullerenes encapsulated inside of nanotubes could significantly affect the physical properties relative to the empty nanotube. Questions naturally arise about the geometries that the hybrid NT/fullerene systems adopt, the resulting elastic properties and energetics, and about certain issues related to the tribology of the system. In addition to achieving fundamental understanding, it is interesting to study such peapod structure as potential functional devices e.g., as nano-pistons, nano-bearing systems, nano-writing devices, or nano-capsule.

A modeling study of nanotubes filled with fullerenes is presented, with focus on a new approach. The objective is to investigate the mechanics of this nanotube/fullerene system, such as the elastic properties, energetics, and certain tribological issues. We have proposed a combined molecular dynamics/continuum approach with account of both the non-bonded and bonded interactions. The derived elastic properties from the Tersoff-Brenner potential are consistent with experimental observations. We have computed cases in which C60 is "injected" at high velocity into different nanotubes. Several important conclusions are made based on the simulations. In addition, we have computed the case where the C60 has zero velocity and is "sucked" into the (10,10) and (9,9) tubes from the sharp force gradient present from the deep attractive potential at the inlet of these tubes. Remarkably, the C60 oscillates back and forth in these nanotube pipes, because at the other end it experiences the same attractive force pulling it back into the tube so that it cannot escape, even though reaching a high velocity at the center of the tube. (The online animations are located at http://tam.mech.nwu.edu/wkl/c60_in_tube). The detail of the computational implementation is presented in a forthcoming paper (Qian, Liu and Ruoff, 2001).

7 CONCLUSIONS

The following goals have been achieved

1. A wavelet-based meshfree particle method is developed to form the foundation of multi-scale analysis.
2. Methodologies for linking micro-scale dislocations to the meso-scale damage models in multi-scale multi-physics methods is developed.
3. Enrichment methods based on length scale dependent theories have been incorporated into the proposed computational models.
4. Computational multi-physics methods will be developed to improve the predictions made for strain-softening behavior that generally coincide with most damage analyses.
5. Approaches to efficiently distribute computations onto large numbers of parallel processors will be concurrently developed with the foregoing meshfree methods.
6. A meshfree continuum approach for the analysis of nano-scale system has been proposed and successfully implemented.

8 ACKNOWLEDGEMENT

This work is supported by the grants from National Science Foundation, the Army Research Office, and Tull Family Endowment.

9 REFERENCES

Belytschko, T. and Tabbara, T., 1996, Dynamic fracture using EFG. *International Journal for Numerical Methods in Engineering*, **39**, p. 3.

Chen, J.S., Pan, C., Wu, C.T. and Liu, W.K., 1996, Reproducing kernel particle Methods for large deformation analysis of nonlinear structures. *Computer Methods in Applied Mechanics and Engineering,* **139**, pp. 195-227.

Hao, S., Liu, W.K., Klein, P., Qian, D., 2001, Multi-scale damage model. manuscript to be submitted.

Hao, S., Park, H., and Liu, W.K., 2001, Moving Particle Finite Element Method. submitted.

Jun, S., Liu, W. K. and Belytschko T., 1998, Explicit Reproducing Kernel Particle Methods for Large Deformation Problems. *International Journal for Numerical Methods in Engineering,* **41**, pp. 137-166.

Kalthoff, J.F., and Winkler, S., 1987a, Failure mode transition at high rates of shear loading. In *Impact Loading and Dynamic Behavior of Materials,* edited by Chiem, C.Y., Kunze, H.D. and Meyer, L.W., Vol.1, pp. 185-195.

Klein, P.A., 1999, Technical Report, Sandia National Laboratories.

Li, S., Liu, W.K., Qian, D., Guduru, P., and Rosakies, R., 2001, Dynamic Shear Band Propogation and Micro-structure of Adiabatic Shear Band. Submitted to *Computer Methods in Applied Mechanics and Engineering.*

Liu, W.K. and Chen, Y., 1995, Wavelet and multiple scale reproducing kernel methods. *International Journal for Numerical Methods in Fluids,* **21**, pp. 901-931.

Liu, W. K. and Jun, S., 1998, Multiple Scale Reproducing Kernel Particle Methods for Large Deformation Problems. *International Journal for Numerical Methods in Engineering,* **41**, pp. 1339-1362.

Liu, W.K., Jun, S., Zhang, Y.F., 1995, Reproducing Kernel Particle Methods. *International Journal for Numerical Methods in Engineering,* **20**, pp. 1081-1106.

Liu, W.K., Jun, S., Li, S., Adee, J., Belytschko, T., 1995, Reproducing Kernel Particle Methods for Structural Dynamics. *International Journal for Numerical Methods in Engineering,* **38**, pp. 1655-1680.

Liu, W.K., Chen, Y., Jun, S., Chen, J.S., Belytschko, T., Pan, C., Uras, R.A. and Chang, C.T., 1996, Overview and Applications of the Reproducing Kernel Particle Methods. *Archives of Computational Methods in Engineering: State of the Art Reviews.* Vol.3, pp. 3-80.

Mason, J.J., Rosakis, A.J., and Ravinchandran, G., 1994, Full field measurement of the dynamic deformation field around a growing adiabatic shear band at the tip of a dynamically loaded crack or notch. *Journal of Mechanics and Physics of Solids,* **42**, pp. 1679-1697.

Qian, D., Liu, W.K. and Ruoff, R.S., 2001, Mechanics of Nanotubes filled with Fullerenes. in preparation.

Ravi-Chandar, K., 1995, On the failure mode transition in polycarbonate dynamic mixed-mode loading. *International Journal of Solids and Structures*, **32**, pp. 925-938.

Rosakis, A.J., 2000, private communication.

Sloan, J., Dunin-Borkowski, R.E., Hutchison, J.L., Coleman, K.S., Williams, V.C., Claridge, J.B., York, A.P.E., Xu, C., Bailey, S.R., Brown, G., Fridrichs, S., Green, M.L.H., 2000. The size distribution, imaging and obstructing properties of C60 and higher fullerenes formed within arc-growth single walled carbon nanotubes. *Chemical Physics Letters*, **316**, 191-198,

Smith, B. W., Monthoux, M. and Luzzi, D.E., 1998, Encapsulated C60 in carbon nanotubes. *Nature*, **336**, p. 323.

Zhang, Y., Iijima, S., Shi, Z. and Gu, Z., 1999, Defects in arc-discharg-produced single-walled carbon nanotubes. *Philosophical Magazine Letters*, **79**, N.7, pp. 473-479.

Zhou, M., Rosakis, A.J., and Ravichandran, G., 1996a, Dynamically propagating shear bands in impact-loaded prenotched plates -i, experimental investigations of temperature signatures and propagation speed. *Journal of Mechanics of Physics and Solids*, **44**, pp. 981-1006.

Zhou, M., Rosakis, A.J., and Ravichandran, G., 1996b, Dynamically propagating shear bands in impact-loaded prenotched plates -ii, numerical simulations. *Journal of Mechanics of Physics and Solids*, **44**, pp. 1007-1032.

Modeling of Joints and Interfaces

Xue Yue[1] and K. C. Park[1]

1. Introduction

Modeling of joints or interfaces in structural systems pose several difficulties. First, both in analytical models and experiment design it is difficult to isolate the intrinsic joint behavior without attaching joints to the interfacing structural elements or substructures. Therefore, incorporation of joints in the form of standard finite elements becomes problematic.

Second, joints usually exhibit complex nonlinearities that are dependent on the loads that they are subjected to as well as directionality of the loads as observed by O'Donnel and Grawley(1985). This makes the implementation of a joint model into a stand-alone structural library a difficult task.

Third, although a host of joint experiments are available in the literature, their effective utilization for the construction of joint models that can be reliably used in model-based simulations has not been adequately addressed.

Fourth, most joint models are based on the coulomb friction law which is time-invariant (Dohner, 2000), thus often inadequate to capture the complex history-dependent multidimensional nonlinearities exhibited in joints. As a consequence, uncertainties in the source of nonlinearities (Beards, 1979) compound the task of developing high-fidelity constitutive relations for joints.

Finally, joint stiffnesses are often radically different from those of the substructures to which they are connected. Due to prevalent stiffness mismatches, standard simulation procedures can either fail to capture the dominant physics or may lead to inaccuracies as well as substantially increased computational effort.

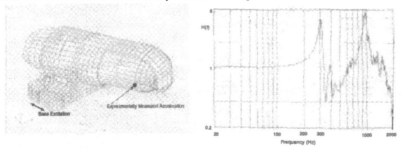

Figure 1. A Sandia Structure for Present Joint Modeling Study

In practice, most structures often consist of a large number of components whose responses are dominated by joint and interface mechanics. But due to joint modeling difficulties, simulation results may poorly predict the accurate response of the structures with joints or interfaces. For example, Fig. 1 shows a horizontal acceleration

---

[1] University of Colorado at Boulder

record measured at the dotted location when the system is subjected to a horizontal excitation. The structure is bolted down to the foundation, which is the main focus of the present analysis. An elaborate finite element model was constructed and two cases have been simulated: an ideal joint (rigid support) and flexible joints. The results of these two cases carried out at Sandia National Laboratories are shown in Figure 2. Clearly, two simulation results do not correlate well with the experimental result shown in Fig. 1.

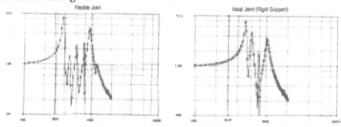

Figure 2. FEM Analysis Results Provided by Sandia National Laboratories

The foregoing difficulty in correlating simulations with experimental results has motivated us to develop an approach for constructing predictive models of structures with joints and interfaces.

2. Modeling of Basic Joint Hysteresis

A significant deviation between simulation models and response data obtained from experiments as discussed in the preceding section calls for a modeling approach that can capture the dynamic behavior of the structures with joints and interfaces which exhibit complex hysteresis behavior. This section introduces a modified multidimensional form of Wen's hysteresis model in (Wen, 1976) and further developed by Fliente(1995). Under dynamic loading, the load-displacement trace produces hysteresis loops caused by damping and/or inelastic deformation as well as friction. It will be further developed to the joint element in later sections. A 2-DOF mass-damper-spring system is employed to investigate the different joint behaviors and the sensitivity of the system response to the variation of the model parameter changes. It is shown that the response of the system with joints or interfaces is strongly affected by the joint behavior, which is determined by the joint parameters. The desired frequency response of the system under dynamic loading can be obtained by adjusting the joint parameters.

2.1 JOINT BEHAVIOR

The mass-damper-spring system is shown in Figure 3 Notice the nonlinear spring connecting mass m1 and m2. It is the representation of the hysteresis joint between the two masses. It provides the nonlinear force to the two masses based on the relative movement of the two masses. The equation of motion of this system can be expressed as:

$$\mathbf{M}\ddot{\mathbf{x}} + \mathbf{D}\dot{\mathbf{x}} + \mathbf{K}\mathbf{x} + \mathbf{f}_n = \mathbf{f}_a \qquad (2.1)$$

where $\mathbf{M}$, $\mathbf{D}$ and $\mathbf{K}$ are the mass, damping and stiffness matrixes, and the nonlinear joint force vector $\mathbf{f}_n = (\, f_n \quad -f_n \,)^T$ is modeled as (Wen, 1976)

$$\dot{\mathbf{f}}_n = \mathbf{A}\, \Delta\dot{\mathbf{x}} - \gamma\, |\mathbf{f}_n|\Delta\dot{\mathbf{x}} - \delta\, |\Delta\dot{\mathbf{x}}|\mathbf{f}_n, \qquad \Delta\dot{X} = \dot{x}_1 - \dot{x}_2 \qquad (2.2)$$

where $\mathbf{A}, \gamma, \delta$ characterize the joint stiffness, the width of hysteresis loop, and the softening and hardening ranges of the hysteresis curve, respectively.

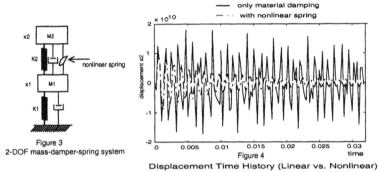

Figure 3
2-DOF mass-damper-spring system

Figure 4
Displacement Time History (Linear vs. Nonlinear)

In order to demonstrate the joint parameters, (2.1) and (2.2) are numerically integrated using the Runge-Kutter fourth-order method. Figure 4 shows the response comparison between the linear model with and without the nonlinear joint. Notice that the displacement with the nonlinear spring decreases faster when compared to the linear model with only material damping, showing that the joint is the primary source of damping.

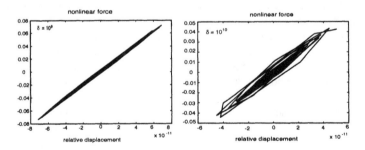

Figure 5. Nonlinear Spring Characteristics corresponding to Different δ

Figure 5 illustrates the role that parameter δ has on the joint force vs. displacement. While keeping $\mathbf{A}$ and γ unchanged, the width of the hysteresis loop increases as δ increases. This translates into higher damping as shown in Figure 6.

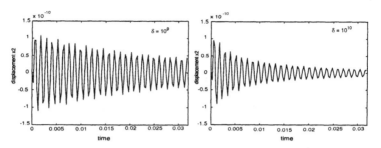

Figure 6. Displacement Time History corresponding to Different δ

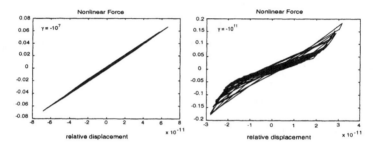

Figure 7. Nonlinear Springg Characteristics Corresponding to Different γ

Figure 7 shows the joint behavior when changing parameter γ only. It can be seen that γ mainly affects the curvature of the hysteresis loop or the hardening and softening of the joint nonlinearity.

Finally, joint parameter **A** determines the initial slope of the hysteresis loop. This means that the choice of **A** would be primarily responsible for any significant role in system frequency changes for jointed structures.

2.2 SYSTEM FREQUENCY RESPONSE

The estimate of the system frequency response function is obtained by Fast Fourier Transform(FFT)-based extraction method (see, e.g., (Juang, 1994)), which can be realized through the following equations:

$$H(\omega_k) = \bar{S}_{fx}(\omega_k) \cdot \bar{S}_{ff}(\omega_k)^{-1} \tag{2.3}$$

where $\bar{S}_{fx}$ and $\bar{S}_{xx}$ are the frequency domain cross- and auto-spectral densities. At a discrete frequency ω_k, they are defined as:

$$\bar{S}_{fx} = \hat{x}(\omega_k) \cdot \hat{f}^*(\omega_k) \tag{2.4}$$

$$\bar{S}_{xx} = \hat{f}(\omega_k) \cdot \hat{f}^*(\omega_k) \tag{2.5}$$

where * denotes complex conjugate.

Figure 8 shows the influence of joint parameters δ and γ on the system frequency response. It is observed that the magnitudes of the two FRF peaks decreases as δ increases. This again proves that δ plays the role of damping. The change in the magnitude ratio of the two peaks is also observed as δ changes. This shows that adjusting the value of δ is one way to obtain the desired system frequency response. On the other hand, γ mainly affects the smoothness and the shape of FRF, it also changes the width of the FRF peak.

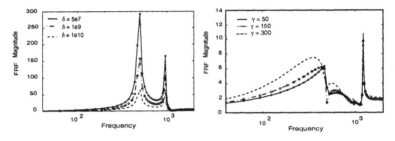

Figure 8. Frequency Response Functions corresponding to Different Joint Parameters

With the foregoing preliminary insight into the present joint model (2.2), we will present the partitioned formulation that is conducive to identify joint properties from experimental data and that facilitates the incorporation of joint elements as a stand-alone module to existing FEM software.

3. Modeling of Nonlinear Structural Joint Via Experimental-Analytical Localized Flexibility Identification

This section presents a procedure for modeling of nonlinear structural joints by utilizing hybrid experimental and analytical localized flexibility. It is assumed that the substructures that are assembled by the structural joints are known a priori either by separate experiments or by validated analytical models. System identification (see, e.g., Alvin and Park, 1994) is then carried out for the assembled structural systems. Localized partitioning procedure developed in Park and Reich(1998), Alvin and Park(1999), and Park and Felippa(1998 and 2000) is then applied to identify the dynamic flexibility of the isolated joints. It is this isolated joint flexibility that is used to identify nonlinear joint parameters. The present joint modeling procedure is applied to simple beam-like structure whose substructures are assembled by a nonlinear joint.

3.1 PARTITIONED FORMULATION OF JOINTED STRUCTURES

Consider two linear substructural systems A and B whose governing semi-discrete equations are given by

$$\mathbf{p}_A\left(\mathbf{x}_A\right) = \mathbf{f}_A, \quad \mathbf{p}_A = \mathbf{M}_A \frac{d^2}{dt^2} + \mathbf{D}_A \frac{d}{dt} + \mathbf{K}_A$$

$$\mathbf{p}_B\left(\mathbf{x}_B\right) = \mathbf{f}_A, \quad \mathbf{p}_B = \mathbf{M}_B \frac{d^2}{dt^2} + \mathbf{D}_B \frac{d}{dt} + \mathbf{K}_B \tag{3.1}$$

Here $\mathbf{x}$ denotes the displacement vector, $\mathbf{M}$ the mass matrix, $\mathbf{D}$ the damping matrix, $\mathbf{K}$ the stiffness matrix, $\mathbf{f}$ the applied force vector, and d/dt means time differentiation. Subscripts A and B identify the two substructures.

If substructures A and B are interfaced directly through Lagrange multipliers, the resulting coupled system is formally obtained as the variational statement

$$\delta\Pi = \delta\Phi_{free} + \delta\pi_c \tag{3.2}$$

$$\Phi_{free} = \delta\mathbf{x}_A^T \left[\mathbf{f}_A - \mathbf{p}_A(\mathbf{x}_A)\right] + \delta\mathbf{x}_B^T \left[\mathbf{f}_B - \mathbf{p}_B(\mathbf{x}_B)\right]$$

$$\delta\pi_c = \delta\lambda_{AB}^T \mathbf{C}_{AB}^T \left(\mathbf{x}_A - \mathbf{x}_B\right) + \left(\delta\mathbf{x}_A - \delta\mathbf{x}_B\right)^T \mathbf{C}_{AB} \lambda_{AB} \tag{3.3}$$

Here $\mathbf{C}_{AB}$ is the (linear) interface matrix that extracts the interface degrees of freedom from both substructures A and B, respectively, and λ_{AB} the corresponding array of Lagrange multipliers. The governing equations are given by the stationarity condition $\delta\Pi = 0$:

$$\begin{bmatrix} \mathbf{p}_A(\cdot) & \mathbf{0} & -\mathbf{C}_{AB} \\ \mathbf{0} & \mathbf{p}_B(\cdot) & \mathbf{C}_{AB} \\ -\mathbf{C}_{AB}^T & \mathbf{C}_{AB}^T & \mathbf{0} \end{bmatrix} \begin{Bmatrix} \mathbf{x}_A \\ \mathbf{x}_B \\ \lambda_{AB} \end{Bmatrix} = \begin{Bmatrix} \mathbf{f}_A \\ \mathbf{f}_B \\ \mathbf{0} \end{Bmatrix} \tag{3.4}$$

A joint is called *ideal* if substructural displacements along the interface boundary satisfy the last of (3.4):

$$\mathbf{C}_{AB}^T \left(\mathbf{x}_A - \mathbf{x}_B\right) = \mathbf{0} \tag{3.5}$$

An ideal joint does not possess mass, stiffness or damping. In physical terms an ideal joint merely connects the interface freedoms of adjacent substructures without reference to its physical properties.

Suppose now that the joint connecting the two substructures is not ideal:

$$\mathbf{C}_{AB}^T \left(\mathbf{x}_A - \mathbf{x}_B\right) \neq \mathbf{0} \tag{3.6}$$

so that the difference accounts for its mechanical properties. A fairly general dynamic model for a joint may be expressed

$$\mathbf{m}_J \ddot{\mathbf{d}} + \bar{\mathbf{q}}_J\left(\dot{\mathbf{d}}, \mathbf{d}\right) = \mathbf{f}_J, \quad \mathbf{d} = \begin{Bmatrix} \mathbf{d}_A \\ \mathbf{d}_B \end{Bmatrix} \tag{3.7}$$

where $\mathbf{m}_J$ is the joint mass, $\bar{\mathbf{q}}_J$ and $\mathbf{f}$ are the joint internal and external forces, respectively, whereas $\mathbf{d}_A$ and $\mathbf{d}_B$ denote the displacements of joint J along the substructural A and B interfaces, respectively, as illustrated in Figure 9.

Consequently, the non-ideal joint condition is replaced by:

$$x_{A_\Gamma} - d_{A_\Gamma} = 0, \quad x_{B_\Gamma} - d_{B_\Gamma} = 0 \tag{3.8}$$

so that $\delta\pi_c$ becomes constraints:

$$\delta\pi_c = \delta\lambda_{AJ}^T \left(C_{AJ}^T x_A - B_{AJ}^T d\right) + \delta\left(C_{AJ}^T x_A - B_{AJ}^T d\right)^T \lambda_{AJ} \\ + \delta\lambda_{BJ}^T \left(C_{BJ}^T x_B - B_{BJ}^T d\right) + \delta\left(C_{BJ}^T x_B - B_{BJ}^T d\right)^T \lambda_{BJ} \tag{3.9}$$

The resulting equations with a non-ideal joint becomes

$$\begin{bmatrix} P_X(.) & 0 & -C \\ 0 & q_J(.) & B \\ -C^T & B^T & 0 \end{bmatrix} \begin{Bmatrix} x \\ d \\ \lambda \end{Bmatrix} = \begin{Bmatrix} f_X \\ f_J \\ 0 \end{Bmatrix}$$

with $\quad x = \begin{Bmatrix} x_A \\ x_B \end{Bmatrix}, \quad \lambda = \begin{Bmatrix} \lambda_{AJ} \\ \lambda_{BJ} \end{Bmatrix}$

$$P_X = \begin{Bmatrix} P_A & 0 \\ 0 & P_B \end{Bmatrix}, \quad B = [B_{AJ} \quad B_{BJ}], \quad C = \begin{bmatrix} C_{AJ} & 0 \\ 0 & C_{BJ} \end{bmatrix} \tag{3.10}$$

where q_J is the left-hand-side operator of the joint model (3.7)

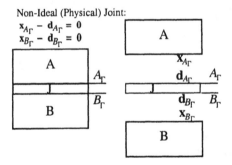

Non-Ideal (Physical) Joint:

Figure 9. Schematic representations of a non-ideal joint.

Equation(3.10) thus provides an element-by-element implementation of joints in simulating of jointed structures regardless of how joints are characterized.

3.2 EXPERIMENTALLY IDENTIFIED STRUCTURAL MODEL AND JOINT CHARACTERIZATION

As stated in Introduction, joints often exhibit complex nonlinearities. Yet, except in limited cases a majority of experimentally determined structural models at best possess linearized stiffness and damping. Suppose that the system-identified modes and mode shapes are expressed in terms of partitioned structural displacements x and the joint displacement d. Then one can express the experimentally identified model as

$$\left\{ \begin{matrix} \mathbf{x} \\ \mathbf{d} \end{matrix} \right\} = \mathbf{H}(\omega)_{exp} \left\{ \begin{matrix} \mathbf{f} \\ \mathbf{f}_J \end{matrix} \right\} \tag{3.11}$$

where $\mathbf{H}(\omega)$ will be called the dynamic partitioned flexibility.

The analytical counterpart to $\mathbf{H}(\omega)_{anal}$ can be expressed from (3.10)

$$\mathbf{H}(\omega)_{anal} = \mathbf{H}_s(\omega) \cdot [\mathbf{I} - \bar{\mathbf{C}}\, \mathbf{F}_B^+ \, \bar{\mathbf{C}}^T \mathbf{H}_s(\omega)]$$
$$\mathbf{H}_s(\omega) = \begin{bmatrix} \mathbf{P}_x^{-1}(\omega) & \mathbf{0} \\ \mathbf{0} & \mathbf{Q}_J^{-1}(\omega) \end{bmatrix} \tag{3.12}$$
$$\bar{\mathbf{C}} = \left\{ \begin{matrix} -\mathbf{C} \\ \mathbf{B} \end{matrix} \right\}, \quad \mathbf{F}_B = \bar{\mathbf{C}}^T \mathbf{H}_s \bar{\mathbf{C}}$$

Note that a comparison of the experimentally determined dynamic flexibility (3.11) and its analytical counterpart (3.12) by assuming that the joint model $\mathbf{q}_J$ are approximated by a linear model:

$$\mathbf{H}(\omega)_{exp} = \mathbf{H}_s(\omega) \cdot [\mathbf{I} - \bar{\mathbf{C}}\, \mathbf{F}_B^+ \, \bar{\mathbf{C}}^T \mathbf{H}_s(\omega)] \tag{3.13}$$

from which $\mathbf{H}_s(\omega)$, for that matter $\mathbf{q}_J$, can be determined by a Riccati-like solution procedure.

However, a general joint model represented by (3.8) are often nonlinear. In order to construct a nonlinear joint model, we first obtain the equivalent linear damping and stiffness of the joint model from

$$\mathbf{q}_{exp}(\omega) = (\mathbf{K}_J + j\omega \mathbf{D}_J - \omega^2 \mathbf{m}_J)\mathbf{d}$$
$$\Downarrow \tag{3.14}$$
$$\mathbf{q}_{exp}(\mathbf{d}, \dot{\mathbf{d}}) = \mathbf{K}_J\, \mathbf{d} + \mathbf{D}_J\, \dot{\mathbf{d}} + \mathbf{m}_J\, \ddot{\mathbf{d}}$$

Based on the equivalent linear joint model obtained in (3.14), we can now proceed to use the nonlinear joint model. We choose the endochronice model introduced in section 2.

3.3 SIMPLIFIED MODELING OF JOINT

In this section we will focus on the modes or mechanisms of joints by employing a simplified model as shown in Figure 10. Note that, unlike the experimental setup wherein the structure is bolted down to a rigid foundation, the foundation of the simplified model consists of two columns and a horizontal beam in order to model the compliance of the lower part of the test structure. Also, the overhanging beam on the left-hand side of the vertical beam is subsequently dropped as it hardly influences the joint characterization.

After some trial and error, the simplified model we have arrived at consists of seven plane beam and one joint element as shown in Figure 10. Note once again that the joint is placed between the flexible foundation and the structure. A large mass is placed on the cantilevered beam that can capture the overall mass of the structure.

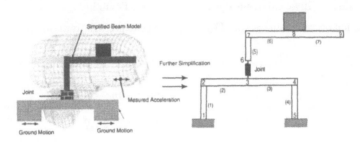

Figure 10. FE Model Simplified from Experimental Setup

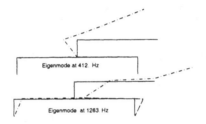

Figure 11. Mode Shapes of First Two Modes of Linear Ideal Model

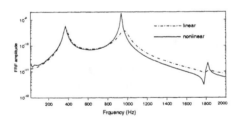

Figure 12. Horizontal FRF Curves at Node 8 for Linear Damped and Nonlinear Joint Model

Examination of the two mode shapes shown in Figure 11 suggests that the rocking modes or the rotational modes at the joint location between nodes 3 and 6 would provide the physical rocking mode of joint and it is this model chosen for subsequent study.

In order to assess the proposed joint identification approach outlined in (3.11) $\sim$ (3.14), the experimentally identified joint characteristics q_{exp} in (3.13) was assumed and the horizontal FRF curve at node 8 of the equivalent linear damped model given as the dotted line in Figure 12. The equivalent linear elemental joint was then identified using (3.12). The endochronic nonlinear joint model (3.14) was then determined by minimizing the difference between the equivalent linear joint model identified from (3.12) and the parameterized nonlinear joint model (3.14).

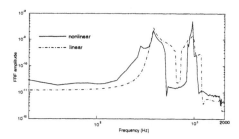

Figure 13. Elemental Joint FRFs

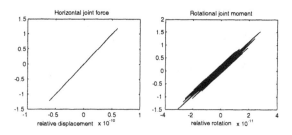

Figure 14. Nonlinear Joint Characteristics

The FRF curves of linear and nonlinear joints after several iterations on the nonlinear joint parameters γ and δ with $p = 1$ are plotted in Figure 14. The corresponding horizontal nonlinear force and rotational nonlinear joint moment are plotted in Figure 13. As expected, the rotational moment curves undergoes substantial hysteresis loops for each transient response period, which accounts for equivalent linear damping.

4. Development of Joint Element and Its Application

We have studied the partitioned joint element's frequency response function in the last section, but left out the details of the joint element construction. This section discusses the general requirements of the joint element, the development of the generic nonlinear joint element restricted to these requirements from the hysteresis model introduced in section 2, and its incorporation into the standard FEM software. This joint element has then been applied to the two different structural systems with interfaces whose substructures are modeled by 2D-beam and continuum elements using ABAQUS.

4.1 JOINT ELEMENT DEVELOPMENT

In general, it is desired that a joint element has following properties: (1) it is scalable from micro to continuum levels; (2) it is inherently multidimensional; (3) it can be easily corroborated with the experiment; and, (4) it should have the basic element attributes such as invariance with respect to rigid-body motions, orientation independent, and passing the patch test.

The tangent matrix of the joint element from the hysteretic model introduced in section 1, as shown following:

$$\mathbf{K}_n = \mathbf{A} - \bar{\gamma} - \bar{\delta} \tag{4.1}$$

where $\bar{\gamma} = \gamma \, |\mathbf{f}_n|$ and $\bar{\delta} = \delta \, [sign(\dot{\mathbf{u}}_j)] \, \mathbf{f}_n$.

Based on above requirements, the tangent matrix of the joint element with rotational degrees of freedom can be expressed as:

$$\mathbf{K}_n = \begin{pmatrix} k_{11} & 0 & 0 & -k_{11} & 0 & 0 \\ 0 & k_{22} & k_{23} & 0 & -k_{22} & k_{23} \\ 0 & k_{23} & k_{33} & 0 & k_{23} & k_{33} \\ -k_{11} & 0 & 0 & k_{11} & 0 & 0 \\ 0 & -k_{22} & -k_{23} & 0 & k_{22} & -k_{23} \\ 0 & k_{23} & k_{33} & 0 & -k_{23} & k_{33} \end{pmatrix} \tag{4.2}$$

with entries subject to:

$$[\, k_{22} \cdot l = 2k_{23} \qquad k_{23} \cdot l = 2k_{33} \,] \tag{4.3}$$

in which l is the characteristic length of the joint element given by

$$k_{ij} = a_{ij} - \gamma_{ij} \cdot |\hat{\mathbf{f}}_{ij}| - \delta_{ij} \, [sign(\dot{\mathbf{u}}_{ij}) \, \hat{\mathbf{f}}_{ij}] \tag{4.4}$$

where $|\hat{\mathbf{f}}_{ij}| = (|f_{ni}| + |f_{nj}|)/2$, and $\hat{\mathbf{f}}_{ij} = (f_{ni} + f_{nj})/2$.

The joint element without the rotational degree of freedom has the matrix format of $\mathbf{K}_n$ as:

$$\mathbf{K}_n = \begin{pmatrix} K_{11} & -K_{11} \\ -K_{11} & K_{11} \end{pmatrix} \quad with \quad K_{11} = \begin{pmatrix} k_{11} & k_{12} \\ k_{12} & k_{22} \end{pmatrix} \tag{4.5}$$

and similarly for k_{ij} in equation(4.4)

4.3 APPLICATIONS OF JOINT ELEMENT

Two structural systems with interfaces are built by the standard FE software ABAQUS. The joint model is implemented as the user defined element. A series of dynamic response simulations corresponding to different joint parameters have been carried out before determining the final joint parameters whose frequency response functions are well correlated with the experimentally observed result.

4.3.1 An Equivalent Beam Model

Figure 15 shows an equivalent spatial beam model to that shown in Figure 1 as we have not been able to access the FEM model generated by Sandia scientists. The equivalent model consists of two substructures, the structure itself and the foundation with 150 two-dimensional beam elements, and the interface is modeled by 11 joint elements.

Notice the joint elements are arranged both in vertical and horizontal directions. The FRF comparison between our joint model and the Sandia flexible joint model shows our equivalent model matches experimental data very well as illustrated in Figure 16.

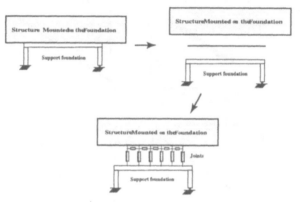

Figure 15. Model Sketch (with joint)

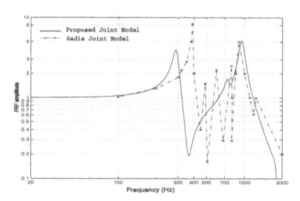

Figure 16. Comparison of Sandia Flexible Joint Model and Proposed Joint Model

The energy dissipation and inelastic deformation accounting for the damping and friction can be observed from the joint element behavior as shown in Figure 17. Figure 18 shows the frequency response difference between the nonlinear structure with joint elements and the linear one without joints. It's easy to see that the damping introduced by the joint model is much higher than pure material damping, and the

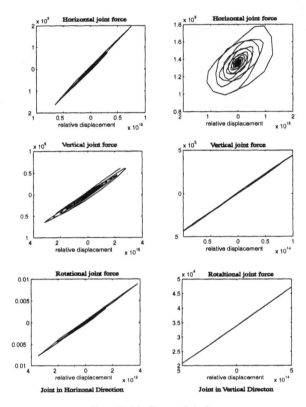

Figure 17. Joint Elements Behavior

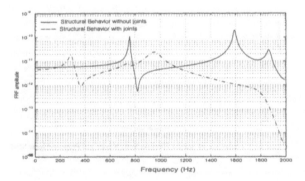

Figure 18. Comparison of Rigid and Flexible Joints for Proposed Model

structural natural frequencies have been greatly reduced due to the softer joint material(implemented through the joint matrix $\mathbf{A}$). This demonstrates that the dynamic responses of the structures are dominated by their joints or interfaces.

4.3.2 An Equivalent Continuum Model

An equivalent continuum model is similar to the equivalent spatial beam model except that the mounted substructure and the foundation are both modeled by two-dimensional continuum elements with 702 plain-stress elements and 10 joint elements.

The linear modal shapes shown in Figure 19 guided us to parameterize the continuum model so that the task for model parameter determination is computationally tractable. Note that while the first mode is primarily rotational, the second mode is a transversal motion.

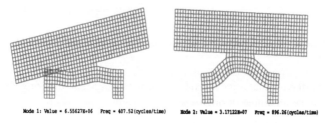

Mode 1: Value = 6.55627E+06   Freq = 407.52(cycles/time)   Mode 2: Value = 3.17122E+07   Freq = 896.26(cycles/time)

Figure 19. The First Two Linear Model Shapes

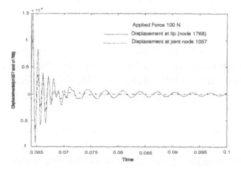

Figure 20. Comparison of Displacement at Tip and Displacement at Joint

Figure 20 shows the displacement time history at the joint node and near the tip of the structure. The displacement at the joint decreases much faster than the one near the tip.

Figure 21 shows a good match of the simulated frequency response at the structure tip with the experimental data. Notice that the second peak becomes much wider than the one produced by the simplified model in section 3.

5. Discussions and Future Work

From the present study on joint modeling approach, we offer the following observations:

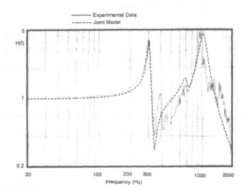

Figure 21. Frequency Response Comparison of Experimental and Proposed Joint Model

1. The overall equivalent linear model on structural system level or substructural level may be identified through a series of careful experiments

2. An accurate determination of joint flexibility from the partitioned flexibility equation (3.13) remains a challenge as the standard Riccati solver is not very suitable. The discrepancy in joint elemental FRFs between the linear model and the Wen model can be attributed to this difficulty.

3. A generic nonlinear analytical structural element is developed for modeling joints and interfaces in structural systems. The hysteretic constitutive law employed in the present joint element incorporates the complex history-dependent multidimensional nonlinearites exhibited in joints. The energy dissipation and inelastic deformation account for damping and friction as manifested in the hysteretic force-displacement loop.

4. It is shown that the proposed joint element satisfies a standard finite element requirements: invariance with respect to rigid-body modes and convergence under constant stress states. Hence, it can be easily incorporated into any standard FEM software.

5. The correlations performed so far using experimental data indicate that the proposed joint element can capture predominant joint behavior, and since the present joint element is independent of interfacing structures, it can be treated as single joint substructure, hence it can be used as an alternative to conventional interface modeling that is difficult to use for non-matching interface meshes and mismatching interface flexibilities using localized Lagrange multiplier method.

6. Joints are not only where structural damages often occur, but do impact how and in what modes the system energy is propagated. Hence, the modes of deformation and their impact on the force transmission mechanisms become a vital information that the structural modeler seeks. The micro-joint modeler and subsequent refined joint modelers must be fully aware of this fact when designing and carrying out micro-joint experiments and subsequent large-order joint simulations.

Encouraged by the present correlations between simulations with joint elements and experimental data, we intend to develop the partitioned version of joint model which can be used in structures with non-matching interface meshes and mismatching interface flexibilities. In addition, we will further refine the joint element that may include local plasticity, friction, and contact-release rattling of joints.

REFERENCES

Alvin, K. F. and Park,K. C., 1999, "Extraction of substructural flexibility from global frequencies and modal shapes," *AIIAA Journal*, **37**(11), 1444–1451.

Alvin, K. F. and Park, K. C. 1994, "A Second-Order Structural Identification Procedure via System Theory-Based Realization," *AIAA Journal*, **32**(2), 397-406.

Bears, C. F., 1979, "Damping in structural joints," *Shock and Vibration Digest*, **11** (9), 35-44.

Dohner, J. et al., 2000, " On the Development of Methodologies for Constructing Predictive Models of Structures with Joints and Interfaces". Sandia National Laboratories, Albuquerque NM.

Foliente, Greg C., 1995," Hysteresis Modeling of Wood Joints and Structural Systems". *Journal of Structural Engineering*, Vol.121, No.6. 1013–1022

Juang, Jer-Nan, 1994, Applied System Identification, Prentice Hall, Englewood Cliffs, NJ, 81-120,

Park, K. C. and Felippa, C. A., 1998, "A variational framework for solution method developments in structural mechanics," *J. Appl. Mech.*, **65**, 242–249.

O'Donnell, K. J. and Crawley, E. F. , 1985, "Identification of nonlinear system parameters in space structure joints using the force-state mapping technique," *MIT SSL Report No. 16-85,*

Park, K. C. and Felippa, C. A. , 1998, "A flexibility-based inverse algorithm for identification of structural joint properties," Proc. *ASME Symposium on Computational Methods on Inverse Problems*, Anaheim, CA., AMD-Vol.228, 11–22.

Park, K. C., Reich, G. W. and Alvin, K. F. 1998, "Structural Damage Detection Using Localized Flexibilities," *Journal of Intelligent Material Systems and Structures*, Vol. 9, No. 11, 911-919.

Park, K. C. and Felippa, C. A., 2000, "A Variational Principle for the Formulation of Partitioned Structural Systems," *International Journal of Numerical Methods in Engineering*, vol. 47, 395-418.

Wen, Y. K. 1976, "Method for Random Vibration of Hysteretic Systems," *Proc. ASCE, Journal of Engineering Mechanics*, 102-EM2, 1976, 249-263

Formulation of a Finite Element Method with Arbitrary Element Geometry

Mark M. Rashid[1] and Philip M. Gullett[2]

1. ABSTRACT

Formulation of the *Variable-Element Topology Finite Element Method* for problems in solid mechanics is described. The method differs from conventional finite element methods in that the shape functions are generated without recourse to an isoparametric (or similar) mapping. Instead, optimized polynomial shape functions are defined on arbitrary polygonal elements. Restrictions relating to element geometry and number and arrangement of nodes are thereby circumvented. The task of automatic mesh generation is a significantly less complex one for the new method than it is for the conventional FEM. Aspects of convergence are discussed for the new method, which is nonconforming in general. Computational results are presented that illustrate the performance of the method in the setting of planar elasticity.

2. INTRODUCTION

In this article, a new general purpose computational method for solid mechanics applications is described. The method, called the *Variable Element Topology Finite Element Method* (Rashid and Gullett, 2000), is finite-element-like in character, and shares with the conventional FEM the essential features that make it so powerful. In addition, VETFEM elements may each possess an essentially arbitrary number of nodes, and are not restricted by a convexity requirement. The task of automatic mesh generation is thereby greatly simplified. The VETFEM is therefore attractive for problems involving very complex geometry and/or evolving

---

[1] Department of Civil and Environmental Engineering, University of California at Davis, Davis, CA 95616. Email: mmrashid@ucdavis.edu
[2] Sandia National Laboratories, Mail Stop 9405, PO Box 0969, Livermore, CA 94551

domain topology, such as fracture and shape optimization problems. Indeed, VETFEM and conventional finite elements may occur in any combination within a single mesh, so that regions of an otherwise conventional mesh that are difficult to discretize – e.g. near the front of an advancing crack – can be selectively treated with VETFEM elements.

In a displacement-based variational setting, the essential task in the construction of any general-purpose approximation method is to devise a means of synthesizing a large number of basis functions on an arbitrary problem domain. These basis functions must be capable of respecting the displacement boundary conditions, and in addition should possess C^0 continuity throughout the domain. (Or, at the least, one must be prepared to deal with the consequences of any departure from this requirement.) The conventional FEM has earned a dominant status in solid mechanics, and a very important position in many other branches of continuum physics, because it possesses these fundamental qualities naturally. In addition, it exhibits other features which make it very convenient to program and use. In particular, the compact support property of finite element basis functions leads to a sparse and banded system of algebraic equations. Also, the support boundaries are defined simply by the element boundaries, so that the evaluation of weak form integrals near boundaries of the domain presents no special difficulty. Finally, the Kronecker delta property of finite element basis functions renders the imposition of essential boundary conditions and interface conditions a trivial matter.

These very powerful properties of the FEM result from the isoparametric formulation of standard elements on a simple parent domain. However, this type of element formulation constrains the mesh generation process rather severely. The motivation therefore exists to explore alternative means of generating suitable families of shape functions for use in Galerkin approximations. Accordingly, the development of approximation methods that do not rely on a conventional mesh was the subject of intensive research throughout the 1990s, and continues to be actively pursued. In these so-called *meshless methods*, interpolation theory is used to construct basis functions of small support from a spatial distribution of nodes only, without the help of explicit connectivity among the nodes. Examples include Smoothed Particle Hydrodynamics (SPH) (Swegle *et al.*, 1995), the Element-Free Galerkin (EFG) method (Belytschko *et al.*, 1994, 1996), the Reproducing Kernel Particle Method (RKPM) (Chen *et al.*, 1996), and the Natural Element Method (NEM) (Sukumar *et al.*, 1998). Although their details differ, all of these methods use some form of interpolation theory to form basis functions. Most of these methods can be considered special cases of the RKPM. Other generalizations of the meshless approach include h-p clouds (Duarte and Oden, 1996), and the partition of unity method (Babuska and Melenek, 1997). In most cases, meshless methods require some form of domain partition to support numerical quadrature (see, e.g., Dolbow and Belytschko, 1999). Because the support boundaries are typically not explicitly known for meshless basis functions, evaluation of the weak-form integrals near boundaries of the domain as well as imposition of displacement boundary conditions can be awkward (Mukherjee and Mukherjee, 1997).

In the spirit of these and other alternatives to the conventional FEM, the VETFEM represents an attempt to explicitly retain the powerful features of the standard FEM, while at the same time reducing the constraints placed by the method on the discretization process. In this connection, the approximation method of Dohrmann and Rashid (2001) is mentioned. This method is similar in some of its aspects to the VETFEM; however, it does not involve explicit construction of basis functions. Instead, approximations to their gradients are directly generated which are, in general, non-integrable. The VETFEM, on the other hand, is based on local polynomial interpolants which are optimized to exhibit favorable properties over arbitrary polygonal patches (i.e. elements). As will be described in section 3, VETFEM basis functions do not, in general, exhibit strict C^0 continuity on inter-element boundaries. As with all nonconforming methods, consistency is therefore a significant issue. A brief account of this and related aspects of the VETFEM is given in section 4. Computational results designed to illuminate the performance of the method are given in section 5.

3. DEVELOPMENT OF THE METHOD

3.1. Governing Boundary-Value Problem

For simplicity, linear elasticity is considered herein, with the understanding that there exists no essential difficulty in employing VETFEM basis functions in a non-linear setting. Also, the development is restricted to two-dimensional elastostatics. The governing BVP is given by

$$\left.\begin{aligned} \nabla \cdot \mathbf{T} + \rho \mathbf{b} &= 0 \\ \mathbf{T} &= \mathbf{C}\mathbf{e} \\ \mathbf{e} &= \nabla_s \mathbf{u} \end{aligned}\right\} \quad \mathbf{x} \in \Omega,$$

$$\mathbf{u} = \bar{\mathbf{u}}, \ \mathbf{x} \in \partial_u \Omega; \quad \mathbf{Tn} = \bar{\mathbf{t}}, \ \mathbf{x} \in \partial_t \Omega. \tag{3.1}$$

Here, $\mathbf{T}(\mathbf{x})$, $\mathbf{e}(\mathbf{x})$, $\mathbf{u}(\mathbf{x})$ are the stress, strain, and displacement fields, respectively, on the domain Ω; $\mathbf{b}$ is the body force, ρ is the density, $\mathbf{C}$ is the rank-four modulus tensor, and $\bar{\mathbf{u}}$, $\bar{\mathbf{t}}$ are the specified displacement and traction acting on their respective parts of the domain boundary. There holds $\partial_u \Omega \cup \partial_t \Omega = \partial \Omega$, $\partial_u \Omega \cap \partial_t \Omega = \varnothing$. Employing a conventional notation, the variational BVP (VBVP) corresponding to (3.1) is given by

find $\mathbf{u} \in V \equiv \{\mathbf{u} \in H^1(\Omega): \mathbf{u} = \bar{\mathbf{u}}, \mathbf{x} \in \partial_u \Omega\}$ such that

$$a(\mathbf{u}, \mathbf{v}) \equiv \int_\Omega \nabla_s \mathbf{v} \cdot \mathbf{C} \nabla_s \mathbf{u} \, da = \int_\Omega \rho \mathbf{b} \cdot \mathbf{v} \, da + \int_{\partial_t \Omega} \bar{\mathbf{t}} \cdot \mathbf{v} \, ds \equiv f(\mathbf{v}) \tag{3.2}$$

for any $\mathbf{v} \in V_0 \equiv \{\mathbf{v} \in H_1(\Omega): \mathbf{v} = 0, \mathbf{x} \in \partial_u \Omega\}$.

The bilinear form $a(\cdot, \cdot)$ is symmetric in its arguments.

The Galerkin approximation method is obtained by restricting V and V_0 to identical finite-dimensional bases spanned by the basis functions $\{\Phi_1, \ldots, \Phi_N\}$, so that

$$\mathbf{u} = \sum_{1 \le a \le N} \mathbf{u}_a \Phi_a + \sum_{N < a \le \bar{N}} \bar{\mathbf{u}}_a \Phi_a , \quad \mathbf{v} = \sum_{1 \le a \le N} \mathbf{v}_a \Phi_a . \tag{3.3}$$

Here, $\Phi_a = 0$ on $\mathbf{x} \in \partial_u \Omega$ for $1 \le a \le N$. Introduction of (3.3) into (3.2) leads to a symmetric linear system of equations for the unknown parameters $\mathbf{u}_a$. If $\Phi_a(\mathbf{x}_b) = \delta_{ab}$ (the Kronecker delta property), then (3.3)$_1$ is an interpolant, the $\mathbf{u}_a$ are nodal values of the displacement, and $\bar{\mathbf{u}}_a = \bar{\mathbf{u}}(\mathbf{x}_a)$.

3.2. VETFEM basis functions

The role of any general-purpose computational method based on (3.2) is to generate an arbitrary number of suitable basis functions Φ_a on a given domain, and then to facilitate evaluation of the integrals in (3.2) on the domain. In the VETFEM, the domain is subdivided into L polygonal elements ω_k such that $\bigcup\limits_{1 \le a \le L} \omega_a = \Omega$ and $\omega_a \cap \omega_b = \varnothing$ if $a \ne b$. Each element possesses nodes at its vertices, as shown in Figure 1. Henceforth, a single representative element ω with n nodes and facets will be considered.

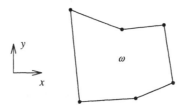

Figure 1. Typical VETFEM element.

As in the conventional FEM, the global basis functions Φ_a are composed of shape functions ϕ_a, $a = 1, \ldots, n$, where $\phi_a = 0$, $\mathbf{x} \notin \omega$. In the VETFEM, these shape functions are written as polynomials in the physical coordinates (x, y):

$$\phi_a = \sum_{1 \le i \le M} G_i^a q_i . \tag{3.4}$$

Here, q_i refers to the ordered list of monomials $\{1, x, y, x^2, xy, y^2, \ldots\}$. Normally, it will be assumed that M corresponds to a complete polynomial through order k, so that $M = (k + 1)(k + 2)/2$. The polynomial coefficients G_i^a are to be set so that the shape functions exhibit certain desirable properties, as described below.

3.2.1. Node point constraints

In order that the Kronecker delta property of the standard finite element method be preserved, the shape function coefficients in (3.4) must be set so that $\phi_a(\mathbf{x}_b) = \delta_{ab}$.

This condition constrains the coefficients by

$$\sum_{1 \le i \le M} G_i^a \, q_i(\mathbf{x}_b) = \delta_{ab} \,. \tag{3.5}$$

These constraints are always linearly independent, and are compatible for sufficiently large M.

3.2.2. Consistency constraints

Whereas the Kronecker delta property is desirable from the standpoint of convenience and utility, certain properties at the inter-element interfaces are *required* if convergence of the overall method is to be assured. In the conventional FEM, the basis functions are continuous at these interfaces, and consistency is automatic. In the VETFEM, however, slight discontinuities on the interfaces away from the nodes may arise due to the inability of finite order polynomials to exactly reproduce a specified variation on element facets (piecewise linear in the current implementation). The origin of these requirements is discussed in the next section. For the present purposes, it suffices to appeal to the intuitive notion that arbitrary incompatibilities between elements is unlikely to result in a convergent method. In fact, it turns out that the requisite level of compatibility is achieved if the average value on each facet of the jump vanishes for all basis functions. This condition is expressed by

$$\int_{\Gamma_b} \phi_a \, ds = \frac{1}{2}(|\Gamma_a|\delta_{ab} + |\Gamma_c|\delta_{cb}) \text{ (no sum)}, \quad 1 \le a, b \le n. \tag{3.6}$$

Here, Γ_a is the facet of $\partial \omega$ defined by nodes a and $mod(a + 1, n)$, and $c = 1 + mod(n - 2 + a, n)$. Equation (3.6) states that the integral of shape function a on a facet containing node a must be equal to half the length of the facet, and zero for other facets. This condition is trivially satisfied if the shape functions exactly reproduce the requisite piecewise linear variation on $\partial \omega$; a circumstance which occurs for sufficiently simple shapes and sufficiently high polynomial order. The integrals in (3.6) may be economically evaluated using Gaussian quadrature.

The constraints (3.6) represent a stricter notion of consistency than do the analogous constraints in the original exposition of the method (Rashid and Gullett, 2000). In that work, the constraints were based on the requirement that a version of the engineering patch test be passed for all element configurations. Equation (3.6), on the other hand, allows for an explicit proof of convergence.

As with the node point constraints (3.5), the constraints (3.6) are linearly independent and consistent for element shapes and polynomial orders of practical interest. The two sets of constraints can be combined into the single expression

$$\mathbf{CG}^a = \mathbf{b}^a \,, \tag{3.7}$$

in which

$$C_{a,i} = q_i(\mathbf{x}_a) , \quad C_{a+n,i} = \int_{\Gamma_a} q_i \, ds ,$$

$$b_a^b = \delta_{ab} , \quad b_{a+n}^b = \tfrac{1}{2}(|\Gamma_a|\delta_{ab} + |\Gamma_c|\delta_{cb})$$

$$\left. \begin{array}{l} 1 \le a \le n \\ 1 \le b \le n \\ 1 \le i \le m \\ c = 1 + mod(n - 2 + a, n). \end{array} \right\} \quad (3.8)$$

The $2n \times m$ matrix $\mathbf{C}$ is defined entirely in terms of the geometry of the element, and must be factored only once in the course of solving for the coefficients for all n shape functions. Clearly, it is necessary that the polynomial order k of the shape functions be chosen sufficiently large so that $m \ge 2n$.

3.2.3. Compatibility and smoothness optimization

Following imposition of the node point and consistency constraints, there will remain, in general, some freedom in the shape function coefficients $\mathbf{G}^a$. This remaining freedom may be put to good effect by attempting to optimize both the inter-element compatibility, as well as the smoothness of the shape functions on the interior of the element. In relation to the former, the degree to which the shape function ϕ_a achieves the desired piecewise linear variation on the element boundary may be assessed with the positive semi-definite functional

$$F_1^a = \int_{\partial\omega} [\nabla\phi_a \cdot \nabla\phi_a - (\mathbf{n} \cdot \nabla\phi_a)^2] \, ds \quad \text{(no sum)} , \qquad (3.9)$$

in which $\mathbf{n}$ is the outward unit normal to $\partial\omega$. The form (3.9) is convenient for application to both 2D and 3D elements. A measure of interior smoothness, on the other hand, is given by

$$F_2^a = \int_{\omega} \nabla\phi_a \cdot \nabla\phi_a \, da \quad \text{(no sum)} , \qquad (3.10)$$

which is also positive semi-definite. A blend of favorable inter-element compatibility and interior smoothness can be achieved by considering the composite functional $F^a = \alpha|\partial\omega|F_1^a + (1 - \alpha)F_2^a$, where $\alpha \in [0, 1)$ is an adjustable parameter. The length of the element boundary is introduced into the first term in order to render both terms dimensionless. Using (3.4), the discrete form of this functional is

$$F^a = \mathbf{G}^{aT}\mathbf{A}\mathbf{G}^a ,$$

$$A_{i,j} = \alpha|\partial\omega| \int_{\partial\omega} [\nabla q_i \cdot \nabla q_j - (\mathbf{n} \cdot \nabla q_i)(\mathbf{n} \cdot \nabla q_j)] \, ds + \qquad (3.11)$$

$$(1 - \alpha) \int_{\omega} \nabla q_i \cdot \nabla q_j \, da .$$

Both integrals in (3.11) may be easily evaluated using Gaussian quadrature; the second integral after an application of Green's theorem.

The algorithmic problem for the determination of the shape function coefficients G_i^a may now be stated as follows: minimize F^a as defined in (3.11), subject to the constraints (3.7). The matrix $\mathbf{A}$ in (3.11) is symmetric and positive, but only semi-definite (and in fact possesses a single vanishing eigenvalue). Accordingly, the constraints (3.7) are first factored to the extent possible by applying a generalized Gaussian elimination algorithm to (3.7) (see Rashid and Gullett, 2000), from which the representation

$$\mathbf{G}^a = \mathbf{B}\mathbf{g}^a + \mathbf{d}^a \tag{3.12}$$

emerges. In (3.12), $\mathbf{B}$ is $m \times (m - 2n)$ and each $\mathbf{d}^a$ is $m \times 1$; these are the outputs of the generalized elimination algorithm. Then, (3.12) is used in $(3.11)_1$, and each F_a is minimized with respect to the $(m - 2n)$ corresponding free parameters $\mathbf{g}^a$ to obtain the linear system

$$\mathbf{B}^T \mathbf{A} \mathbf{B}\mathbf{g}^a = -\mathbf{B}^T \mathbf{A} \mathbf{d}^a . \tag{3.13}$$

The system (3.13) is square, with dimension $(m - 2n)$, and is guaranteed to be of full rank. The coefficient matrix is independent of the shape function number and so must be factored only once for the element. The shape function coefficients are then recovered by repeated back substitution to obtain $\mathbf{g}^a$, followed by use of (3.12).

In summary, the shape functions are given by polynomials in the physical coordinates which interpolate the nodal data, and which also satisfy inter-element compatibility *in the mean* on each facet. In addition, the polynomials are optimized to achieve a blend of minimum incompatibility at inter-element interfaces, and maximum smoothness on the element interior. This strategy differs in some of its details from that originally proposed by Rashid and Gullett (2000). In particular, the compatibility constraints have been strengthened as mentioned previously. Also, in the original formulation the compatibility and smoothness optimizations were performed sequentially, rather than in weighted combination as described here. In extensive numerical tests, the latter has been found to result in better control over the shape function quality.

In the current implementation of the VETFEM, the polynomial order for the shape functions is taken as simply $k = n + 1$. This results in 36 polynomial coefficients and 12 constraints for a six-node element. With regard to the weighting factor α (see equation (3.11)), the final system of equations (3.13) is guaranteed to be of full rank for any value in the range [0, 1). A value of 0.2 has been found to give favorable overall shape function quality under all circumstances.

Figure 2 shows a square VETFEM element containing six nodes, along with surface plots of three of its shape functions. Whereas isoparametric element formulations containing midside nodes are common in the conventional FEM, it should be borne in mind that the element of Figure 2 attempts *linear*, and not quadratic, variation between each node. From the figure, it may be seen that strict piecewise linearity on all facets is not maintained for some of the shape functions, as may be expected due to the finite order of the shape function polynomials.

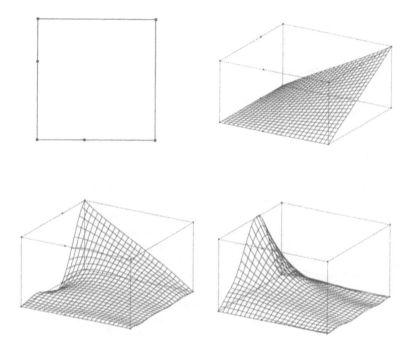

Figure 2. VETFEM element with six nodes, and surface plots of three of its shape functions.

3.3. Implementation

Computer implementation of the VETFEM is straightforward, and follows along the lines of the conventional FEM. The VETFEM implementation used herein employs a standard frontal solution scheme in which only minor modifications were made to accommodate a variable number of nodes per element. The primary difference between VETFEM and FEM algorithms comes in the element formulation, culminating in the calculation of the shape function gradients. The VETFEM shape function polynomial coefficients are computed prior to the assembly and solution phases of the calculation. Formation of the element stiffness matrix is entirely standard. Owing to the fact that the integrands are polynomials, the stiffness matrix integrals may be evaluated exactly. On the other hand, for nonlinear applications stress sampling points and numerical integration must be used, as in the conventional FEM. See Rashid (1998), Rashid and Gullett (2000), and Evans (1993) for discussions of various aspects of numerical quadrature on regions of arbitrary shape.

VETFEM meshes may be generated using a variety of methods. The Voronoi diagram of a domain with respect to a random distribution of nuclei makes an entirely satisfactory VETFEM mesh. Of course, conventional FEM meshes can be used in the VETFEM as well. Alternatively, a simple rule based recursive subdivision strategy has been employed to generate the results presented in section 5.

In this strategy, a region of the domain is selected for subdivision into either two or three subregions, based on a small set of simple rules. In the present 2D setting, the perimeter of the domain serves to define the initial region, and subdivision halts when a user prescribed target is reached for the mesh parameter h. The subdivision algorithm allows for arbitrary mesh gradation to be included in a simple way, although the results reported herein were generated using homogeneous meshes for purposes of comparison with uniform-mesh FEM results.

4. CONVERGENCE

This section contains a brief account of some aspects of convergence of the VET-FEM. No claims of mathematical rigor are made; rather, the intention is to provide an outline of the issues involved in proving convergence of the VETFEM. In particular, questions of existence and uniqueness of solutions, basic to any conventional discussion of convergence, are largely bypassed here. A fuller account is to appear elsewhere. To provide a point of departure, convergence of conforming methods is addressed first.

In what follows, homogeneous essential boundary conditions ($\bar{\mathbf{u}} = \mathbf{0}$) will be assumed for economy of presentation, so that the sets V and V_0 are identical (see equation (3.2)). This restriction engenders no essential loss in generality. In standard (conforming) Galerkin approximations to elliptic VBVPs, the fundamental error bound is provided by Céa's Lemma (see Strang and Fix, 1973), which states:

$$\|\mathbf{u} - \mathbf{u}_h\| \leq C \inf_{\mathbf{v}_h \in V_h} \|\mathbf{u} - \mathbf{v}_h\| \,. \tag{4.1}$$

Here, $\mathbf{u} \in V$ is the exact solution to the VBVP; its existence and uniqueness is assured if the bilinear form $a(\cdot, \cdot)$ satisfies a suitable definition of ellipticity. $\mathbf{u}_h$, on the other hand, is the approximate solution, and belongs to a finite-dimensional function space V_h spanned by $\{\Phi_1, \ldots, \Phi_N\}$ (see equation (3.3)). In (4.1), h is a mesh parameter which may be taken as the diameter of the largest element, C is a constant independent of the mesh, and $\|\cdot\| = \sqrt{a(\cdot, \cdot)}$ is the *energy norm* induced by the bilinear form. Céa's Lemma asserts that the approximate solution minimizes the energy norm of the error. The standard Galerkin method is therefore seen to be optimal, and the approximate solution may be viewed as the projection of the exact solution onto V_h.

In view of (4.1), convergence of the standard Galerkin method is seen to hinge on the relationship between V and V_h in the limit $h \to 0$. Specifically, if this limit results in an approximation space which is dense in V under the norm $\|\cdot\|$, then convergence is implied. Intuitively, this occurs if arbitrary polynomials of suitable order can be represented on each element. For second order differential operators such as (3.1), this so-called approximability condition (or *completeness* in the engineering literature) requires that arbitrary linear functions be representable on each element. Isoparametric elements of standard type all meet this condition, and convergence is therefore assured.

Turning now to nonconforming methods such as the VETFEM, most of the results (e.g. Céa's Lemma) relating to convergence of Galerkin approximations are unavailable, and in fact very few results of a general nature can be proven.

Nonconforming basis functions give rise to an approximation space V_h which is not contained by the actual solution space; i.e. $V_h \not\subset V$. The approximation is therefore termed *exterior*. Furthermore, the bilinear form $a(\cdot, \cdot)$ is not defined on V_h, because the discontinuities at inter-element interfaces lead to products of Dirac delta functions in the definition of the bilinear form. It is therefore necessary to modify the VBVP itself by replacing the bilinear form with an approximate one $a_h(\cdot, \cdot)$. In nonconforming finite element methods, this is typically done by replacing the integral over Ω in $a(\cdot, \cdot)$ with an assemblage of element contributions, thus ignoring the singularities on the element boundaries. With this replacement, the approximate problem becomes

$$\text{find } \mathbf{u}_h \in V_h \text{ such that } a_h(\mathbf{u}_h, \mathbf{v}_h) = f(\mathbf{v}_h) \text{ for any } \mathbf{v}_h \in V_h, \qquad (4.2)$$

where it is noted that the linear form $f(\cdot)$ is retained from the original VBVP, and is well-defined on V_h.

Convergence evaluations for nonconforming methods are complicated by the fact that the approximate problem (4.2) differs from the original VBVP. As with the original VBVP, completeness (i.e. the approximability condition) requires that arbitrary linear functions on elements be representable. VETFEM interpolants are readily seen to satisfy this requirement, based on the manner in which their polynomial shape functions are constructed. Satisfaction of the approximability condition means that the expanding approximation space V_h comes continuously closer, under mesh refinement, to containing the exact solution. Clearly this is a necessity if the method is to be convergent. However, in the absence of Céa's Lemma (4.1), there remains the question of convergence of the approximate solution $\mathbf{u}_h \in V_h$ to the exact solution, and not to some other function, as the approximation space expands. Or, equivalently, does the exact solution satisfy the approximate VBVP (4.2) in the limit as $h \to 0$? This issue can be referred to as *consistency*, in somewhat loose analogy to the precisely defined notion of consistency in the finite difference method. So, it might be said that completeness relates to the *availability* of the exact solution (or increasingly close approximations thereto) in the approximation space V_h, whereas consistency refers to the ability of the method to *select* the right function from V_h.

Given the completeness of the approximation space V_h, the basic tool for evaluation of consistency is provided by the following result (Strang and Fix, 1973; Ciarlet, 1978):

$$\|\mathbf{u} - \mathbf{u}_h\|_h \leq C \left[\inf_{\mathbf{v}_h \in V_h} \|\mathbf{u} - \mathbf{v}_h\|_h + \sup_{\mathbf{v}_h \in V_h} \frac{|a_h(\mathbf{u}, \mathbf{v}_h) - f(\mathbf{v}_h)|}{\|\mathbf{v}_h\|_h} \right]. \qquad (4.3)$$

This error bound is applicable to nonconforming methods with elliptic bilinear forms, and has come to be known as "Strang's Lemma." In (4.3), $\|\cdot\|_h$ is the energy norm induced by the approximate bilinear form, and C is a constant that is independent of the mesh. In the event that the approximation space $V_h \subset V$ (i.e. is conforming), the second term vanishes, and Strang's Lemma reduces to Céa's Lemma. It is therefore seen that the former is a nonconforming generalization of the latter. The second term in (4.3) is often called the "consistency term."

Application of Strang's Lemma to conclusively show convergence is not at all trivial, and usually must be linked to specific BVPs. The *patch test* (Irons and Loikkanen, 1983; Taylor *et al.*, 1986; Zienkiewicz and Taylor, 1997) is a practical way in which to gain insight into both completeness and consistency for particular types of boundary data, and has been used by the engineering community since the 1960s. In the patch test, an arbitrarily discretized domain is subjected to boundary conditions that implicate a linearly varying solution. The test is passed if the solution at all degrees of freedom in the mesh agrees with the underlying linear solution. The patch test is typically carried out as a matter of numerical experimentation, with a strong implication of general good behavior emerging from a sufficiently large body of numerical evidence. On the other hand, Strang (1972) (see also Strang and Fix, 1973) has introduced an analytic version of the patch test, which, while somewhat more difficult to perform than the numerical one, removes the specificity to particular mesh configurations. The patch test assesses completeness and consistency for the particular boundary data used in the test, and its passage is therefore clearly necessary for convergence more generally. However, its sufficiency remains in dispute. Arguments in favor of its sufficiency rest upon the assumption that a certain correspondence between $a(\cdot, \cdot)$ and $a_h(\cdot, \cdot)$ for a globally linear solution implies correspondence for all locally linear solutions. This conclusion, although intuitive and reasonable, apparently has never been proven. In any case, it bears mention that the VETFEM has been seen to pass the numerical patch test in extensive experimentation, and in addition can be shown to pass Strang's analytic patch test as well (Gullett, 2001).

In an important related development, Stummel (1979) has put forth what he calls the "generalized patch test," which is shown by him to constitute a necessary and sufficient condition for convergence of a nonconforming method that is applied to an elliptic bilinear form. Stummel's theorem considers the behavior of a suitably defined quantity for a sequence of meshes whose mesh parameters h_i decrease monotonically to zero as $i \to \infty$. Specifically, in the present context a tensor $\mathbf{S}_i$ is defined by

$$\mathbf{S}_i = \sum_{1 \le a \le L_i} \int_{\partial \Omega_a} \psi \, \mathbf{u}_{h_i} \otimes \mathbf{n} \, \mathrm{d}s \tag{4.4}$$

for each mesh in the sequence, where $\psi \in C_0^\infty(\Omega)$ is an arbitrary test function, homogeneous on $\partial \Omega$. A necessary and sufficient condition for convergence of the approximate solution to the exact one is that $\mathbf{S}_i \to 0$ as $i \to \infty$. This result holds trivially for a conforming method, because the integrands in (4.4) cancel pairwise on all internal element interfaces in this case. For nonconforming methods it must be proven. In the case of the VETFEM, the result (4.4) can be shown to hold with some effort, and depends crucially on the consistency constraints described in subsection 3.2.2.

5. COMPUTATIONAL EXAMPLES

A simple elastostatic problem is considered for the purpose of illustrating the convergence of VETFEM solutions. The L-shaped domain shown in Figure 3 is loaded by a uniform normal traction on the left-hand edge, and is constrained against horizontal displacement on the right-hand edge. Additionally, the upper-most node on the right-hand edge is constrained against vertical displacement as well. The material properties were taken to be $E = 10^5$ psi, $v = 0.3$, whereas the normal traction on the left-hand edge was $0.01E$. The solution was carried out under plane strain conditions. The figure shows both a conventional FEM mesh consisting of standard four node isoparametric elements, and a VETFEM mesh with a similar mesh parameter h. Four different sets of analyses were carried out, with a distinct mesh parameter h corresponding to each set. In each set, 10 completely different VETFEM mesh realizations were used, along with one conventional FEM mesh for reference. The VETFEM meshes were generated using the subdivision strategy mentioned above. The present tactic of reporting results for several different meshes of similar h values is intended to convey a practically useful sense of both the rate of convergence with h, and the accuracy of the solution for a given level of mesh refinement.

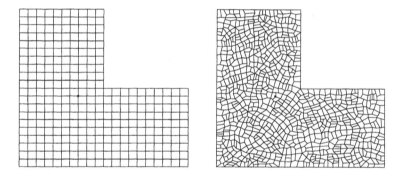

Figure 3. FEM (left) and VETFEM meshes corresponding to similar mesh parameters. Stress and displacement results are reported in Figure 4 for the location indicated with a dot.

Representative analysis results are shown in Figure 4. The figure shows the the vertical displacement component and the maximum principal stress at the location $(0.7, 0.9)$ as functions of the mesh parameter. The results of all 40 VETFEM analyses as well as of FEM runs corresponding to each of the four analysis sets appear in the graphs. The displacement values were normalized by the magnitude of a reference displacement, whereas the stresses were normalized by a reference maximum principal stress. These reference values, in turn, were obtained by averaging the 10 VETFEM results with the highest level of mesh refinement. Displacement and stress values were computed directly from the VETFEM or FEM interpolant at the sampling point; i.e. no smoothing from nodal or other values was performed. Whereas evaluation of the stress at an arbitrary point within an element is not normally regarded as the most accurate way to access a finite element solution, this approach was taken here in order to remove extraneous "post-processing" aspects from the presentation of results.

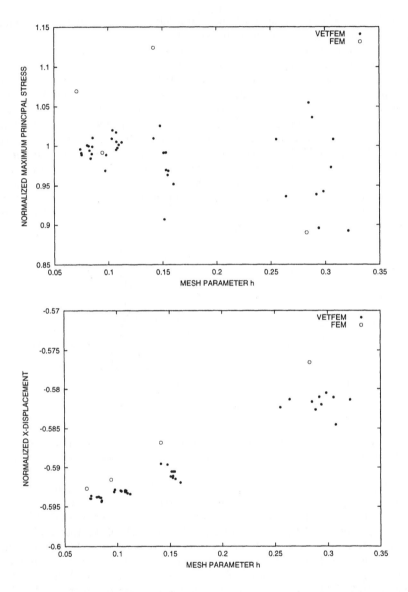

Figure 4. VETFEM and FEM results for the maximum principal stresses (top) and
vertical displacement component at location (0.7, 0.9). The ordinate values
were normalized as described in the text.

Examination of Figure 4 suggest an $O(h)$ rate of convergence for both the
stresses and displacements; obviously so in the latter case. This is the expected
result based on the well known performance of conventional bilinear quadrilateral
elements. Also, it is noted that the VETFEM results appear to compare favorably
with the conventional FEM results at a given level of mesh refinement, at least for

this specific problem. However, it bears mention that the numbers of VETFEM elements and nodes required to achieve a given value of *h* was typically 2.5 to 3 times greater than the corresponding numbers for the FEM meshes. This is largely due to the fact that the simple domain geometry allowed for square finite elements of uniform size in this particular problem, whereas the element diameter in the VETFEM meshes typically ranged over an interval roughly equal to the average element diameter. Finally, it is mentioned that the VETFEM mesh generation algorithm ensures that all elements have no fewer than four nor more than eight nodes. Most elements contain four, five, or six nodes, with a typical median of 4.6 nodes per element.

6. CONCLUSION

The VETFEM is attractive in relation to the conventional FEM chiefly because it engenders fewer and less restrictive constraints on the mesh generation process. By employing methods familiar in nonconforming approximation theory, the VET-FEM can be shown to be a convergent method. Also, the example problem studied herein suggests that the method's performance in practice is competitive with that of the standard FEM. In terms of implementation architecture, the VETFEM is virtually identical to the FEM. The price of the added meshing flexibility is that the computational expense of generating the shape functions is greater than for the conventional method. However, this cost increases only linearly with the problem size, and has been seen in practice to account for only a small fraction of the total processor time in all but the smallest problems.

Extension of the VETFEM to 3D is currently being pursued, and will be reported on in the future. The most straightforward extension engenders a restriction to planar element facets. Even with this constraint, however, automatic generation of 3D VETFEM meshes is considerably simpler than for high quality conventional meshes, particularly if hexahedral elements are required.

7. ACKNOWLEDGEMENT

It is a pleasure to acknowledge the support of the joint NSF/Sandia LCE Program under NSF grant CMS-9732331 with the University of California, Davis. Also, MMR is grateful for a visiting appointment at the Technical University of Denmark, during which the manuscript was prepared.

8. REFERENCES

Babuska, I., and Melenek, J. M., 1997, The partition of unity method. *International Journal for Numerical Methods in Engineering,* **40**, pp. 727-758.
Belytschko, T., Lu, Y. Y., Gu, L., 1994, Element-free Galerkin methods. *International Journal for Numerical Methods in Engineering,* **37**, pp. 229-256.
Belytschko, T., Krongauz, Y., Organ, D., Fleming, M., and Krysl, P., 1996, Meshless methods: An overview and recent developments. *Computer Methods in Applied Mechanics and Engineering,* **139**, pp. 3-47.

Chen, J. S., Pan, C., Wu, C. T., and Liu, W. K., 1996, Reproducing kernel particle methods for large deformation analysis of nonlinear structures. *Computer Methods in Applied Mechanics and Engineering,* **139**, pp. 195-227.

Ciarlet, P. G., 1978, *The Finite Element Method for Elliptic Problems,* volume 4 of: *Studies in Mathematics and its Applications,* (Amsterdam: North Holland).

Dohrmann, C. R., and Rashid, M. M., 2001, Polynomial approximation of shape function gradients from element geometries. *International Journal for Numerical Methods in Engineering,* in press.

Dolbow, J., and Belytschko, T., 1999, Numerical integration of the Galerkin weak form in meshfree methods. *Computational Mechanics* **23**, pp. 219-230.

Duarte, C. A., and Oden, J. T., 1996, An h-p adaptive method using clouds. *Computer Methods in Applied Mechanics and Engineering,* **139**, pp. 237-262.

Evans, G., 1993, *Practical Numerical Integration,* (Chichester: John Wiley and Sons).

Gullett, P. M., 2001, *The Variable Element Topology Finite Element Method,,* PhD dissertation, University of California at Davis.

Irons, B. M., and Loikkanen, M., 1983, An engineer's defense of the patch test. *International Journal for Numerical Methods in Engineering,* **19**, pp. 1391-1401.

Mukherjee, Y. X., and Mukherjee, S., 1997, On boundary conditions in the element-free Galerkin method. *Computational Mechanics,* **19**, pp. 264-270.

Rashid, M. M., 1998, The arbitrary local mesh replacement method: An alternative to remeshing for crack propagation analysis. *Computer Methods in Applied Mechanics and Engineering,* **154**, pp. 133-150.

Rashid, M. M., and Gullett, P. M., 2000, On a finite element method with variable element topology. *Computer Methods in Applied Mechanics and Engineering,* **190**, pp. 1509-1527.

Strang, G., 1972, Variational crimes in the finite element method. In *The Mathematical Foundations of the Finite Element Method with Applications to Partial Differential Equations,* edited by Aziz, A. K., (New York: Academic Press), p. 797.

Strang, G., and Fix, G. J., 1973, *An Analysis of the Finite Element Method,* (Englewood Cliffs, NJ: Prentice Hall).

Stummel, F., 1979, The generalized patch test. *SIAM Journal on Numerical Analysis,* **16**, pp. 449-471.

Sukumar, N., Moran, B., and Belytschko, T., 1998, The natural element method in solid mechanics. *International Journal for Numerical Methods in Engineering,* **43**, pp. 839-888.

Swegle, J. W., Hicks, D. L., and Attaway, S. W., 1995, Smoothed particle hydrodynamics stability analysis. *Journal of Computational Physics,* **116**, pp. 123-134.

Taylor, R. L., Simo, J. C., Zienkiewicz, O. C., and Chan, A. C., 1986, The patch test: A condition for assessing finite element convergence. *International Journal for Numerical Methods in Engineering,* **22**, pp. 39-62.

Zienkiewicz, O. C., and Taylor, R. L., 1997, The finite element patch test revisited: A computer test for convergence, validation, and error estimates. *Computer Methods in Applied Mechanics and Engineering,* **149**, pp. 223-254.

Meshless Methods for Life Cycle Engineering Simulation: Natural Neighbour Methods

Brian Moran[1] and Jeong Yoo[1]

1 INTRODUCTION

Meshless or meshfree methods have been developed in recent years to circumvent restrictions imposed by formal element connectivity requirements in traditional finite element methods. A broad class of methods such as the element-free Galerkin method (Belytschko, Lu and Gu, 1994) is based on moving least squares They show tremendous promise for problems involving large deformations as well as moving interfaces or discontinuities. An overview of developments in meshfree methods, was given by Belytschko et al. (1996). Other promising recent methods include the Reproducing Kernel Particle Method (Liu, Jun and Zhang, 1995), h-p clouds (Duarte and Oden, 1996), the partition of unity finite element method (Melenk and Babuska, 1996), and the meshless local Petrov-Galerkin method (Atluri and Zhu, 1998), to mention just a few. In this paper, a class of meshfree methods called *natural neighbour* methods are reviewed and their potential for application to life cycle engineering simulations is discussed.

Natural neighbour methods are based on so-called natural neighbour coordinates which are well-known interpolants for multivariate data fitting (here, the term coordinates should be interpreted as akin to shape functions). The appeal of natural neighbour coordinates is that they are interpolants for irregular nodal configurations. This same feature makes them highly attractive for meshless, Galerkin type schemes for solving partial differential equations. Natural neighbour coordinates are based on, and developed via, the Voronoi tessellation of a set of nodes. They were first introduced by Sibson (1980). The natural neighbours of a point are the nodes. Braun and Sambridge (1995) introduced the Sibson natural neighbuor interpolants for the solution of partial differential equations in a scheme termed the Natural Element Method (NEM). Sukumar, Moran and Belytschko (1998) demonstrated the usefulness of the NEM for problems in solid mechanics, with emphasis on linear elasticity. Sukumar and Moran (1999) showed that the method could be extended to problems requiring C^1 continuity (such as plate bending) thorugh the use of Bezier-splines in conjunction with the Sibson shape

---

[1] Northwestern University

functions.   Beuche, Sukumar and Moran (2000) investigated the dispersive properties of the NEM and showed that the overall behavior of NEM in elastodynamics was much improved over finite elements. In that work, both consistent mass matrices and a lumped mass matrix based on the nodal Voronoi cell were used. Reeves and Moran (2000) showed the utility of the NEM for contaminant transport problems.

A key feature of the natural neighbour coordinates is that they are interpolants and that they are precisely linear on convex boundaries (alternatively stated, the shape functions on convex boundaries depend only on nodes on the boundary and not on interior nodes). This means that essential (Dirichlet) boundary conditions may be applied exactly as in finite elements, without modification. On non-convex boundaries, sufficient discretization is required to assure that deviations from linearity (or deviations from conformal behavior at an interior interface) are slight. Sukumar et al. (1998) showed that excellent results are obtained even with non-convex boundaries using this approach.   Cueto, Doblare and Gracia (2000) introduced a modification of the NEM based on the concept of α-shapes which essentially renders the Sibson shape functions conforming on non-convex boundaries. Belikov et al. (1997) showed that natural neighbour coordinates or interpolants are not unique and they proposed a new non-Sibsonian interpolant. Sukumar, Moran, Semenov and Belikov (2001) introduced the non-Sibsonian interpolant into a Natural Neighbour Galerkin method. They showed that the non-Sibsonian interpolant is precisely linear on non-convex boundaries and interfaces and thus essential boundary conditions may be exactly imposed on any type of boundary without correction of the shape functions. The non-Sibsonian shape functions are also more efficient to compute than their Sibsonian counterparts.

All of the above Galerkin schemes require integration of the weak form of the governing equations. This is typically accomplished using Gaussian quadrature on background cells. The topological dual of the Voronoi tessellation is a Delaunay triangulation and quadrature over these triangles was used for integration of the weak form. This lead to poor satisfaction of the patch test, with errors in the third or fourth significant digit as opposed to near machine precison. Yoo, Moran and Chen (2001) developed a Nodal Natural Neighbour Method. In the Nodal Natural Neighbour method, the stabilized conforming nodal integration (SCNI) scheme of Chen et al. (2001) was introduced into the natural neighbour method in conjunction with the non-Sibsonian coordinates. Key features of the Nodal Natural Neighbour method are the ease and efficiency of implementation, significant improvement in the patch test (near machine precision), significantly improved behavior for near-incompressible problems, and accurate computation of strain gradients.

In the remainder of the paper, the above features of natural neighbour methods are described in more detail. Numerical examples are presented to illustrate the performance of the method, with particular emphasis on the Nodal Natural

Neighbour method. Some concluding remarks address future developments of the method.

The outline of the paper is as follow. In the following section, Sibsonian and non-Sibsonian interpolants are described and key properties highlighted. In section 3, the nodal integration scheme for the weak form is described and its key properties are discussed. Some properties of the nodal integration scheme including possible instability associated with the choice of numerical integration of the Voronoi cell boundary integrals as well as the treatment of near incompressibility are presented in section 4. In section 5 numerical examples illustrating the behavior of the method in patch tests and several problems in elastostatics including near incompressibility are presented. Concluding remarks are made in Section 7.

2 NATURAL NEIGHBOUR COORDINATES

Sibson (1980) developed the concept of natural neighbour interpolation as a means for multivariate data fitting. The Voronoi diagram is the basis for the development of natural neighbour coordinates. The nodal placement is arbitrary and there is no formal connectivity requirements between the nodes. Rather the connectivity emerges in the development of the shape functions through the natural neighbour concept. In Figure 1(a), the Voronoi diagram for a set of five nodes is presented. The Voronoi diagram (also called the 1st order Voronoi diagram) is defined through

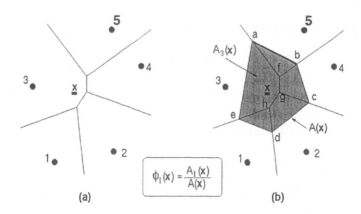

Figure 1 (a) Voronoi diagram for a set of five nodes in the plane. (b) Second order Voronoi polygons for definition of Sibson natural neighbour coordinates, ϕ_I.

the division of the plane in Figure 1(a) into regions T_I (Voronoi polygons) associated with each node I. The Voronoi polygon is given by

$$T_I = \left\{ \mathbf{x} \in R^2 : d(x, x_I) < d(x, x_J) \forall J \neq I \right\} \tag{1}$$

where $a(\mathbf{x}, \mathbf{x}_J)$ is the Euclidean distance between points $\mathbf{x}$ and $\mathbf{x}_J$.

The next step in defining natural neighbour coordinates at a point $\mathbf{x}$ is to introduce that point into the Voronoi tessellation. If $\mathbf{x}$ is considered as a node along with the other nodes in the set, the Voronoi diagram in Figure 1(b) is obtained. The nodes which share a facet of the Voroni polygon $T_{\mathbf{x}}$ (*abcde*) about $\mathbf{x}$ are the natural neighbours of $\mathbf{x}$ (in this case all five nodes are natural neighbours of $\mathbf{x}$). Different geometric features of the original tessellation and the tessellation incorporating the point $\mathbf{x}$ are used to define the different natural neighbour coordinates or shape functions for representing the primary variables in Galerkin methods for solving partial differential equations. For example, the displacement field, $\mathbf{u}$, in elasticity is written as a linear combination of the shape functions ϕ_I as

$$\mathbf{u}(\mathbf{x}) = \sum_{I=1}^{n} \phi_I(\mathbf{x}) \mathbf{u}_I \tag{2}$$

where $\mathbf{u}_I$ are the nodal data or coefficients and n is the number of natural neighbours. The procedures for Sibson and non-Sibsonian coordinates are described below.

2.1 Sibson Interpolation

Sibson natural neighbour coordinates are defined through a partition of unity based on the second order Voronoi polygons about $\mathbf{x}$. The second order polygons about $\mathbf{x}$ are the intersections of $T_{\mathbf{x}}$ with the original polygons T_I (see Sukumar et al., (2001) for more details). The coordinates are defined as the ratio of the intersection area to the area of $T_{\mathbf{x}}$, i.e.,

$$\phi_I(\mathbf{x}) = \frac{A_I(\mathbf{x})}{A(\mathbf{x})} \tag{3}$$

where $A(\mathbf{x})$ is the area of $T_{\mathbf{x}}$ and $A_I(\mathbf{x})$ is the area of intersection of $T_{\mathbf{x}}$ and T_I, e.g., for node 3, $A_3(\mathbf{x})$ is the area of the polygon *afghe* in Figure 1(b).

The Sibson natural neighbour coordinates (or shape functions) have the following properties of positivity, interpolation and partition of unity, respectively (Sukmar et al, 1998; Sukumar et al., 2001)

$$0 \le \phi_I(\mathbf{x}) \le 1, \qquad \phi_I(\mathbf{x}_J) = \delta_{IJ}, \qquad \sum_{I=1}^{n} \phi_I(\mathbf{x}) = 1 \qquad (4)$$

where n is the number of natural neighbours. They also have the linear completeness property

$$\mathbf{x} = \sum_{I=1}^{n} \phi_I(\mathbf{x})\mathbf{x}_I \qquad (5)$$

A detailed description of these properties is given in Sukumar et al. (1998).

2.2 Non-Sibsonian Interpolation

To define the non-Sibsonian interpolants, consider again the Voronoi polygon of the point $\mathbf{x}$ of Figure 1(b). For illustrative purposes, a similar figure is shown in Figure.2, to which we now refer. The interpolant at the point $\mathbf{x}$, referred to here as the site p, is obtained by considering p as a node in the tessellation. The Voronoi polygon of p is denoted as $V(p)$ and its closure is denoted as $\overline{V}(p)$. The Voronoi polygons of the natural neighbours of p are denoted by $V(p_I)$ and their closure by $\overline{V}(p_I)$. Define $\partial V(p, p_J)$, which may be an empty set, as

$$\partial V(p, p_J) = \left\{ p : p \in \overline{V}(p_I) \cap \overline{V}(p_J), J \ne I \right\} \qquad (6)$$

For example, $\partial V(p, p_5)$ in Figure 2 is the common facet of the Voroni cells for site p and node p_5. The non-Sibsonian interpolant is defined as

$$\phi_I(p) = \frac{\dfrac{|\partial V(p, p_I)|}{d(p, p_I)}}{\sum_J \dfrac{|\partial V(p, p_J)|}{d(p, p_J)}} \qquad (7)$$

where $|\cdot|$ denotes the Lebesgue measure, which is the facet length here. The quantity $d(p, p_J)$ is the Euclidean distance between site p and the node p_J. The non-Sibsonian interpolants also have the properties (4) and (5).

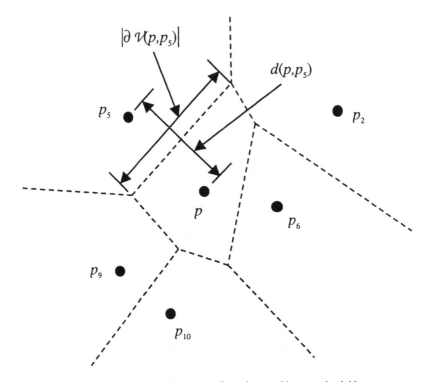

Figure 2 Voronoi polygons for a site p and its natural neighbours

2.3 Treatment of Essential Boundary Conditions

Both the Sibson and non-Sibsonian coordinates (or shape functions) described above are interpolants. This means that the nodal coefficients are the actual values of the nodal variables and this has implications for the application of essential or Dirichlet boundary conditions in Galerkin schemes. The Sibson interpolant is linear on convex (including straight line) boundaries. Together with the interpolant property, this means that, unlike other meshless methods which require correction procedures to enforce essential boundary conditions, essential boundary conditions (on convex boundaries) can be applied exactly as in finite elements. However, the Sibson interpolant is not precisely linear on non-convex boundaries. Sukumar et al.

(1998) show that, with sufficient refinement on the boundary, deviations from linearity (which is essentially non-conforming behavior) on non-convex boundaries or material interfaces is slight. Cueto et al. (2000) introduced a method based on α–shapes to correct the behavior on non-convex boundaries, i.e., to render the Sibson interpolant conforming.

With the non-Sibsoninan interpolant, however, the behavior is precisely linear on both convex and non-convex boundaries or interfaces. No correction procedure or modification of the interpolant is required beyond subdomain interpolation (see Sukumar et al, (2001) for details). Non-Sibsoninian interpolants between two points *A* and *B* on the boundary of an inclusion are shown in Figure 3 where it can be seen that the interpolants are precisely linear whether viewed from the inclusion (convex) or matrix (non-convex) sides. This also implies that coupling of the natural neighbour method based on non-Sibsonian shape functions to linear finite elements can be accomplished directly without need for correction or transition functions (Sukumar et al., 2001).

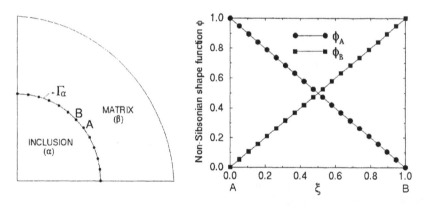

Figure 3 Linear approximation along a non-convex boundary (non-Sibsonian interpolant)

3 IMPLEMENTATION OF THE WEAK FORM

The governing equations (strong form) for linear elasticity are given by

$$\sigma_{ij,j} + f_{,i} = 0 \quad \text{in } \Omega$$

$$\varepsilon_{ij} = u_{(i,j)}$$

$$\sigma_{ij} = C_{ijkl}\varepsilon_{kl} \tag{8}$$

$$u_i = \bar{u}_i \qquad \text{on } \partial\Omega_u$$

$$\sigma_{ij}n_j = \bar{t}_i \qquad \text{on } \partial\Omega_\sigma$$

The corresponding weak or variational form (principle of virtual work) is

$$\int_\Omega \sigma_{ij}\delta\varepsilon_{ij}\,d\Omega = \int_\Omega f_i\delta u_i\,d\Omega \quad \forall\,\delta u_i \in V \tag{9}$$

Here, V is the space of kinematically admissible functions (satisfying homogeneous boundary conditions).

3.1 Standard Gradient Operator and Gaussian Quadrature

To implement the weak form, a gradient operator can be defined in the usual fashion by differentiating (2). Thus the strains are given by

$$\varepsilon_{ij} = u_{(i,j)} = \sum_I \tfrac{1}{2}\left(\phi_{I,i}u_{Ii} + \phi_{I,j}u_{Ij}\right) \qquad \{\varepsilon\} = \begin{Bmatrix} \varepsilon_{11} \\ \varepsilon_{22} \\ 2\varepsilon_{12} \end{Bmatrix} = \sum_I \mathbf{B}_I\mathbf{d}_I \tag{10}$$

where $\mathbf{d}_I$ are the nodal displacements (recall that all the shape functions we are dealing with are interpolants) and the symmetric gradient operator is given by the usual $\mathbf{B}$ matrix with nodal submatrices

$$\mathbf{B}_I = \begin{bmatrix} \phi_{,1} & 0 \\ 0 & \phi_{,2} \\ \phi_{,2} & \phi_{,1} \end{bmatrix} \tag{11}$$

Details on how to obtain expressions for the derivatives of the Sibson shape functions are given in Sukumar et al. (1998). A similar procedure is used for the gradients of the non-Sibsonian shape functions. With the strain thus defined, the implementation of the waek form is completed through choice of a suitable quadrature scheme. In implementations involving Gaussian integration (Sukumar, et al., 1998; Sukumar et al., 2001; Beuche et al., 2000; Sukumar and Moran, 1999) quadrature was carried out over background cells generated by the Delaunay triangulation (the topological dual of the Voronoi diagram). While this approach was generally successful, poor results were sometimes obtained in patch tests with errors in the third or fourth significant digit. Many meshless method exhibit this kind of behavior which is a result of the use of Gaussian quadrature, in conjunction with the non-polynomial nature of the functions to be integrated, and the lack of coincidence of the integration domains with the shape function supports. Dolbow and Belytschko (1999) developed a strategy for the resolution of this problem in the context of Gaussian integration and the element-free Galerkin method.

3.2 Strain Smoothed Gradient Operator and Nodal Integration

A stabilized conforming nodal integration (SCNI) scheme was developed by Chen et al. (2001) as an alternative to Gaussian quadrature on background cells. The method uses a stabilized smoothed strain field, defined through a weighted integal around the boundary of the Voronoi cell associated with each node. The nodal integration scheme gave improved performance over Gaussian quadrature when used for integration of the weak form in the reproducing kernel particle method. Because natural neighbour interpolants are based on the Voroni digram to begin with, this nodal integration scheme is ideally suited for natural neighbor methods. Yoo, Moran and Chen (2001) introduced a Nodal Natural Neighbour method based on the above-mentioned nodal integration scheme. In Yoo et. al. (2001), the strain is defined as (see Figure 4)

$$\tilde{\varepsilon}_{ij}\left(\mathbf{x}_L\right)=\frac{1}{A_L}\int_{\Gamma_L}\frac{1}{2}\left(u_i n_j + u_j n_i\right)\!d\Gamma \tag{12}$$

where Γ_L is the boundary of the Voroni cell associated with node L, possibly cut by the boundary of the domain or material subdomain of interest. In Figure 4, the integration contour over the six faces of Ω_{10} ($V\!\left(p_{10}\right)$ cut by the boundary Γ^h) is shown.

Introducing the interpolation (2) into (12) leads to the following expressions for the strain matrix and the nodal sub-matrices of the symmetric gradient operator (evaluated at each node $\mathbf{x}_L$)

$$\left\{\tilde{\varepsilon}\right\}=\left\{\begin{array}{c}\tilde{\varepsilon}_{11}\\ \tilde{\varepsilon}_{22}\\ 2\tilde{\varepsilon}_{12}\end{array}\right\}=\sum_I \mathbf{B}_I \mathbf{d}_I, \qquad \tilde{\mathbf{B}}_I\left(\mathbf{x}_L\right)=\frac{1}{A_L}\int_{\Gamma_L}\left(\begin{array}{cc}\phi_I n_1 & 0\\ 0 & \phi_I n_2\\ \phi_I n_2 & \phi_I n_1\end{array}\right)d\Gamma \tag{13}$$

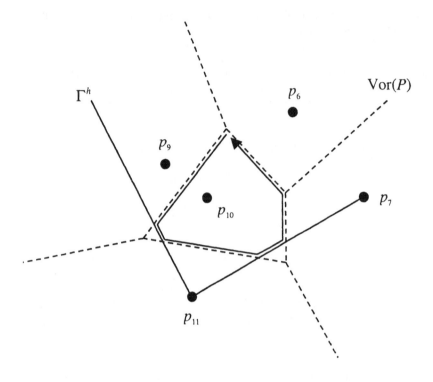

Figure 4 Contour integration over the boundary of the Voronoi cell (cut by the domain boundary)

Chen et al. (2001) evaluated the integrals around Γ_L using a trapezoidal rule. However, Yoo et al. (2001) noted that due to the intrinsic properties of the non-Sibsonian interpolant, the trapezoidal rule can give rise to spurious mechanisms. Such mechanisms are eliminated by using Gaussian quadrature on each facet. Note that this use of Gaussian quadrature is solely for purposes of evaluating the strain (or other gradients as needed). The integration of the weak form is achieved through nodal integration using the nodal strains as evaluated in (13). The resulting expression for the nodal stiffness matrices is given by a simple sum over the nodes of the integrand weighted by the area of the nodal Voronoi cell (with straightforward adjustments made for cells which are intersected by a boundary)

$$\mathbf{K}_{IJ} = \int_{\Omega^h} \tilde{\mathbf{B}}_I^T \mathbf{C} \tilde{\mathbf{B}}_J \, d\Omega = \sum_L \tilde{\mathbf{B}}_I^T (\mathbf{x}_L) \mathbf{C} \tilde{\mathbf{B}}_J (\mathbf{x}_L) A_L \qquad (14)$$

See Yoo et al. (2001) for additional details and properties of the nodal integration scheme in the nodal natural neighbour method as well details of the integration mechanism and its avoidance.

In Yoo et al. (2001), non-Sibsonian interpolants were used for all applications of the nodal natural neighbour scheme because of the advantages of this interpolant for essential boundary and interface conditions. Sibson coordinates could also be used where non-convex boundaries or interfaces are not an issue (or where the lack of conformity of the Sibson interpolant is made sufficiently negligible through use of refined nodal spacing on the boundaries). It would be interesting to compare accuracy when the Sibson interpolant is used in nodal natural neighbour method instead of the non-Sibsonian interpolant.

3.5 Strain Gradients

By repeated application of the nodal gradient operator procedure of (10) an (11), strain gradients may be readily calculated. This makes the nodal natural neighbour method a promising candidate for problems involving higher gradients. As a check on the accuracy of this approach, Yoo et al. (2001) calculated strain gradients in a 2D linear elasticity simulation of a beam bending problem and excellent results were obtained.

3.4 Near-Incompressible Problems

In Yoo et al. (2001), an interesting byproduct of the nodal integration scheme was observed, namely that in the solution of a linear elasticity problem (plate with a hole) for near-incompressible material behavior (Poisson's ratio as large as 0.499999999999 with no significant deterioration in the relative error over results for Poisson's ratio of 0.25; only slight oscillations in stress are observed) no modification of the method was required to avoid locking. The reason for this can be linked to the number of constraints involved: in the nodal integration scheme there is a single incompressibility constraint at each node and two degrees of freedom per node for an optimal constraint ratio of 2. The method thus appear very promising for near-incompressible problems such as in large deformation plasticity or slightly compressible hydrodynamics. Further investigation of this is warranted.

4 NUMERICAL EXAMPLES

Two numerical examples are presented to illustrate the performance of the nodal natural neighbour method. The patch test for various irregular nodal configurations serves to illustrate the improved accuracy of the nodal integration scheme over Gaussian quadrature. An example consisting of a plate with an inclusion indicates the accuracy of the method in handling problems with material interfaces and non-convex boundaries. Two-point Gaussian quadrature was used in the evaluation of the contour integral in the gradient operator in (13).

4.1 Patch Test

For the natural element method with Gaussian quadrature, machine precision is not achieved on patch tests. This is due to the non-polynomial nature of the shape functions and the lack of coincidence of the integration domains (background cells) and the shape function supports (which are unions of the circumcircles of the Delaunay triangles generated by the node and its natural neighbours). With the direct nodal integration scheme in the nodal natural neighbour method, however, excellent performance on patch tests is observed. In Table 1, patch test results for the three nodal configurations shown in Figure 5 are given. The results for the nodal natural neighbour method are significantly more accurate than the results for the natural element method which uses Gaussian quadrature.

Figure 5 Nodal configurations for patch tests with 8, 25 and 70 nodes, respectively

Table 1 Relative errors in $L^2(\Omega)$ norm in patch tests

Patch	NEM	NNNM
8 nodes	9.3E-3	8.6E-16
25 nodes	7.5 E-4	2.4E-16
70 nodes	4.4E-3	1.2E-15

4.2 Plate with Inclusion

The problem of a plate with an inclusion undergoing a transformation strain incorporates several interesting features of the nodal natural neighbour method.

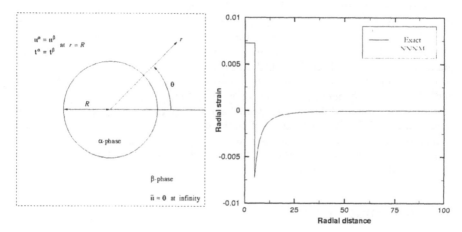

Figure 6 Inclusion in an infinite matrix. Results for radial strain as a function of radial distance.

Firstly, it involves Dirichlet boundary conditions and a material interface. The latter is a non-convex boundary from the viewpoint of the matrix. Secondly, because of the properties of the non-Sibsonian shape functions discussed above, namely precise linearity on convex or non-convex boundaries, the essential boundary and interface conditions are implemented exactly as in finite elements with no modification apart from defining the shape functions over the appropriate material subdomain of the body. Nodes on the interface belong to both subdomains and the shape function properties assure that the shape functions are perfectly conforming on the interface. Finally, the nodal integration is also carried out over the two subdomains which accounts for the discontinuity in strain and material properties across the interface (Yoo et al., 2001). In Figure 6, the results for the radial strain as a function of radial distance from the center of the inclusion are seen to be in excellent agreement with the exact solution (for an infinite domain).

5 CONCLUDING REMARKS

A summary of natural neighbour methods has been presented with emphasis on the recently developed nodal natural neighbour method (Yoo et al., 2001). For integration of the weak form of the governing equations, a nodal integration scheme is used. The nodal integration scheme is based on the same Voronoi tessellation used to compute the shape functions. The shape functions used are the so-called non-Sibsonian interpolants and thus essential boundary and interface conditions are implemented with no correction of the shape functions required. Patch test results for the nodal natural neighbour method are significantly improved over previous natural neighbour Galerkin implementations based on Gaussian quadrature. The nodal natural neighbour method shows excellent promise for problems with near incompressible material behavior. Also, the gradient operator used in the nodal integration scheme yields nodal values of strains, i.e., the nodes are the integration points and also the points at which strains and (in history dependent problems) internal variables are defined. Higher order gradients are computed through a repeated application of the gradient operator. For these reasons the nodal natural neighbour method is particularly promising for problems involving large deformations and/or higher order gradients. Our current emphasis is on the implementation of the method for small and large deformation classical and gradient plasticity.

ACKNOWLEDGEMENTS

The authors are grateful for the support of the National Science Foundation through grant CMS-9732319 to Northwestern University and for the opportunity to participate in the joint program on Life-Cycle Engineering Simulation with Sandia National Laboratories. Helpful discussions with Ted Belytschko, Northwestern University and Natarajan Sukumar, University of California at Davis are gratefully acknowledged.

6 REFERENCES

Atluri, S. and Zhu, T., 1998, A new meshless local Petrov-Galerkin (MLPG) approach. *Computational Mechanics*, 22, pp.117-127.

Belikov, V., Ivanov, V., Kontorovich, V., Korytnik, S., and Semenov, A., 1997, The non-Sibsonian interpolation: A new method of interpolation of the values of a function on an arbitrary set of points. *Computational Mathematics and Mathematical Physics*, 37(1), pp. 9-15.

Belytschko, T., Krongauz, Y., Organ, D., Fleming, M., and Krysl, P., 1996, Meshless methods: An overview and recent developments. *Computer Methods in Applied Mechanics and Engineering*, 139, pp. 3-47.

Belytschko, T., Lu, Y., and Gu, L., 1994, Element-free Galerkin methods. *International Journal for Numerical Methods in Engineering*, 37, pp. 229-256.

Beuche, D., Sukumar, N., and Moran, B., 2000, Dispersive properties of the natural element method. *Computational Mechanics*, 25, pp. 207-219.

Braun, J. and Sambridge, M., 1995, A numerical method for solving partial differential equations on highly irregular evolving grids. *Nature*, 376, pp. 655-660.

Chen, J.S., Wu, C. T., Yoon, S., and You, Y., 2001, A stabilized conforming nodal integration for Galerkin meshfree methods. *International Journal of Numerical Methods in Engineering*, 50, pp. 435-466.

Cueto, E., Doblare, M., and Gracia, L., 2000, Imposing essential boundary conditions in the natural element method by means of density-scaled •-shapes. *International Journal for Numerical Methods in Engineering*, 49(4), pp. 519-546.

Dolbow, J. and Belytschko, T., 1999 Numerical integration of the Galerkin weak form in meshfree methods. *Computational Mechanics*, 23, pp. 219-230.

Duarte, C. and Oden, J., 1996, A h-p adaptive method using clouds. *Computer Methods in Applied Mechanics and Engineering*, 139, pp. 237-262.

Liu, W., Jun, S., and Zhang, Y., 1995, Reproducing kernel particle methods. *International Journal for Numerical Methods in Fluids*, 20, pp. 1081-1106.

Melenk, J. and Babuska, I.,1996, The partition of unity finite element method: basic theory and applications. *Computer Methods in Applied Mechanics and Engineering*, 139, pp. 289-314.

Reeves, H. and Moran, B., 2000, Meshless methods in contaminant hydrology. Computational Methods in Water Resources XIII, pp. 713-718.

Sibson, R., 1980, A vector identity for the Dirichlet tessellation. *Mathematical Proceedings of the Cambridge Philosophical Society*, 87, pp. 151-155.

Sukumar, N., Moran, B. and Belytschko, T. B. 1998, The natural element method in solid mechanics. *International Journal of Numerical Methods in Engineering*, 43, pp. 839-887.

Sukumar, N. and Moran, B., 1999, C^1 natural neighbor interpolant for partial differential equations. *Nuerical Methods for Partial Differential Equations*, 15, pp. 417–447.

Sukumar,N., Moran, B., Semenov, Y., and Belikov, B., 2001, Natural neighbuor Galerkin methods. *International Journal of Numerical Methods in Engineering*, 50 pp. 1-27.

Yoo, J., Moran, B. and Chen J.S., 2001, A nodal natural neighbour method, submitted for publication

Aspects of Lifecycle Engineering for Reliable Microelectromechanical Systems Containing Multilayer Material Systems

Martin L. Dunn[1], Ken Gall[1], Yanhang Zhang[1] and Brian Corff[1]

1 INTRODUCTION

Microelectromechanical systems (MEMS) technologies have matured dramatically over the last decade. While they began as an offspring of silicon microelectronics technology, initially capitalizing on the microelectronics infrastructure, it has grown to be much broader, expanding to include additional materials and manufacturing processes. As such, MEMS abound with applications that require rich and direct connections between system performance and materials behavior issues. To date these connections are perhaps less open to design freedom than traditional mechanical systems because of a limited palette of material choices and limited microfabrication techniques that derive from compatibility with CMOS technology (see Spearing (2000) and Ehrfeld et al. (1999) for reviews on materials issues in MEMS and Madou (1997) for a review of microfabrication technologies). The simultaneous application of new materials and development of new fabrication techniques is growing rapidly; the result is and will continue to be increased flexibility for designers who are looking toward MEMS technologies for innovative solutions to more and more complex problems.

Of paramount importance to the development of reliable MEMS devices capable of performing sensing, actuation, and computing functions in a robust manner is a solid understanding of the thermomechanical behavior of MEMS materials and structures at length scales relevant to MEMS. The present understanding of damage development of MEMS materials at the appropriate length scales and the relationships between damage mechanisms and processing conditions, geometry, thermomechanical loading, and environmental conditions is quite limited. With an understanding of the microstructural damage mechanisms, one can in principal abstract this detailed information into a form that can be

---

[1] University of Colorado at Boulder

incorporated into material damage models, perhaps in terms of a set of internal state variables, that can then be used in finite element codes to create robust design tools for lifecycle engineering of MEMS.

In this chapter we describe a number of issues that arise in the lifecycle engineering of MEMS composed of multilayer material systems. Such material systems arise naturally in many of the existing fabrication technologies for MEMS which rely on the patterning, deposition, and etching of thin film materials. We address only some lifecycle engineering concerns particular for the use of multilayer films, recognizing that additional significant lifecycle engineering concerns exist that are not discussed here. For example, stiction (adhesion) between micromechanical structures and the substrate is a significant issue for many surface micromachined structures.

2 LIFECYCLE OF MEMS MULTILAYERS

Multilayer material systems abound in microelectromechanical systems (MEMS) applications, serving both active and passive structural roles. In these many applications dimensional control is a critical issue. Surface micromachined mirrors, for example, require optically flat surfaces. Holographic data storage and optical beam steering for display applications need reflective surfaces free of optical phase distortion to increase the signal-to-noise ratio, minimize cross-talk, and keep the system in focus. The necessary level of optical flatness is difficult to achieve due to curvature commonly seen in micromachined mirrors composed of several different material layers. For other applications, such as RF MEMS, cantilever and bridge like structures are used to make electrostatic switches. It is important to control warpage of these large area, thin actuator, structures in order to achieve desired deflection versus voltage relationships, on/off switching times, and RF frequency response. An example of a gold/polysilicon bimaterial cantilever beam that serves as the actuatable component in a micromechanical switch array is shown in Fig. 1 (Miller et al., 2001). An inherent characteristic of such multilayer material structures is that misfit strains between the layers (for example, due to intrinsic processing stresses or thermal expansion mismatch between the materials upon a temperature change) lead to stresses in the layers and deformation of the structures. These thermomechanical phenomena act over the lifecycle of a MEMS component, driven by various factors during various stages of the component lifetime including processing, post-processing steps such as packaging, and the intended use environment.

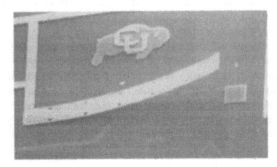

Figure 1 A 425 μm long gold/polysilicon actuatable microconnector. The initial curvature is caused by misfit strains; the beam deflects approximately 18 μm above the substrate.

3   THERMOMECHANICAL BEHAVIOR OF MEMS MULTILAYERS

As MEMS technologies largely derived from the microelectronics industry, so has the companion understanding of the thermomechanical behavior of multilayer film systems.   While for MEMS applications, the basic phenomena of stress and curvature development in multilayers is the same as in microelectronics applications, significant differences exist and must be well understood to design reliable MEMS. These include:

- In MEMS applications, the thicknesses of the layers are on the order of microns, and usually comparable. This leads to much larger deflections, relative to the thickness of structures, than are observed in microelectronics applications where films of thickness on the order of a micron are deposited on substrates several hundred microns thick. This can make it necessary to include geometric nonlinearity in order to accurately model deformation. Furthermore, the geometric nonlinearity can lead to bifurcations in the deformation behavior. These can be detrimental when dimensional stability is a requirement, or can be beneficial for actuator applications.
- Since the layers are of comparable thicknesses, stresses can vary appreciably through the thickness of the layers. As a result it may be necessary to consider not only stress, but also stress gradient effects on damage initiation processes. Furthermore, from a practical manner the average stress in the layer may not be suitable to characterize film stresses as it is in many microelectronics applications (for example via wafer curvature measurements and the use of the Stoney (1909) equation.
- Tradeoffs between stress and curvature exist. For example, for a given metal film thickness in a metal/polysilicon microstructure, varying the polysilicon thickness can reduce the stress in the metal, but at the expense of increasing the curvature. The significance of this is obvious as many MEMS applications

have strict deformation requirements, perhaps more severe than stress requirements.

- The patterning geometry of metal films on polysilicon (or other materials) may be arbitrary, as opposed to fully blanketed or thin line patterns.
- The curvature, and thus stresses, may vary significantly over the in-plane dimensions of the structure; thus, the average curvature may be insufficient to adequately describe the deformation state of the structure.
- MEMS microstructures are often subjected to many thermal and/or mechanical cycles since devices are often driven at resonance.

In terms of the understanding of thermomechanical behavior of multilayer films for MEMS, we make the following broad observations:

- Quite a bit of data has been accumulated for metal films, driven by microelectronics, but the results are not necessarily directly applicable to MEMS for the reasons described above. For example, extensive work has been performed to connect stress voiding to local crystallographic texture (see for example, Keller et al., 1999). They showed that void initiation sites were correlated with sites of local grain orientations that differ from a <100> texture. We note that the stress state in these studies is basically equibiaxial (in the plane of the film) and uniform through the thickness because it sits on a thick substrate. Similar issues are expected with regard to MEMS technologies, but the details may differ because the stress states are not equibiaxial and uniform.
- Modelling strategies for stress analysis of MEMS have been and are being developed. Regarding mechanical behavior, they typically consider only linear elastic material behavior (some consider geometric nonlinearity) and thus do not include important lifecycle issues such as material microstructure evolution and damage initiation and propagation.

In the remainder of this section we describe some of the phenomena we have studied in our laboratory over the past few years with regard to these issues. In order to organize the discussion, we break the phenomena into four general areas: deformation behavior after processing/fabrication, deformation behavior in the intended service environment, monotonic deformation behavior over time, and cyclic deformation behavior over time. We emphasize that this discussion is far from exhaustive regarding thermomechanical deformation as many interesting and still not well-understood phenomena exist.

In order to study the thermomechanical deformation of MEMS multilayers, we designed a series of beam and plate microstructures. The beams were up to 300 μm long x 50 μm wide. We recognize that the behavior of these microstructures really requires the use of plate, not beam, theory, but for convenience we will refer to them as beams throughout this paper. We also designed a series of circular gold/polysilicon plate microstructures with diameters ranging from 150 μm – 300 μm. Both the beams and plates were fabricated using the MUMPS surface micromachining process (Koester et al., 2001). This yielded a polysilicon layer either 1.5 or 3.5 μm thick, covered by a 0.5 μm thick gold layer.   The microstructures are supported on the substrate by a 16 μm diameter polysilicon post to enable them to rest as freely as possible. Scanning electron micrographs of typical beam and plate microstructures are shown in Fig. 2.

Full-field measurements of the out-of-plane displacement $w(x,y)$ of the microstructures were made using an interferometric microscope with scanning white light interferometry as the temperature was changed in a prescribed manner in a temperature chamber. From the measured $w(x,y)$ data, the curvature was computed by fitting $w(x,y)$ to a 6th-order polynomial in x and y and then differentiating this analytically to determine $\kappa_x(x,y)$ and $\kappa_y(x,y)$. From these pointwise values of curvature, the average curvature over the plate was computed.

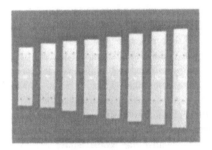

Figure 2 An array of gold/polysilicon beams (the longest is 300 μm). and a 300 μm diameter gold/polysilicon microstructure.

3.1 Deformation Behavior After Fabrication

Figure 3 shows the curvature vs. temperature behavior of two 300 μm x 50 μm gold/polysilicon beams for six and two heating and cooling cycles after release. Upon heating, the response appears to be theromoelastic up to a certain temperature, after which the curvature varies little with the increasing temperature. Upon cooling, the response is thermoelastic. The curvature upon return to room temperature is greater than that at the beginning of the cycle. Subsequent cycles are similar: the ΔT required before yield-like behavior increases (like following a stress-strain curve in the elastic-plastic region), and the subsequent room temperature curvature is successively larger.

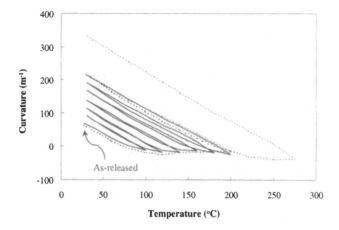

Figure 3 Curvature vs. temperature during thermal cycles of two beam microstructures conducted after release.

The nonlinearity in the curvature-temperature response is most likely caused by microstructural changes in the gold film, such as grain growth, which have been widely reported in the case of thin metal films on thick substrates (see for example, Nix, 1989; Keller et al., 1999). Upon heating for a sufficient period of time the microstructure will presumably stabilize. Here, neither the temperature nor time necessary has been studied. The impact of this deformation behavior after processing is significant for the lifecycle engineering of MEMS. As an example, consider the bilayer beam of Fig. 1 that serves as the active mechanical element in a switch array. A key design specification is the curvature of the beam after packaging. Suppose a packaging process exposes the beam to a temperature of 150°C. If the beam is designed to have the as-released curvature, the packaging process will substantially raise the curvature, resulting in device failure. If one understands this deformation response, though, it can be used in the design process to facilitate high device yield.

3.2 Deformation Behavior in Service

Numerous studies have elucidated the basic thermomechanical response of layered plates when subjected to temperature changes or other sources of misfit strains between the layers. These have come in the context of many technological applications, the most common being structural composite materials (e.g., Hyer, 1981) and thin film/substrate systems for microelectronics (e.g., Masters and Salamon, 1993; Finot and Suresh, 1996; Finot et al., 1997; Freund, 2000). When such a layered plate is subjected to a temperature change, two key aspects of deformation occur: straining of the midplane and bending. When the transverse deflections due to bending are of prime importance, one way to broadly characterize the deformation response, especially for plates with relatively large in-plane dimensions as compared to their thickness, is in terms of the average curvature developed as a function of temperature change. Formally, the curvature is a second-rank tensor, and for the type of layered plate problems considered here it can be wholly described by the two principal curvature components, e.g., in the x- and y- directions, κ_x and κ_y. The curvature is a pointwise quantity meaning it varies from point to point over the in-plane dimensions of the plate. To illustrate the nature of the deformation, consider the seemingly simple case of a plate with total thickness much less than the in-plane dimensions of the plate, composed of two isotropic layers with different material properties (elastic modulus and thermal expansion), subjected to a temperature change. In terms of the average curvature variation as a function of temperature change, three deformation regimes have been identified (Masters and Salamon, 1993; Finot and Suresh, 1996; Finot et al., 1997; Freund, 2000). The first regime, *I*, consists of a linear relation between the average curvature and temperature change where $\kappa_x = \kappa_y$, i.e., the average curvature is spherically symmetric. This deformation regime is characterized by both small transverse displacements and rotations and so conventional thin-plate theory adequately describes the deformation. The second regime, *II*, consists of a nonlinear relation between the average curvature and temperature, but again $\kappa_x = \kappa_y$. The behavior is due to geometric nonlinearity that results when the deflections become excessively large relative to the plate thickness and they contribute significantly to the in-plane strains. In these two regimes the symmetric deformation modes are stable. The second regime ends at a point when the deformation response bifurcates from a spherical to an ellipsoidal shape, i.e., $\kappa_x \neq \kappa_y$. At this point, the beginning of regime *III*, it becomes energetically favorable for the plate to assume the ellipsoidal shape because of the increased midplane straining required to maintain the spherical deformation under an increasing temperature change. After the bifurcation, the curvature in one direction increases while that perpendicular to it decreases; the plate tends toward a state of cylindrical curvature. This discussion has been cast in the context of linear material behavior. Additional deformation regimes result if material nonlinearity is present, for example, yielding (see Finot and Suresh, 1996, for a discussion of some of these issues).

These phenomena are illustrated in Fig. 4 which shows contour plots of the measured and predicted displacement field $w(x,y)$ at room temperature for the circular gold/polysilicon plates of all four sizes. The $D = 150$ µm samples deform in a spherically symmetric manner; contours of constant transverse displacement $w(x,y)$ are nearly circles. This is also the case as the size increases to $D = 200$ and 250 µm, although the displacements increase as the plate size increases. At $D = 300$ µm, though, the transverse displacement contours are not circular, but elliptical, indicating that the deformation is no longer spherically symmetric and that the bifurcation described previously has occurred. It is apparent that when subjected to the same temperature change, both the magnitude and deformation mode of the different size plates depend on the plate size. In Fig. 5 the average curvatures in the x- and y-directions, along the paths y = 0 and x = 0, respectively, are plotted as a function of temperature during cooling. Measurements are shown with symbols (the lines connecting them are used simply to aid viewing), and finite element predictions are shown with solid lines. The x- and y- directions are taken to be aligned with the principal curvatures after bifurcation. Before bifurcation, the response is spherically symmetric and so the x- and y-directions are arbitrary and indistinguishable. The measurements and predictions show the three regimes of deformation as discussed previously. In regime *I*, the curvature-temperature response is independent of plate size. In regime *II*, though, there is a strong dependence on plate size. Although only one set of data exists in regime *III*, calculations and measurements that are not shown demonstrate a dependence on plate size. Good agreement between the measurements and predictions exists in all three deformation regimes. The major discrepancy is the sharpness of the bifurcation for the D = 300 μm plates; it is quite sharp in the predictions but much more gradual in the measurements. Additional detail regarding these results can be found in Dunn et al., (2000, 2001).

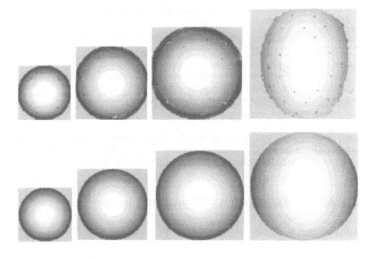

Figure 4 Contour plots of the measured (top), and predicted (bottom) transverse displacements $w(x,y)$ at room temperature for four gold/polysilicon circular plates: D = 150, 200, 250, and 300 μm from left to right. Each contour band represents a displacement of 0.13, 0.22, 0.29, and 0.48 μm for the D = 150, 200, 250, and 300 μm microstructures, respectively.

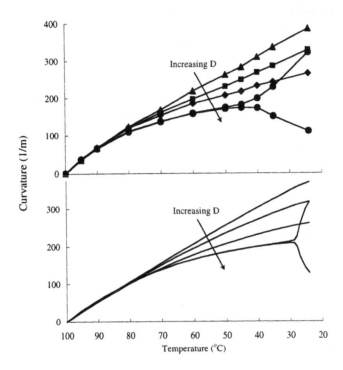

Figure 5 Average measured (top) and predicted (bottom) curvature as a function of temperature change upon cooling from 100°C to room temperature. The curves from top to bottom are for the $D =$ 150, 200, 250, and 300 μm microstructures, respectively.

3.3 Monotonic Time-Dependent Deformation Behavior

Figure 6 shows the results of measurements made on similar microstructures to demonstrate the effects of stress relaxation in the metal films used in microstructures. For a series of gold/polysilicon beams and plates, we first cycled them between room temperature and 190°C a couple of times to stabilize the microstructure (as described in Section 3.1), at least for the subsequent tests. From 190°C the microstructures were cooled to 120°C and the curvature was measured at 20°C increments. The microstructures were then held at 120°C for nearly a month and the curvature was measured periodically during this time. Figure 6 illustrate a number of interesting phenomena of both structural and material origin.

The deformation during the temperature drop results from thermoelastic phenomena described in Section 3.2. The results for multiplayer plates show both linear and geometric nonlinear behavior, while those from the beams are completely linear. This is because the beams are able to approach a cylindrical bending configuration without stretching the midplane. Consistent with the size effects described in Section 3.2, the 200 μm plate does not bifurcate, while the 300 μm plate does. During the isothermal hold, stress relaxation is observed in all of the microstructures. We presume the relaxation to be predominately in the gold film, although this is not totally conclusive at this point. Both beam microstructures show significant stress relaxation. The beam with the 3.5 μm polysilicon film (open square symbols) appears to have fully relaxed while the one with the 1.5 μm polysilicon film (closed square symbols) has not. Note that the stresses in the former are larger than those in the latter, but the stress gradient is less. Both plate microstructures also experience substantial stress relaxation with the 200 μm plate exhibiting behavior similar to the beams. Interestingly, for microstructures that have buckled during the thermoelastic deformations, the stress relaxation can cause them to unbuckle. We have attempted to model the phenomena shown in Fig. 6 assuming power law creep in the gold, however, we have not been able to describe all of the data in a self-consistent manner (Zhang and Dunn, 2001). The reasons why are presently unclear, but may be due to the difference in stress gradients in the gold film in the different cases.

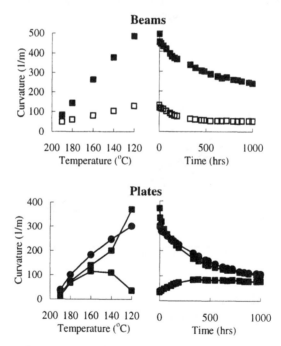

Figure 6 Curvature vs. the temperature during the cooling process from 190°C to 120°C, and curvature vs. time during the isothermal hold period at T = 120°C of gold (0.5 μm thick)/polysilicon (filled squares for 1.5 μm thick and open squares for 3.5 μm thick) 300μm x 50μm beams; and 200 μm (filled circles) and 300 μm (filled squares) diameter plate microstructures.

3.4 Cyclic Deformation Behavior

The effects of repeated thermomechanical cycles on multilayer MEMS components have been scarcely studied. In fact to our knowledge, the corresponding problem in the microelectronics context has been the subject of only limited study itself (see for example, Thouless et al., 1993; Shen and Suresh, 1995; Koike et al., 1998; Keller et al., 1999). Figure 7 shows some preliminary results in this direction (see Gall et al., 2001 for more detail). Gold/polysilicon beams were thermal cycled over a range of $\Delta T = 250$ °C and a mean temperature of $T_m = 75$ °C, where $\Delta T = T_{max} - T_{min}$ and $T_m = (T_{max} + T_{min})/2$ for over 8000 cycles. After the first cycle, the curvature vs. temperature response is linear and unchanged over the first 100 cycles. After 100 cycles, the thermo-mechanical response shows an appreciable shift towards smaller curvatures, although the thermoelastic curvature-temperature response appears to be unchanged. Microstructural observations and modelling are needed to fully understand this evolutionary behavior during cycling. Moreover, further experimentation is necessary to clarify and quantify additional variables such as the role of the temperature range versus the mean temperature on the cyclic response. With the aforementioned microstructural and experimental information, mechanism-based models of the cyclic response can be developed as a step towards new multilayer MEMS design tools.

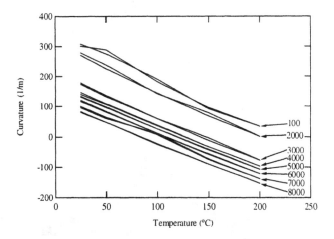

Figure 7 Curvature vs. temperature during thermal cycling ($\Delta T = 175$ °C, $T_m = 112.5$ °C) of a 300 μm by 50 μm gold/polysilicon beam.

4 SOME LIFECYCLE ENGINEERING NEEDS FOR RELIABLE MEMS

To design functional and reliable multilayer MEMS structures, it is necessary to understand and predict their thermomechanical response for a variety of possible loading paths and structure geometries. The results of the previous section have highlighted some aspects of this response for gold/polysilicon multilayers. In order to efficiently address these phenomena during design, reliable models incorporating microstructural evolution must be developed and implemented in design tools. In general, this is a Herculean task.

Specifically regarding multilayer films for MEMS, although an understanding of the film-level stress state is in place, there is a need to understand how this nominal stress state impacts grain-level stress states in the textured polycrystalline films. In particular, it is desirable to understand the effects of different grain misorientations on local stress states. In the context of microelectronics this would entail passing from the biaxial uniform film stress to grain-level microstresses. In MEMS, though, the effects of through-thickness stress gradients must also be considered.

An understanding of damage mechanisms and their evolution under these nonuniform local stress states is also needed. This should include the effects of stress gradients, both in-plane and through the thickness. The understanding of such small-scale damage mechanisms may be aided by the use of appropriate atomistic simulations. For example, modified embedded atom method (MEAM) atomistic simulations can be used to guide, and also support and help solidify conclusions from companion SEM and TEM studies. Material parameters can be altered rapidly in simulations, allowing the systematic identification of nanostructural features that may play a first order role in damage development. The features can then be the focus of TEM and SEM studies to ultimately confirm their mechanistic role. With regard to multilayer MEMS, such simulations will be helpful in identifying the dependence of damage initiation processes on relative orientations of Au and Si grains; this includes Au/Au and Si/Si grain boundaries along with Au/Si grain boundaries including the effects of defects (for example voids which are common as a result of the deposition process) at interfaces. We have carried out preliminary simulations of this type regarding the deformation of perfect [100]Al‖[100]Si interfaces and attempted to connect MEAM simulations and continuum-based traction separation laws for the decohesion of aluminum/silicon interfaces (Gall et al., 2000).

Additional MEMS test structures should be designed, fabricated, and tested to study the effects of various multiaxial stress states, stress gradients through the thickness, and monotonic and cyclic loading. Concurrent with such testing, extensive microstructural characterization is needed to identify damage modes and microstructural evolution. Ultimately this type of data can be incorporated into various facets of lifing codes (for example into continuum parameters in traction-separation laws for interfaces or evolution equations for internal state variables) as has previously been done for the lifecycle engineering of cast aluminum alloys (Gall et al., 2000).

5 REFERENCES

Dunn, M. L., Zhang, Y., and Bright, V. M., 2000, "Linear and Geometrically Nonlinear Behavior of Metal/Polysilicon Plate Microstructures Subjected to Temperature Changes," in *Proceedings of the MEMS Symposium, ASME International Mechanical Engineering Congress and Exposition*, A. P. Lee, A. P. Malshe, F. K. Forster, R. S. Keynton, and Q. Tan, MEMS-Vol. 2, pp. 207-213.

Dunn, M. L., Zhang, Y., and Bright, V., 2001, "Deformation and Structural Stability of Layered Plate Microstructures Subjected to Thermal Loading," submitted for publication.

Ehrfeld, W., Hessel, V., Lowe, H., Schulz, C., and Weber, L., 1999, "Materials of LIGA Technology," *Microsystem Technologies*, Vol. 5, pp. 105-112.

Finot, M. and Suresh, S., 1996, Small and Large Deformation of Thick and Thin Film Multilayers: Effect of layer Geometry, Plasticity, and Compositional Gradients. *J. Mech. Phys. Solids* 44, 683.

Finot, M., Blech, I. A., Suresh, S., and Fujimoto, H., 1997, "Large Deformation and Geometric Instability of Substrates with Thin Film Deposits," *J. Appl. Phys.*, Vol. 81, pp. 3457-3464.

Freund, L. B., 2000, Substrate Curvature due to Thin Film Mismatch Strain in the Nonlinear deformation Range. *J. Mech. Phys. Solids* 48, 1159.

Gall, K., Dunn, M. L., Zhang, Y., and Corff, B., 2001, "Thermal Cycling Response of Layered Gold/Silicon MEMS Structures," submitted for publication.

Gall, K., Horstemeyer, M. F., Baskes, M. I., and Van Schilfgaarde, M., 2000, "Atomistic Simulations on the Tensile Debonding of an Aluminum-Silicon Interface," *Journal of the Mechanics and Physics of Solids*, Vol. 48, pp. 2183-2212.

Hyer, M. W., 1981, "Some Observations on the Cured Shape of Thin Unsymmetric Laminates," *J. Compos. Mater.*, Vol. 15, pp. 175-194.

Keller, R. M., Baker, S. P., and Arzt, E., 1999, "Stress-Temperature Behavior of Unpassivated Thin Copper Films," *Acta Mater.*, Vol. 47, pp. 415-426.

Keller, R. R., Kalnas, C. E., and Phelps, J. M., 1999, "Local Crystallography and Stress Voiding in Al-Si-Cu Versus Copper Interconnects," *J. Appl. Phys.*, Vol. 86, pp., 1167-1169.

Koester, D. A., Mahadevan, R., Hardy, B., and Markus, K. W., 2001, *MUMPs™ Design Rules*, Cronos Integrated Microsystems, A JDS Uniphase Company, http://www.memsrus.com/cronos/svcsrules.html.

Koike, J., Utsunomiya, S, Shimoyama, Y., Maruyama, K., and Oikawa, H., 1998, Thermal Cycling Fatigue and Deformation Mechanism in Aluminum Alloy Thin Films on Silicon. *J. Mater. Res.* 13, 3256.

Madou, M., 1997, *Fundamentals of Microfabrication*, CRC Press.

Masters, C. B. and Salamon, N. J., 1993, "Geometrically Nonlinear Stress-Deflection Relations for Thin Film/Substrate Systems," *Int. J. Engng. Sci.*, Vol. 31, pp. 915-925.

Miller, D. C., Zhang, W., and Bright, V. M., 2001, "Micromachined, Flip-Chip

Assembled, Actuatable Contacts for Use in High Density Interconnection in Electronics Packaging," *Sensors and Actuators A*, Vol. 89, pp. 76-87.

Nix, W. D., 1989, Mechanical Properties of Thin Films. *Metall. Trans.* 20A, 2217.

Shen, Y.-L. and Suresh, S., 1995, Thermal Cycling and Stress Relaxation Response of Si-Al and Si-Al-SiO$_2$ Layered thin films. *Acta. Metall. Mater.* 43, 3915.

Spearing, S. M., 2000, "Materials Issues in Microelectromechanical Systems (MEMS), *Acta Materialia, Vol.* 48, pp. 179-196.

Stoney, G. G., 1909, The Tension of Metallic Films Deposited by Electrolysis. *Proc. Roy. Soc. Lon.* A82, 172.

Thouless, M. D., Cupta, J., and Harper, J. M. E., 1993, "Stress Development and Relaxation in Copper Films During Thermal Cycling," *J. Mater. Res.*, Vol. 8, pp. 1845-1852.

Zhang, Y. and Dunn, M. L., 2001, Stress Relaxation and Creep of Gold/Polysilicon Layered MEMS Microstructures Subjected to Thermal Loading. *Proceedings of the 2000 International Mechanical Engineering Congress and Exposition (IMECE 2001)*, in press.

Variational Multiscale Methods to Embed Macromechanical Formulations with Fine Scale Physics

Krishna Garikipati[1]

1.1 INTRODUCTION

As deformation phenomena approach the length scales of grains (1μm and lower in most polycrystalline solids), the interactions between defects (dislocations, microvoids and microcracks) and the effects of their elastic fields play increasingly important roles. Conventional macroscopic continuum theories of inelasticity—such as plasticity or damage—fail at resolving this effect. A treatment of interactions between individual defects is, however, ruled out; their densities are too high to warrant such an approach even at these finer scales. On the other hand, the hardening effects at crack tips, results of nanoindentation tests (Nix, 1989; McElhaney *et al.*, 1998), microtorsional and microbending experiments (Fleck *et al.*, 1994; Stolken & Evans, 1998) can be explained with notable success by various classes of strain gradient plasticity theories (Fleck & Hutchinson, 1997; Gao *et al.*, 1999; Arsenlis & Parks, 1999). The observed finite width of shear bands is also represented by these theories since they introduce a length scale to the problem. A separate approach to incorporating the influence of microstructure is seen in the formal treatment of configurational microforces (Gurtin, 1995). A related work is (Gurtin, 2000), which treats crystal plasticity in the framework of microforces and encompasses plastic strain gradient effects. A third class of models operating at microscopic length scales consists of traction-displacement laws specified on internal surfaces. Included in this category are laws specifying the evolution of tangential traction driven by slip on shear bands (Armero & Garikipati, 1996) and embedded fracture zone models [see (Tvergaard & Hutchinson, 1992), for example].

In application to boundary-value problems, each class of models outlined above prevails in domains whose dimensions are in the near micron and submicron-range. While using numerical techniques, the finest scales of the discretization must then be in the submicron range. This does not pose a large

---

[1] University of Michigan

computational cost if the size of the solid domain over which the boundary-value problem is to be solved is also in the micron-range. On the other hand, a wide range of applications require the inclusion of fine scale physics associated with such microscopic features in otherwise macroscopic domains. *Yet, the overall physical dimensions of the problems remain on the order of 1–10 meters and computational efficiency cannot be forsaken in employing the microstructural theories.* This difficulty would seem to suggest the approach of embedding micromechanical models in macroscopic ones as one alternative. As another option, a microforce theory may be used. In either case the approach should give rise to mathematical models in which deformations at different scales are coupled. Computational efficiency is served if such multiscale models can then be posed solely in terms of the macroscopic coarse scale deformations.

The approach described here is termed the Variational Multiscale Method. The method has previously been applied to the localization of deformation by the author and coworkers (Garikipati & Hughes, 1998; Garikipati & Hughes, 2000a). Subsequently, the approach was extended to embed a simple micromechanical law in the macromechanical formulation (Garikipati & Hughes, 2000b; Garikipati, 2001). The method assumes a decomposition of the displacement into coarse and fine scale components. The fine scale component is regarded as being driven by the micromechanical description, while the coarse scale depends upon the macromechanical model. The micromechanical model is used to express the fine scale displacement—analytically or numerically—in terms of the coarse scale, and it is substituted back into the macroscopic model. The result is a multiscale formulation posed entirely in terms of the coarse scale, and with the micromechanical model embedded.

The variational basis of this multiscale method results in formulations to which finite element or meshless methods can be applied.

The rest of this chapter focuses upon the technical approach, beginning in Section 1.2 with the application of the method to a chosen model for fine scale physics. Earlier work is discussed in Section 1.3, showing how the variational multiscale approach has already been applied to embedding a chosen micromechanical law. Numerical examples are also included in this section, thereby establishing the viability of the proposed method. The long term scope of this approach are outlined in Section 1.4.

1.2 MULTISCALE TREATMENT OF A THEORY GOVERNING FINE SCALE PHYSICS

1.2.1 Fleck-Hutchinson strain gradient plasticity theory

The development in this subsection assumes small deformations. At the outset we introduce a scale separation of the displacement, u, as $u = \bar{u} + u'$, into coarse and fine scale components respectively. In a variational setting, a displacement variation, w, arises, and is also decomposed as: $w = \bar{w} + w'$. Using the coarse scale variation $\bar{w}$, the stress σ, the body force f, and the traction t, the weak form of the classical (non-gradient) theory is,

$$\int_{\Omega} (\bar{w}_{i,j}\sigma_{ij})\,dV = \int_{\Omega} \bar{w}_k f_k\,dV + \int_{\partial\Omega} \bar{w}_k t_k\,dS, \tag{1}$$

where the domain of integration, Ω, is to be thought of as a macroscopic solid body. The coarse scale displacement, $\bar{u}$, and its variation, $\bar{w}$, are taken to vary on macroscopic scales; say, greater than $10^2\mu$m. Thus, Equation (1) can be interpreted as a projection of the macroscopic balance law (stress equilibrium) on these coarse scale functions. We will refer to this equation as the coarse scale weak form. The fine scale fields, u' and w', however, vary on much smaller scales in the micron and submicron range. Thus, if Equation (1) were rewritten with the fine scale variation, w', replacing its coarse scale counterpart, $\bar{w}$, it would admit the obvious interpretation of projecting the macroscopic balance law on these fine scale functions. Suppose now that there exists a subdomain, $\Omega' \subset \Omega$, in which it is of interest to take the fine scale view. Furthermore, let the microstructural nature of Ω' be such that the fine scale physics discussed in Section 1.1 comes into play in this subdomain. It would then be of interest to jettison the classical continuum theory in favor of one that can resolve this fine scale physics. For the purposes of the present discussion we would replace the classical theory with the Fleck-Hutchinson strain gradient plasticity theory. On Ω', therefore, we use the following fine scale weak form, obtained from the Fleck-Hutchinson strain gradient plasticity theory:

$$\int_{\Omega'} (w'_{i,j}\sigma_{ij} + w'_{k,ij}\beta_{ijk})\,dV = \int_{\Omega'} w'_k f_k\,dV + \int_{\partial\Omega'} w'_k t_k\,dS$$

$$+ \int_{\partial\Omega'} (Dw'_k)r_k\,dS + \sum_i \oint_{C'_i} w'_k p_k\,ds. \tag{2}$$

Observe that the fine scale displacement variation, w', is used. In addition to the fields introduced with Equation (1), β is the higher-order stress that is conjugate to strain gradients, $\eta_{ijk} = u_{k,ij}$ (to be introduced below) and r is the couple stress traction. The normal surface gradient operator is denoted $D(\bullet)$. The last term on the right hand-side arises when the surface $\partial\Omega'$ has edges denoted by closed curves C_i', $i = 1, \ldots, n$. In this setting, p represents a line load resulting from the higher order stress, β. The strain gradient, η, its conjugate higher-order stress, β, the couple stress traction, r and the line load, p, are viewed as micromechanical quantities since they appear only with the micromechanical strain gradient plasticity theory and their influence diminishes if the deformation varies at larger scales (say, on the order of 10^{-5} m). Further details and background (Fleck & Hutchinson, 1997) are omitted here; instead, attention is turned toward the variational multiscale approach for this theory.

Given the two weak forms, we consider the following set of constitutive relations for σ_{ij} and β_{ijk}:

$$\sigma_{ij} = C_{ijkl}\varepsilon_{kl}^{e}, \qquad \beta_{ijk} = J_{ijklmn}\eta_{lmn}^{e},$$

$$\varepsilon_{kl} = \varepsilon_{kl}^{e} + \varepsilon_{kl}^{p}, \qquad \eta_{lmn} = \eta_{lmn}^{e} + \eta_{lmn}^{p}, \qquad (3)$$

$$f(\boldsymbol{\sigma}, \boldsymbol{\beta}) \leq 0, \qquad \dot{\varepsilon}^{p} = \lambda\frac{\partial f}{\partial\boldsymbol{\sigma}}, \qquad \dot{\eta}^{p} = \lambda\frac{\partial f}{\partial\boldsymbol{\beta}}.$$

In Equation (3), ε_{kl}^{e} and η_{lmn}^{e} are the elastic strain and strain gradient respectively. The usual fourth-order elasticity tensor is denoted C_{ijkl}, while J_{ijklmn} is a sixth-order tensor: a material constant, with dimensions of stress$\times$length^{2}, relating the higher-order stress, β_{ijk} and elastic strain gradient, η_{lmn}^{e}. A length scale thus needs to be introduced as a material property. A strain gradient *flow theory* of plasticity is assumed. The third line of Equation (3) shows the yield criterion and the associative flow rules for rates of plastic strain, $\dot{\varepsilon}^{p}$, and plastic strain gradient, $\dot{\eta}^{p}$, respectively.

Up to this stage the decompositions, $u = \bar{u} + u'$ and $w = \bar{w} + w'$ are somewhat general other than for the scale separation. They are made more precise by requiring that u' and w' vanish outside Ω'. This gives:

$$\varepsilon_{ij} := \frac{1}{2}(u_{i,j} + u_{j,i}) = \begin{cases} \frac{1}{2}(\bar{u}_{i,j} + \bar{u}_{j,i}) & : \text{ in } \Omega - \Omega' \\ \frac{1}{2}(\bar{u}_{i,j} + \bar{u}_{j,i}) + \frac{1}{2}(u_{i,j}' + u_{j,i}') & : \text{ in } \Omega', \end{cases}$$

$$(4)$$

$$\eta_{ijk} := u_{k,ij} = \begin{cases} \bar{u}_{k,ij} & : \text{ in } \Omega - \Omega' \\ \bar{u}_{k,ij} + u'_{k,ij} & : \text{ in } \Omega'. \end{cases}$$

Since the strain gradient theory is applied only over the microstructural sub-domain, Ω', it is implied that in $\Omega - \Omega'$, the classical, non-gradient theory provides a sufficiently accurate description of deformation phenomena and material response, and that strain gradients play no role. As seen from the kinematics associated with η_{ijk}, $\bar{u}$ might produce a contribution to strain gradients in $\Omega - \Omega'$; however, we will assume that these terms play no role in the material response over this subdomain.

The weak form of the fine scale problem [Equation (2) in this case], will be used to express u' in terms of the remaining fields in Ω'. The yield condition and flow rule in Equation (3) render the stress, σ, and higher-order stress, β, nonlinear and history-dependent in ε and η respectively. In order to have ε and η (alternatively, the first and second displacement gradients) appear explicitly in Equation (2), σ and β must be expanded up to terms of first order in $\Delta\varepsilon$ and $\Delta\eta$, where $\Delta(\bullet)$ denotes an incremental quantity. The following substitutions are made in Equation (2): $\sigma_{ij} = \sigma^0_{ij} + C^{ep}_{ijkl}\Delta\varepsilon_{kl}$, $\beta_{ijk} = \beta^0_{ijk} + J^{ep}_{ijklmn}\Delta\eta_{lmn}$. The terms σ^0_{ij} and β^0_{ijk} are of zeroth order while C^{ep}_{ijkl} and J^{ep}_{ijklmn} are the associated elastoplastic tangents, which upon contraction with $\Delta\varepsilon_{kl}$ and $\Delta\eta_{lmn}$ provide the corresponding first-order corrections. Employing Equation (4) the following integral relation arises for the *incremental* fine scale displacement:

$$\int_{\Omega'} (w'_{i,j}C^{ep}_{ijkl}\Delta u'_{k,l} + w'_{k,ij}J^{ep}_{ijklmn}\Delta u'_{n,lm})dV = \int_{\Omega'} w'_k f_k dV + \int_{\partial\Omega'} w'_k t_k dS$$

$$+ \int_{\partial\Omega'} (Dw'_k)r_k dS + \sum_i \oint_{C'_i} w'_k p_k ds$$

$$- \int_{\Omega'} \left(w'_{i,j}\sigma^0_{ij} + w'_{k,ij}\beta^0_{ijk} \right) dV$$

$$- \int_{\Omega'} \left(w'_{i,j}C^{ep}_{ijkl}\Delta\bar{u}_{k,l} + w'_{k,ij}J^{ep}_{ijklmn}\Delta\bar{u}_{n,lm} \right) dV \quad (5)$$

Observe that (5) expresses $\Delta u'$ in terms of the coarse scale field, $\Delta\bar{u}$, and the remaining micro and macromechanical fields. As is the practice in computational plasticity, an iterative scheme based upon some variant of the Newton-Raphson Method is involved in determining the updated fields:

$$u'_{i+1} = u'_i + \Delta u', \quad \bar{u}_{i+1} = \bar{u}_i + \Delta \bar{u}, \quad i = i + 1, \tag{6}$$

until convergence is obtained (see REMARK 1). [In Equation (6), i is the iteration number.] In order to solve Equation (5) for $\Delta u'$, we adopt finite-dimensional approximations of $u', \bar{u}$ and the corresponding variations:

$$u'^h = \sum_A N'^A \alpha_A, \; w'^h = \sum_A N'^A \theta_A; \; \bar{u}^h = \sum_A \bar{N}^A d_A, \; \bar{w}^h = \sum_A \bar{N}^A c_A, \tag{7}$$

where a superscript $(\bullet)^h$ denotes a finite-dimensional approximation of the corresponding field. Substituting (7) in (5), the latter equation can be solved for $\Delta \alpha$, and a functional expressing $\Delta u'^h$ in terms of the other micro and macromechanical terms is obtained:

$$\Delta u'^h = \mathcal{U}'^h[\Delta \bar{u}^h, \sigma^{h^0}, \beta^{h^0}, C, J, t, r, p]. \tag{8}$$

Using $\sigma^h_{ij} = \sigma^{h^0}_{ij} + C^{\text{ep}}_{ijkl}[\text{sym}(\Delta \bar{u}^h_{k,l}) + \text{sym}(\Delta u'^h_{k,l})]$ over Ω' in the coarse scale weak form, Equation (1), and substituting the above functional [Equation (8)] for $\Delta u'^h$, the fine scale field is eliminated from the variational formulation. As before, $\sigma^{h^0}_{ij}$ is the zeroth-order term and $C^{\text{ep}}_{ijkl}[\text{sym}(\Delta \bar{u}^h_{k,l}) + \text{sym}(\Delta u'^h_{k,l})]$ is the first order correction. *This will result in a single weak form for the multiscale method with the strain gradient theory (which governs the fine scale physics) embedded in the classical macromechanical continuum formulation.* The resulting weak form of the variational multiscale method is specific to this particular micromechanical law.

Since a partial differential equation [the fine scale weak form, Equation (2)] is solved to obtain $\Delta u'^h = \mathcal{U}'^h[\Delta \bar{u}^h, \sigma^{h^0}, \beta^{h^0}, C, J, t, r, p]$, the functional, $\mathcal{U}'^h$, is of integral form. On eliminating $\Delta u'^h$ from the coarse scale weak form, Equation (1), the resulting multiscale weak form is therefore nonlocal in $\Delta \bar{u}^h$.

REMARK 1. A typical check for convergence would be $|\Delta \bar{u}^h + \Delta u'^h| < \delta$, where δ is a chosen numerical tolerance related to the machine precision. Another could be specified by requiring that the magnitude of the residual in Equation (5) be less than δ, where the residual consists of all terms not involving first order corrections.

REMARK 2. The incremental solution scheme outlined above requires integration algorithms for the strain gradient plasticity model. Such algorithms have been formulated by the author and possess the features of classical return mapping algorithms for phenomenological plasticity. The multiscale method and these integration algorithms are also applicable, with minor changes, to the recent improvements proposed in a mechanism-based strain gradient plasticity theory (Gao *et al.*, 1999; Huang *et al.*, 2000).

REMARK 3. The development in this section has been presented as an enhancement of the macroscopic continuum formulation by embedding a microscale material law, thus leading to more accurate coarse scale solutions. It is however of some importance to note that the fine scale solution can also be recovered from the coarse scale: Given the solution $\bar{u}^h$, use of the functional, $\mathcal{U}'$, returns u'^h. This is a simple "post processing" step in the numerical setting.

1.2.2 Numerical example

To demonstrate the embedding of this strain gradient theory a bimaterial shear layer is considered. We wish to embed, in turn, a soft material and hard material within a "matrix". The strain gradient plasticity theory will be applied to the neighborhood, Ω', of the embedded layer in an attempt to better represent the fine scale physics. The embedded layer is contained in Ω'. The domain and boundary value problem are shown schematically in Figure 1. Assuming the domain to be infinite in the vertical direction allows the problem to be reduced to one dimension. Of interest are the vertical displacement, u_2, the corresponding shear strain, $\varepsilon_{12} = \frac{1}{2}u_{2,1}$, the strain gradient, $u_{2,11}$, and the shear stress, σ_{12}.

The embedded layer in each case is 2×10^{-6}m in width. The material length scale which enters the constitutive law for the strain gradient plasticity theory has been specified to be 1×10^{-6}m. Figure 2a is a plot of the strain when the embedded layer is subject to plastic softening. As might be expected, a localization of strain is observed within the softening layer. The width of the localized band is related to the material length scale. In this example it is $\approx 2 \times 10^{-6}$m in width. The shear stress distribution for a hardening embedded layer appears in Figure 2b. Attention is drawn to the stress concentrations located at the edges of the hardening layer. It is pointed out that, with a classical (non-gradient) plasticity theory, the shear stress remains constant over the domain. This is shown by the points labelled "C". For the strain

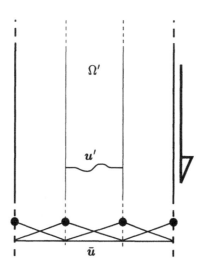

Figure 1: Schematic of shear layer with the fine scale region.

gradient plasticity theory, however, this is not the case. The equilibrium equation involves the conventional stress σ and the higher-order stress β, and in the presence of a hard inclusion, a stress concentration and intensification is seen. What appears as a discontinuity in the shear stress for the strain gradient plasticity theory (the points labelled "SG") is actually a high gradient at the boundary of the hard layer. The shear stress diminishes away from the boundaries; both within the embedded layer and in the matrix.

Both characteristics represented in this numerical example: a softening band width determined by the material length scale, and the stress intensification near a hard inclusion, are hallmarks of a strain gradient theory. The fine scale physics is represented in the multiscale model which was constructed by embedding the fine scale strain gradient theory in the macroscopic formulation. Importantly, this was done without refining the mesh size down to the scales of the embedded layer. A uniform mesh size of 1m was used throughout. The region Ω' was taken to be the element containing the embedded layer (of width 2×10^{-6}m) and the strain gradient plasticity theory specified as the fine scale physics model over all of Ω'.

Microforce theories of the class mentioned in Section 1.1 can also be treated in this framework. The fine scale equation would be the microforce balance law written in weak form, incorporating corresponding constitutive relations. The variational multiscale method provides a formalism to ensure tight coupling

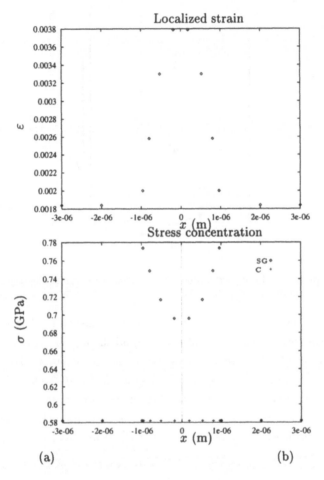

Figure 2: (a) Localized strain in a softening fine scale region. (b) Stress concentrations at the edge of a hardening fine scale region.

between fine (micro) and coarse (macro) scales. Details will be presented elsewhere. In order to demonstrate the viability of the variational multiscale approch, an earlier application (Garikipati, 2001) is presented in Section 1.3.

1.3 PAST WORK: THE METHOD APPLIED TO TRACTION-DISPLACEMENT LAWS

1.3.1 Multiscale formulation

A macroscopic body, Ω, is assumed to possess an internal surface, Γ, upon which it is of interest to specify a class of traction-displacement laws governing shear and normal tractions. *The finite strain theory is adopted in this case.*

The microstructural finescale subdomain, Ω', is such that $\Gamma \subset \Omega' \subset \Omega$.

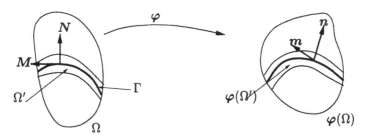

Figure 3: The internal surface, $\Gamma \subset \Omega'$, its tangent M and normal N in the reference configuration, Ω, mapped forward by the deformation, φ, to the spatial configuration $\varphi(\Omega)$.

The weak form of the fine scale problem can be manipulated in this case to the strong requirement of traction continuity on Γ:

$$T_m m + T_n n = \tau n \Big|_{\Gamma^-}, \qquad m = FM, \qquad n = F^{-T}N, \qquad (9)$$

where M is the tangent and N is the normal to Γ in the reference configuration of the body, F is the deformation gradient, and T_m and T_n are traction components along m and n respectively in the current configuration (Figure 3). The macroscopic Kirchhoff stress is denoted τ and is obtained from the stored energy function, ψ, as $\tau = (\partial \psi / \partial F^e) F^{eT}$.

A micromechanical constitutive model is introduced in the form of traction-displacement laws on Γ:

$$|T_m| = T_{m_0} - \mathcal{H}_m \xi_m, \quad |T_n| = T_{n_0} - \mathcal{H}_n \xi_n, \quad \mathcal{H}_m, \mathcal{H}_n > 0, \quad \xi_m, \xi_n \geq 0. \quad (10)$$

The traction components T_m and T_n decay from their peak values as ξ_m, the slip on Γ, and ξ_n, the normal displacement on Γ, increase. This introduces a displacement jump, $[\![u]\!] = \xi_m FM + \xi_n FN$, on Γ. Of central importance is the fact that $[\![u]\!]$ gives rise to the fine scale displacement, u', as shown below.

The deformation field includes a discontinuous displacement, $[\![u]\!]H_\Gamma$, on the surface Γ, as follows:

$$\varphi(X) = X + \underbrace{\hat{u}(X)}_{\text{continuous}} + \underbrace{F\left(\xi_m M + \xi_n N\right) H_\Gamma}_{[\![u]\!]H_\Gamma, \text{ discontinuous}} \qquad (11)$$

The Heaviside function, H_Γ, enforces the discontinuity via the displacement jump, $[\![u]\!]$. Turning to the one-dimensional setting of Figure 4 for motivation, it is observed that for a displacement field admitting a discontinuity, a coarse scale approximation, $\bar{u}$, can be constructed via the linear interpolation, $\bar{N}(X)$, depicted as a broken line. The difference between the actual field and $\bar{u}$ appears in the center of the figure. Such a field can be approximated via the interpolation $N'(X)$ shown on the right of the figure.

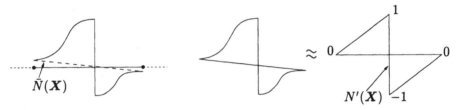

Figure 4: Coarse and fine scale interpolations $\bar{N}(X)$ and $N'(X)$ in a one-dimensional setting.

In a discretized setting where the various displacement fields are replaced by their finite-dimensional approximations, the coarse and fine scale fields are defined via the interpolations:

$$\bar{u}^h(X) = \sum_A \bar{N}^A(X)d_A, \quad u'^h(X) = N'(X)\underbrace{F^h\left(\xi_m^h M + \xi_n^h N\right)}_{[\![u]\!]^h}, \quad (12)$$

where d_A are nodal values of $\bar{u}^h$. Such an additive decomposition of the displacement leads to a multiplicative decomposition of the deformation gradient as

$$F^h = \underbrace{\left(1 + \sum_A d_A \otimes \frac{\partial \bar{N}^A}{\partial X}\right)}_{\bar{F}^h} \underbrace{\left(1 - \left(\xi_m^h M + \xi_n^h N\right) \otimes \frac{\partial N'}{\partial X}\right)^{-1}}_{F'^h}, \quad (13)$$

where $\bar{F}^h$ and F'^h are, respectively, the coarse scale and fine scale deformation gradients.

With this elucidation of multiscale kinematics, we turn to the embedding of this model of fine scale physics: Equation (9a) enforces consistency between the micromechanical and macromechanical descriptions. With Equation (10), it relates the discontinuous displacement components ξ_m and ξ_n to the macromechanical fields in $\Omega - \Gamma$. Since ξ_m and ξ_n determine the fine scale through Equation (12b), it follows that (9a), (10) and (12b), taken together, express the fine scale field, u'^h, in terms of $\bar{u}^h$ and the remaining micro and macromechanical fields. Formally, a functional is obtained: $u'^h = \mathcal{U}'^h \left[\bar{u}^h, \tau^h, f, t, T_m^h, T_n^h, \xi_m^h, \xi_n^h \right]$. As in Section 1.2 this functional is to be substituted for the fine scale displacement in the weak form of the coarse scale problem, which is merely the weak form for macroscopic stress equilibrium, under finite strains, using coarse scale displacement variations, $\bar{w}$. The fine scale displacement field is eliminated, resulting in a single weak form: the weak form of the multiscale problem. The discussion at the begining of this paragraph explains how this leads to an embedding of the micromechanical traction-displacement law.

1.3.2 Numerical example with the traction-displacement law

The macromechanical formulation of multiplicative finite strain plasticity is applied to a hyperelastic, isotropic material. The stored energy function is quadratic in logarithmic principal stretches. The example assumes plane strain conditions. When a bifurcation condition is met (in this case, loss of strong ellipticity of the tangent modulus tensor), the micromechanical surface, Γ, forms as depicted in Figure 5. The embedded micromechanical law becomes active. In the example below it reduces to the tangential component of traction driven by the slip on Γ. The problem is loaded by the downward movement of the rigid block.

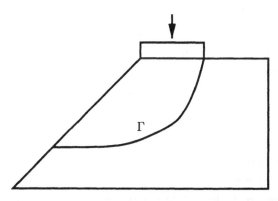

Figure 5: The micromechanical law is imposed on a curved surface, Γ.

Figure 6 shows the deformed mesh. The surface, Γ, is contained in the shaded microstructural subdomain, Ω'. Limit analysis indicates that the shape of Γ could be either a logarithmic spiral (which is roughly the shape obtained in Figure 6) or a circle (Lubliner, 1990). The sharply resolved deformation is evident as is the rotation of the block and material beneath it.

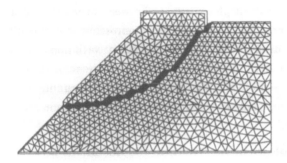

Figure 6: Deformed mesh and microstructural subdomain, Ω' (shaded).

The outline and example in this section demonstrate the embedding of the chosen micromechanical model in the macroscopic formulation using the variational multiscale method. For this particular case, the embedding in the continuum macromechanical formulation is in an analytic form. The appropriate macroscopic response is obtained, reflective of the traction displacement law on the internal surface. A tight coupling between fine and coarse displacements is maintained, and in fact, the discontinuous displacements can be recovered from the fine scale solution. A decrease in support reactions is expected with the decrease in traction along such slip surfaces. Such an evolution of support reaction is observed without any mesh size or orientation sensitivity (results not shown here) even with the rather coarse meshes used here.

1.4 LONG TERM SCOPE OF THIS WORK

The treatment of fine scale physics envisaged in this chapter encompasses micromechanical theories that prevail at length scales of a micron. The long term goal of this work is to build a broad formalism by which to embed— potentially any—such theories in macromechanical continuum formulations. As seen with the treatment of traction-separation laws and the Fleck-Hutchinson strain gradient theory, some of these micromechanical models *require* a vehicle like the multiscale methods proposed here to make the notion of this embedding meaningful. Others, such as microforce theories naturally incorporate such coupling between micro and macroscale response. In such cases, the methods proposed here will serve as computational formulations to preserve the tight coupling of scales needed to realize the potential of these theories.

References

ARMERO, F., & GARIKIPATI, K. 1996. An Analysis of Strong Discontinuities in Multiplicative Finite Strain Plasticity and their Relation with the Numerical Simulation of Strain Localization in Solids. *Int. J. Solids and Structures*, **33**, 2863–2885.

ARSENLIS, A., & PARKS, D. M. 1999. Crysallographic Aspects of Geometrically-necessary and Statistically-stored Dislocation Density. *Acta Materialia*, **47**, 1597–1611.

FLECK, N. A., & HUTCHINSON, J. W. 1997. Strain Gradient Plasticity. *Advances in Applied Mechanics*, **33**, 295–361.

FLECK, N. A., MULLER, G. M., ASHBY, M. F., & HUTCHINSON, J. W. 1994. Strain Gradient Plasticity: Theory and Experiment. *Acta Metallurgica et Materialia*, **42**, 475–487.

GAO, H., HUANG, Y., & NIX, W. D. 1999. Mechanism-based Strain Gradient Plasticity–I. Theory. *Journal of the Mechanics and Physics of Solids*, **47**, 1239–1263.

GARIKIPATI, K. 2001. *A Variational Multiscale Method to Embed Micromechanical Surface Laws in the Macromechanical Continuum Formulation*. To appear in *Computer Modelling in Engineering Sciences*.

GARIKIPATI, K., & HUGHES, T. J. R. 1998. A Study of Strain Localization in a Multiple Scale Framework — The One Dimensional Problem. *Comp. Methods in Applied Mech. Engrg.*, **159**, 193–222.

GARIKIPATI, K., & HUGHES, T.J.R. 2000a. A Variational Multiscale Approach to Strain Localization— Formulation for Multidimensional Problems. *Comp. Methods in Applied Mech. Engrg.*, **188**, 39–60.

GARIKIPATI, K., & HUGHES, T.J.R. 2000b. *Embedding Micromechanical Laws in the Continuum Formulation — A Multiscale Approach applied to*

Discontinuous Solutions. To appear in the inaugural issue of *International Journal for Computational Civil and Structural Engineering.*

GURTIN, M. E. 1995. The Nature of Configurational Forces. *Archives for Rational Mechanics and Analysis,* **131**, 67–100.

GURTIN, M. E. 2000. On the Plasticity of Single Crystals: Free Energy, Microforces, Plastic-strain Gradients. *Journal of the Mechanics and Physics of Solids,* **48**, 989–1036.

HUANG, Y., GAO, H., NIX, W. D., & HUTCHINSON, J. W. 2000. Mechanism-based Strain Gradient Plasticity–II. Analysis. *Journal of the Mechanics and Physics of Solids,* **48**, 99–128.

LUBLINER, J. 1990. *Plasticity Theory.* Macmillan Publishing Co.

MCELHANEY, K. W., VLASSAK, J. J., & NIX, W. D. 1998. Determination of Indenter Tip Geometry and Indentation Contact Area for Depth-sensing Indentation Experiments. *Journal of Mat. Res.,* **13**, 1300–1306.

NIX, W. D. 1989. Mechanical Properties of Thin Films. *Metallurgical Transactions,* **20A**, 2217–2245.

STOLKEN, J. S., & EVANS, A. G. 1998. Microbend Test Method for Measuring the Plasticity Length Scale. *Acta Materialia,* **46**, 5109–5115.

TVERGAARD, V., & HUTCHINSON, J. W. 1992. The Relation between Crack Growth Resistance and Fracture Process Parameters in Elastic-plastic Solids. *Journal of the Mechanics and Physics of Solids,* **40**, 1377–1397.

Computational Multiscale Fatigue Analysis and Life Prediction for Composite Structures

Jacob Fish[1] and Qing Yu[1]

ABSTRACT

A multiscale fatigue analysis model is developed for brittle composite materials. A mathematical homogenization theory is generalized to account for multiscale damage effects in heterogeneous media and a closed form expression relating nonlocal microphase fields to the overall strains and damage is derived. The evolution of fatigue damage is approximated by the first order initial value problem with respect to the number of load cycles. A novel accelerating technique is developed for the integration of continuum damage mechanics based fatigue cumulative law. Consistency adjustment procedures are developed to couple fatigue damage cumulation and mechanical responses. Numerical studies are conducted to study the accuracy and computational efficiency of the proposed model for both low-cycle and high-cycle fatigue.

1.0 INTRODUCTION

In recent years several fatigue models have been developed within the framework of continuum damage mechanics (CDM) (Bhattacharya and Ellingwood, 1999; Chaboche and Lesne, 1988; Chow and Wei, 1994; Marigo, 1985; Papa, 1993; Pass *et al.*, 1993). Within this framework, internal state variables are introduced to model the fatigue damage. Degradation of material response under cyclic loading is simulated using constitutive equations which couple damage cumulation and mechanical responses. The microcrack initiation and growth are lumped together in the form of the evolution of damage variables from zero to some critical value. Most of the existing CDM based fatigue damage models are based on the classical (local) continuum damage theory even though it is well known that the accumulation of damage leads to strain softening and loss of ellipticity in elasto-statics and hyperbolicity in elasto-dynamics (Bazant, 1991; Bazant and Pijaudier-Cabot, 1988; Belytschko and Lasry, 1989; Geers, 1997). To alleviate the deficiencies inherent in the local CDM theory, a number of regularization techniques have been devised to limit the

[1]Department of Civil, Mechanical and Aeronautical Engineering
Rensselaer Polytechnic Institute, Troy, NY 12180

size of the strain localization zone, including the nonlocal damage theory (Bazant and Pijaudier-Cabot, 1988) and gradient-dependent models (Geers *et al.*, 1998). Recent advances in CDM based theories (Mazars and Pijaudier-Cabot, 1996; van Vroonhoven and deBorst, 1999) revealed the intrinsic links between the nonlocal CDM theory and fracture mechanics providing a possibility for building a unified framework to simulate crack initiation, propagation and overall structural failure under cyclic loading.

When applying the CDM based fatigue model to the life prediction of engineering systems, the coupling between mechanical response and damage cumulation poses a major computational challenge. This is because the number of cycles to failure, especially for high-cycle fatigue, is usually as high as tens of millions or more, and therefore, it is practically not feasible to carry out a direct cycle-by-cycle simulation for the fully coupled models even with today's powerful computers. The accelerating technique, or the so-called "cycle jump" technique (Chaboche, 1988), for the integration of fatigue damage cumulative law must be developed for the fully coupled fatigue analysis. Several efforts have been made recently as reported by Chow and Wei (1994) and Pass *et al.* (1993).

For composite materials, the fatigue damage mechanisms are very complex primarily due to significant differences in fatigue behavior of microconstituents (Dauskardt and Ritchie, 1991; Suresh, 1991). The proposed fatigue analysis model is based on the multiscale nonlocal damage theory for composites and the fatigue damage cumulative law stated on the smallest scale of interest (Fish *et al.*, 1999). In Section 2, the multiscale nonlocal damage theory based on the mathematical homogenization is briefly reviewed with emphasis on its application to fatigue of composites. Double scale asymptotic expansions of damage and displacements lead to the closed form expressions relating local (microscopic) fields to overall (macroscopic) strains and damage. In Section 3, a novel fatigue damage cumulative law is described by extending the CDM based static damage evolution law (Fish *et al.*, 1999). The integration of fatigue law is approximated by the first order initial value problem with respect to the number of load cycles. Adaptive Modified Euler method in conjunction with the step size control is used to integrate the initial value problem. Consistency adjustment procedures are introduced to ensure that the integration of the initial value problem preserves thermo-mechanical equilibrium, compatibility and constitutive equations. An adaptive integration scheme of the initial value fatigue problem is presented in Section 4. In Section 5, we study the computational efficiency of the proposed multiscale fatigue model and compare its performance to the available test data. Discussion and future research directions conclude the book chapter.

2.0 REVIEW OF THE MULTISCALE NONLOCAL DAMAGE THEORY FOR COMPOSITES

The microstructure of a composite material is assumed to be locally periodic (Y-periodic) with a period defined by the Representative Volume Element (RVE). The size of RVE is assumed to be very small compared to the characteristic length of the macro domain so that the asymptotic homogenization is applicable (Sanchez-Palencia, 1980). Let x be the macroscopic coordinate vector in the macro domain

Ω and $y \equiv x/\varsigma$ be the microscopic position vector in RVE denoted by Θ. ς denotes a very small positive number compared with the dimension of Ω and $y \equiv x/\varsigma$ is a stretched coordinate vector in the microscopic domain. We assume that all quantities have two explicit dependences: one on the macroscopic level x and another on the level of microconstituents y. Using the classical nomenclature, any Y-periodic function f can be represented as

$$f^\varsigma(x) \equiv f(x, y(x)) \tag{2.1}$$

where superscript ς denotes the Y-periodicity of function f. The indirect macroscopic spatial derivatives of f^ς can be calculated by the chain rule as

$$f^\varsigma_{,x_i}(x) = f_{,x_i}(x, y) + \frac{1}{\varsigma} f_{,y_i}(x, y) \tag{2.2}$$

where the comma followed by a subscript variable x_i denotes a partial derivative with respect to the subscript variable (i.e. $f_{,x_i} \equiv \partial f / \partial x_i$). Summation convention for repeated subscripts is employed, except for the subscripts x and y. Attention is restricted to small deformation theory.

To model fatigue damage, we define a scalar damage variable ω^ς as a function of microscopic and macroscopic position vectors, i.e., $\omega^\varsigma = \omega(x, y)$. The constitutive equation on the microscale is derived from the strain-based continuum damage theory. Following Simo and Ju (1987), the free energy density has the form of

$$\Psi(\omega^\varsigma, \varepsilon^\varsigma_{ij}) = (1 - \omega^\varsigma) \Psi_e(\varepsilon^\varsigma_{ij}) \tag{2.3}$$

where $\omega^\varsigma \in [0, 1)$ is a scalar damage variable on the microscale. The elastic free energy density function is given as $\Psi_e(\varepsilon^\varsigma_{ij}) = \frac{1}{2} L_{ijkl} \varepsilon^\varsigma_{ij} \varepsilon^\varsigma_{kl}$. We assume that microconstituents possess homogeneous properties and satisfy

$$\sigma^\varsigma_{ij,x_j} + b_i = 0 \qquad \text{in} \quad \Omega \tag{2.4}$$

$$\sigma^\varsigma_{ij} = (1 - \omega^\varsigma) L_{ijkl} \varepsilon^\varsigma_{kl} \qquad \text{in} \quad \Omega \tag{2.5}$$

$$\varepsilon^\varsigma_{ij} = u^\varsigma_{(i,x_j)} \qquad \text{in} \quad \Omega \tag{2.6}$$

$$u^\varsigma_i = \bar{u}_i \qquad \text{on} \quad \Gamma_u \tag{2.7}$$

$$\sigma^\varsigma_{ij} n_j = \bar{t}_i \qquad \text{on} \quad \Gamma_t \tag{2.8}$$

where σ^ς_{ij} and $\varepsilon^\varsigma_{ij}$ are components of stress and strain tensors; L_{ijkl} denotes the elastic constitutive tensor components; b_i is a body force assumed to be independent of y; u^ς_i denotes components of the displacement vector; Γ_u and Γ_t are the boundary portions where displacements $\bar{u}_i$ and tractions $\bar{t}_i$ are prescribed; n_i denotes the normal vector component on the boundary. We assume that the interface between the phases is perfectly bonded, i.e. $[\sigma^\varsigma_{ij} \hat{n}_j] = 0$ and $[u^\varsigma_i] = 0$ at the interface, Γ_{int}, where $\hat{n}_i$ denotes the normal vector to Γ_{int} and $[\bullet]$ is a jump operator; the subscript pairs with parenthesizes denote the symmetric gradients defined as $u^\varsigma_{(i,x_j)} \equiv (u^\varsigma_{i,x_j} + u^\varsigma_{j,x_i})/2$

Since the discretization with grid spacing on the scale of material constituents is not feasible, mathematical homogenization theory is employed to account for microstructural effects without explicitly representing the details of the microstructure in the global analysis. This is accomplished by approximating the displacement field, $u_i^{\varsigma}(x) = u_i(x, y)$ and the damage variable, $\omega^{\varsigma}(x) = \omega(x, y)$, in terms of the double-scale asymptotic expansions on $\Omega \times \Theta$:

$$u_i(x, y) \approx u_i^0(x, y) + \varsigma u_i^1(x, y) + \ldots \tag{2.9}$$

$$\omega(x, y) \approx \omega^0(x, y) + \varsigma \omega^1(x, y) + \ldots \tag{2.10}$$

where the superscripts denote the order of terms in the asymptotic expansion. Using these expansions, a non-local damage theory for brittle composites has been developed by Fish *et al.* (1999). In the remaining of this section we briefly summarize the main results relevant to the fatigue problem which is the focus of this book chapter. For more details on non-local damage model we refer to (Fish *et al.*, 1999).

By inserting the expansion (2.9) and (2.10) into the boundary value problem (2.4)-(2.8), we can obtain a set of equilibrium equations for various orders of ς starting from $O(\varsigma^{-2})$. The successive solutions to these equations lead to the asymptotic expansion of the microscopic strain field in RVE:

$$\varepsilon_{ij}^{\varsigma}(x, y) = A_{ijmn}(y)\bar{\varepsilon}_{mn}(x) + G_{ijmn}(y)d_{mn}^{\omega}(x) + O(\varsigma) \tag{2.11}$$

where $\bar{\varepsilon}_{mn}(x)$ is the elastic strain in the macro domain and $d_{mn}^{\omega}(x)$ is a damage-induced macroscopic strain; $G_{ijmn}(y)$ is the local distribution function of damage-induced strain, which can be obtained by solving a linear boundary value problem in Θ with Y-periodic boundary conditions, i.e.

$$\{L_{ijkl}(I_{klmn} + H_{(k, y_l)mn})\}_{,y_j} = 0 \tag{2.12}$$

where $H_{imn}(y)$ is a Y-periodic third rank tensor with symmetry $H_{imn} = H_{inm}$, and $G_{ijmn}(y) = H_{(i, y_j)mn}(y)$; $A_{ijmn}(y)$ is the elastic strain concentration function given by

$$A_{ijmn} = I_{ijmn} + G_{ijmn} ; \quad I_{ijmn} = \frac{1}{2}(\delta_{im}\delta_{jn} + \delta_{in}\delta_{jm}) \tag{2.13}$$

where δ_{mk} is Kronecker delta. The damage-induced strain $d_{mn}^{\omega}(x)$ can be related to the macroscopic strain via a fourth rank tensor such that

$$d_{kl}^{\omega}(x) = D_{klmn}(x)\bar{\varepsilon}_{mn}(x) \tag{2.14}$$

where $D_{klmn}(x)$ represents the influence of the fatigue damage cumulation on the macroscopic response

$$D_{klmn}(x) = \left(I_{klst} - \sum_{\eta=1}^{n} B_{klst}^{(\eta)}\omega^{(\eta)}(x)\right)^{-1}\left(\sum_{\eta=1}^{n} C_{stmn}^{(\eta)}\omega^{(\eta)}(x)\right) \tag{2.15}$$

where η denotes different phases in RVE such that $\bigcup\limits_{\eta=1}^{n} \Theta^{(\eta)} = \Theta$; $\omega^{(\eta)}(x)$ represents the phase average damage; $B_{ijkl}^{(\eta)}$ and $C_{ijkl}^{(\eta)}$ are given by

$$B_{ijkl}^{(\eta)} = \frac{1}{|\Theta|}(\tilde{L}_{ijmn} - \bar{L}_{ijmn})^{-1}\int_{\Theta} G_{stmn}L_{stpq}G_{pqkl}\,d\Theta \tag{2.16}$$

$$C^{(\eta)}_{ijkl} = \frac{1}{|\Theta|}(\tilde{L}_{ijmn} - \bar{L}_{ijmn})^{-1}\int_{\Theta} G_{stmn}L_{stpq}A_{pqkl}\,d\Theta \tag{2.17}$$

$$\tilde{L}_{ijmn} = \frac{1}{|\Theta|}\int_{\Theta} L_{ijmn}\,d\Theta \quad \text{and} \quad \bar{L}_{ijkl} = \frac{1}{|\Theta|}\int_{\Theta} L_{ijmn}A_{mnkl}\,d\Theta \tag{2.18}$$

where $|\Theta|$ is the RVE volume and $\bar{L}_{ijkl}$ the homogenized elastic stiffness tensor. Based on (2.14) and (2.15), it can be seen that the damage-induced strain $d^{\omega}_{mn}(x)$ vanishes when the microstructure is damage free.

After solving for the local strain field, the homogenized strain field can be obtained by integrating the first order term in (2.11) over the phase volume $\Theta^{(\eta)}$, which yields

$$\varepsilon^{(\eta)}_{ij} = \frac{1}{|\Theta^{(\eta)}|}\int_{\Theta^{(\eta)}} \varepsilon^0_{ij}\,d\Theta = A^{(\eta)}_{ijkl}\bar{\varepsilon}_{kl} + G^{(\eta)}_{ijkl}D_{klmn}\bar{\varepsilon}_{mn} \tag{2.19}$$

where

$$A^{(\eta)}_{ijkl} = \frac{1}{|\Theta^{(\eta)}|}\int_{\Theta^{(\eta)}} A_{ijkl}\,d\Theta \quad \text{and} \quad G^{(\eta)}_{ijkl} = \frac{1}{|\Theta^{(\eta)}|}\int_{\Theta^{(\eta)}} G_{ijkl}\,d\Theta \tag{2.20}$$

The resulting constitutive equation for the phase average field can be expressed as

$$\sigma^{(\eta)}_{ij} = (1 - \omega^{(\eta)})L^{(\eta)}_{ijmn}\varepsilon^{(\eta)}_{mn} \tag{2.21}$$

where $\sigma^{(\eta)}_{ij}$ is the phase average stress. The overall homogenized stress field is given as

$$\bar{\sigma}_{ij} = \sum_{\eta=1}^{n} \nu^{(\eta)}\sigma^{(\eta)}_{ij} \tag{2.22}$$

where $\nu^{(\eta)}$ is the volume fractions for phase $\Theta^{(\eta)}$ in the RVE satisfying $\sum_{\eta=1}^{n}\nu^{(\eta)} = 1$

Finally, the phase free energy density corresponding to the nonlocal constitutive equation (2.21) is given as

$$\Psi^{(\eta)}(\omega^{(\eta)}, \varepsilon^{(\eta)}_{ij}) = \frac{1}{2}(1 - \omega^{(\eta)})L^{(\eta)}_{ijmn}\varepsilon^{(\eta)}_{mn}\varepsilon^{(\eta)}_{ij} \tag{2.23}$$

The energy release rate and the energy dissipation inequality (Chaboche, 1988; Simo and Ju, 1987) applied to the phase average fields can be expressed as

$$Y^{(\eta)} = -\frac{\partial\Psi^{(\eta)}}{\partial\omega^{(\eta)}} = \frac{1}{2}L^{(\eta)}_{ijmn}\varepsilon^{(\eta)}_{mn}\varepsilon^{(\eta)}_{ij} \quad \text{and} \quad Y^{(\eta)}\dot{\omega}^{(\eta)} \geq 0 \tag{2.24}$$

Since $L^{(\eta)}_{ijmn}$ is assumed to be a positive definite fourth order tensor it follows that $Y^{(\eta)} \geq 0$ and consequently, $\dot{\omega}^{(\eta)} \geq 0$ must hold

In the next two sections, we will focus on the definition of the fatigue damage cumulation law for the multi-phase composites and on the adaptive integration scheme for the initial value problem corresponding to the fatigue damage cumulation.

3.0 FATIGUE DAMAGE CUMULATIVE LAW

Following Marigo (1985), the fatigue damage cumulative law for each material phase can be expressed along the lines of the power law in viscoplasticity (Odqvist, 1974):

$$
\dot{\omega}^{(\eta)}(x,t) = \begin{cases} 0 & \bar{\vartheta}^{(\eta)} < \bar{\vartheta}_{ini}^{(\eta)} \\ \left[\dfrac{\Phi^{(\eta)}}{\omega^{(\eta)}}\right]^{\gamma^{(\eta)}} \dfrac{\partial \Phi^{(\eta)}}{\partial \bar{\vartheta}^{(\eta)}} < \dot{\bar{\vartheta}}^{(\eta)} >_{+} & \bar{\vartheta}^{(\eta)} \geq \bar{\vartheta}_{ini}^{(\eta)} \end{cases}
\tag{3.1}
$$

where $\omega^{(\eta)} \in [0,1)$ represents the phase average damage; the heaveside operator $< >_{+}$ denotes $< \bullet >_{+} = sup\{0, \bullet\}$; $\Phi^{(\eta)}$ can be expressed as:

$$
\Phi^{(\eta)} = \frac{atan\left[\alpha^{(\eta)}\left(\dfrac{< \bar{\vartheta}^{(\eta)} - \bar{\vartheta}_{ini}^{(\eta)} >_{+}}{\bar{\vartheta}_{0}^{(\eta)}}\right) - \beta^{(\eta)}\right] + atan(\beta^{(\eta)})}{\dfrac{\pi}{2} + atan(\beta^{(\eta)})}
\tag{3.2}
$$

where $\alpha^{(\eta)}$, $\beta^{(\eta)}$, and $\bar{\vartheta}_{0}^{(\eta)}$ are phase material constants which can be calibrated by static uniaxial loading test for each phase.

$\bar{\vartheta}_{ini}^{(\eta)}$ in (3.1) is usually interpreted as an "endurance domain" within which the change of strain state along any loading path does not lead to the growth of damage.

$\bar{\vartheta}^{(\eta)}$, often termed as the damage equivalent strain, is defined as the square root of the phase damage energy release rate (2.24) (Simo and Ju, 1987) along with the weighting tensor

$$
\bar{\vartheta}^{(\eta)} = \sqrt{\frac{1}{2}\hat{L}_{ijkl}^{(\eta)}\hat{\varepsilon}_{mn}^{(\eta)}F_{ijmn}^{(\eta)}\hat{\varepsilon}_{st}^{(\eta)}F_{klst}^{(\eta)}}
\tag{3.3}
$$

where $\hat{L}_{ijkl}^{(\eta)}$ and $\hat{\varepsilon}_{ij}^{(\eta)}$ represent the elastic phase constitutive tensor and nonlocal phase strain tensor, respectively, both expressed in principal directions; $F_{ijkl}^{(\eta)}$ denotes the phase strain weighting tensor accounting for the different behavior in tension and compression caused by cycling loading. In matrix notation equation (3.3) is given as

$$
\bar{\vartheta}^{(\eta)} = \sqrt{\frac{1}{2}(F^{(\eta)}\hat{\varepsilon}^{(\eta)})^{T}\hat{L}^{(\eta)}(F^{(\eta)}\hat{\varepsilon}^{(\eta)})}
\tag{3.4}
$$

where the superscript T represents the transpose; the principal nonlocal phase strain can be written as $\hat{\varepsilon}^{(\eta)} = [\varepsilon_1^{(\eta)}, \varepsilon_2^{(\eta)}, \varepsilon_3^{(\eta)}]^{T}$ and the phase strain weighting matrix is heuristically defined as

$$
F^{(\eta)} = \begin{bmatrix} h_1^{(\eta)} & 0 & 0 \\ 0 & h_2^{(\eta)} & 0 \\ 0 & 0 & h_3^{(\eta)} \end{bmatrix}
\tag{3.5}
$$

$$h_\xi^{(\eta)} \equiv h(\varepsilon_\xi^{(\eta)}) = \frac{1}{2} + \frac{1}{\pi} \text{atan}[a_1(\varepsilon_\xi^{(\eta)} - a_2)], \quad \xi = 1, 2, 3 \tag{3.6}$$

where a_1 and a_2 are material constants. In the limit, as $a_1 \to \infty$ and $a_2 = 0$, the weight function reduces to $h(\varepsilon_\xi^{(\eta)}) = <\varepsilon_\xi^{(\eta)}>_+/\varepsilon_\xi^{(\eta)}$, which corresponds to the case where compression does not promote damage.

The exponent $\gamma^{(\eta)}$ in (3.1) is a stress-dependent parameter for each phase material. When $\gamma^{(\eta)} \to \infty$ the fatigue damage power law reduces to the rate-independent form in the sense that the damage evolution in this case is controlled by the value of $\Phi^{(\eta)}/\omega^{(\eta)}$ (Marigo, 1985). By assuming constant exponent in the power law (Pass *et al.*, 1993), fatigue damage cumulation law in the form of (3.1) can be reduced to the modified Palmgreen-Miner's model (Fatemi and Yang, 1998). More elaborate fatigue damage model can be defined by making the exponent to be stress dependent (function of the maximum and the mean stress values (Chaboche and Lesne, 1988)). Here, we assume that parameter $\gamma^{(\eta)}$ depends on the mean and the maximum values of principal phase stresses, i.e.

$$\gamma^{(\eta)} = g(c_i^{(\eta)}, \bar\sigma_{max}^{(\eta)}, \bar\sigma_{mean}^{(\eta)}) \tag{3.7}$$

where $c_i^{(\eta)}$ is a set of phase material constants; $\bar\sigma_{max}^{(\eta)}$ and $\bar\sigma_{mean}^{(\eta)}$ are dimensionless quantities defined as

$$\bar\sigma_{max}^{(\eta)} = \frac{\hat\sigma_{1max}^{(\eta)}}{2\sigma_u^{(\eta)}} \quad \text{and} \quad \bar\sigma_{mean}^{(\eta)} = \frac{\hat\sigma_{1max}^{(\eta)} + \hat\sigma_{1min}^{(\eta)}}{2\sigma_u^{(\eta)}} \tag{3.8}$$

where $\hat\sigma_{1max}^{(\eta)}$ and $\hat\sigma_{1min}^{(\eta)}$ are the maximal and minimal principal phase average stresses at a given global position; $\sigma_u^{(\eta)}$ is the ultimate strength of the phase material. A further discussion on the calibration of (3.7) is given in Section 5.2.

4.0 INTEGRATION OF NONLOCAL FATIGUE DAMAGE CUMULATIVE LAW

To develop an efficient acceleration technique for the integration of fatigue damage cumulative law, a constant amplitude cyclic loading history is typically subdivided into a series of load cycle blocks, each block consisting of several load cycles. One of the fatigue damage models developed in (Pass *et al.*, 1993) assumes that in each block of cycles, the mechanical response is independent of fatigue damage cumulation until the local rupture occurs. In another model developed for homogeneous materials (Chow and Wei, 1994), the first cycle in the block in which the damage increment is caused by inelastic deformation, is used to compute a constant rate of fatigue damage growth in that block. The major shortcomings of these models are threefold: (i) the deviation from the equilibrium path caused by the integration of fatigue damage cumulation law, (ii) the difficulty in estimating an adequate block size, especially in the initial and near-rupture loading stages where the growth of fatigue damage is very rapid, and (iii) applicability to heterogeneous materials.

In what follows the nonlocal damage cumulative law will be approximated by the first order initial value problem with respect to the number of load cycles and subsequently solved using the Adaptive Modified Euler's method with the maximum damage increment control.

Let us return to the fatigue damage cumulative law defined in (3.1). Since this fatigue law is stated in the rate form, it is necessary to integrate it along the loading path to obtain the current damage state. The nonlocal phase damage increment in one load cycle can be expressed as

$$\int_{t}^{(t+\tau_0)} \dot{\omega}^{(\eta)} dt = \int_{t}^{(t+\tau_0)} \left[\frac{\Phi^{(\eta)}}{\omega^{(\eta)}}\right]^{\gamma^{(\eta)}} \frac{\partial \Phi^{(\eta)}}{\partial \bar{\vartheta}^{(\eta)}} < \dot{\vartheta}^{(\eta)} >_{+} dt \qquad (4.1)$$

where t is the time at the beginning of a load cycle and τ_0 is the period of the cyclic loading. The above integration has to be carried at each Gauss point in the macro domain. Assuming that the increment of phase damage in one load cycle is very small, we can approximate the derivative of the nonlocal damage parameter with respect to the number of load cycles as

$$\frac{d\omega^{(\eta)}}{dN}\bigg|_{K} \approx \int_{t}^{(t+\tau_0)} \dot{\omega}^{(\eta)} dt \equiv \Delta\omega^{(\eta)}\big|_{K} = \omega^{(\eta)}\big|_{K} - \omega^{(\eta)}\big|_{K-1} \qquad (4.2)$$

where N denotes the number of load cycles; $\omega^{(\eta)}\big|_{K}$ is the phase damage at the end of load cycle K which can be obtained by the incremental finite element analysis for this cycle with initial damage $\omega^{(\eta)}\big|_{K-1}$ and the corresponding initial strain/stress conditions.

Using the forward Euler's method the nonlocal phase damage after ΔN_K cycles from the current load cycle K can be approximated by

$$\omega^{(\eta)}\big|_{(K+\Delta N_K;\,\Delta N_K)} = \omega^{(\eta)}\big|_{K} + \Delta N_K \Delta\omega^{(\eta)}\big|_{K} \qquad (4.3)$$

where $\omega^{(\eta)}\big|_{(K+\Delta N_K;\,\Delta N_K)}$ represents the approximate solution of the nonlocal phase damage at the end of load cycle $K + \Delta N_K$ with the block size ΔN_K and the initial nonlocal damage $\omega^{(\eta)}\big|_{K}$.

It is important to note that updating the damage variable while keeping the rest of the fields fixed violates constitutive equations. This inconsistency is subsequently alleviated in two steps: (i) update the nonlocal phase stresses using the overall strain from the end of cycle K, $\bar{\varepsilon}_{ij}\big|_{K}$, (ii) carry out nonlinear finite element analysis to equilibrate discrete equilibrium equations. We will refer to this two-step process as the *consistency adjustment*.

For forward Euler's one-step method the block size ΔN_K should be selected to ensure accuracy. The block size can be adaptively selected by keeping the nonlocal phase damage increment sufficiently small when the damage increases rapidly and vice versa. This can be expressed as follows:

$$\Delta N_K = int\left\{\Delta\omega_a^{(\eta)} / \max_{gauss}(\Delta\omega^{(\eta)})\big|_{K}\right\} \qquad (4.4)$$

where operator $int\{ \bullet \}$ denotes the truncation to the decimal part; $\Delta\omega_a^{(\eta)}$ is a user-defined allowable tolerance of phase damage increment per cycle; $max(\ )$ is computed with respect to all integration points in the macro-problem. There are two major reasons to monitor the value of maximum damage increment. First, is to ensure the existence of the initial value problem, i.e., if the damage growth rate in cycle K in at least one of the Gauss points is very high, the approximation of the initial value problem might be inaccurate, and thus the block size ΔN_K evaluated by (4.4) should be set to zero. In this case, the method reduces to the direct cycle-by-cycle approach. The second reason is to ensure accuracy of the aforementioned consistency adjustment process.

The fatigue life, denoted as N_{max}, can be expressed as

$$N_{max} = n + \sum_{K=1}^{n} \Delta N_K; \qquad \Delta N_K \geq 0 \tag{4.5}$$

where n is the number of the cycle blocks in the loading history which is also the actual number of the cycles carried out in the case of the direct simulation. The maximal value of n is $\text{int}\{1/(\Delta \omega_a^{(\eta)})\}$ provided that the failure occurs when $\omega^{(\eta)}$ reaches one at the critical Gauss point.

To control solution accuracy of the initial value problem we adopt the modified Euler's integrator (Stoer and Bulirsch, 1992) with the initial block size determined by (4.4). The nonlocal phase damage at load cycle $K + \Delta N_K$ (4.3) is then defined as

$$\omega^{(\eta)}\big|_{(K+\Delta N_K;\, \Delta N_K)} = \omega^{(\eta)}\big|_K + \frac{\Delta N_K}{2}(\Delta \omega^{(\eta)}\big|_K + \Delta \omega^{(\eta)}\big|_{K+\Delta N_K}) \tag{4.6}$$

where $\Delta \omega^{(\eta)}\big|_K$ is evaluated by (4.2) while $\Delta \omega^{(\eta)}\big|_{K+\Delta N_K}$ is also obtained by (4.2) after substituting $K + \Delta N_K$ for K; $\omega^{(\eta)}\big|_{K+\Delta N_K}$ is the first order approximation defined in (4.3).

The problem of integration of phase fatigue damage cumulative law can be stated as follows:

<u>Given</u>: the tolerance err; allowable damage increment per load cycle $\Delta \omega_a^{(\eta)}$; initial nonlocal phase damage in each phase $\omega^{(\eta)}\big|_{K-1}$, and the overall strain $\bar{\varepsilon}_{mn}\big|_{K-1}$ at the beginning of load cycle K.

<u>Find</u>: the size of the block ΔN_K; fatigue life N_{max}; nonlocal phase damage and overall strain at the end of cycle $K + \Delta N_K$.

The adaptive scheme is summarized as follows:

i.) Carry out the incremental finite element analysis for one load cycle with initial nonlocal phase damage $\omega^{(m)}\big|_{K-1}$ and the overall strain $\bar{\varepsilon}_{mn}\big|_{K-1}$. Denote the nonlocal phase damage at the end of this cycle as $\omega^{(\eta)}\big|_K$, and the overall strain as $\bar{\varepsilon}_{mn}\big|_K$. At each integration point in the macro domain estimate the rate of nonlocal phase damage in the current load cycle, $\Delta \omega^{(\eta)}\big|_K$, as defined by (4.2).

ii.) Calculate the initial block size using (4.4).

iii.) At each integration point in the macro domain, compute the approximate solution $\omega^{(\eta)}\big|_{(K+\Delta N_K;\, \Delta N_K)}$ with the block size ΔN_K using the modified Euler's method (4.6); then using the block size $\Delta N_K/2$, compute $\omega^{(\eta)}\big|_{(K+\Delta N_K;\, \Delta N_K/2)}$ by two successive uses of (4.6).

vi.) Find the maximum error among all the integration points and check the convergence by:

$$\max_{gauss}\{\big|\omega^{(\eta)}\big|_{(K+\Delta N_K;\, \Delta N_K)} - \omega^{(\eta)}\big|_{(K+\Delta N_K;\, \Delta N_K/2)}\big|\} \leq err \tag{4.7}$$

If (4.7) is false, set $\Delta N_K \leftarrow \Delta N_K/2$ and go back to *iii)*. Otherwise, update the fatigue life by (4.5), i.e. $N_{max}\big|_K = N_{max}\big|_{K-1} + \Delta N_K + 1$; update the approximation of the nonlocal phase damage at the end of load cycle $K + \Delta N_K$, $\omega^{(m)}|_{(K + \Delta N_K; \Delta N_K/2)}$, and the overall strain $\varepsilon_{mn}|_K$. Then, compute the nonlocal phase strains $\varepsilon_{mn}^{(m)}|_K$ and $\varepsilon_{mn}^{(f)}|_K$ using (2.19).

vii.) Perform consistency adjustment: (i) calculate nonlocal phase stresses $\sigma_{ij}^{(m)}|_K$ and $\sigma_{ij}^{(f)}|_K$ by (2.21), and macroscopic stresses $\sigma_{ij}|_K$ by (2.22); (ii) equilibrate discrete solution using nonlinear finite element analysis. Finally, set $K \leftarrow K + 1$ and go to *i.)* for the next block of cycles.

5.0 NUMERICAL EXAMPLES

In this section, we assume that the composite material consists of two phases, matrix ($\eta = m$) and reinforcement ($\eta = f$), denoted by $\Theta^{(m)}$ and $\Theta^{(f)}$ such that $\Theta = \Theta^{(m)} \cup \Theta^{(f)}$. For simplicity, we assume that fatigue damage occurs in the matrix phase only, i.e. $\omega^{(f)} \equiv 0$.

5.1 Qualitative Examples for the Two-Phase Fibrous Composites

The first set of numerical examples investigates the computational efficiency and accuracy of the proposed fatigue model. We consider the classical stress concentration problem - a thin plate with a centered small circular hole, as shown in Figure 5.1. The plate is assumed to be composed of 0/0 ply of fibrous composite. The plate is subjected to uniaxial tension perpendicular to the fiber direction. The two transverse directions coincide with the X and Y axes. The fiber direction is aligned along the Y axis. The properties of the two micro-phases are as follows:

Matrix: $v^{(m)} = 0.733$, $E^{(m)} = 69\text{GPa}$, $\mu^{(m)} = 0.33$

Fiber: $v^{(f)} = 0.267$, $E^{(f)} = 379\text{GPa}$, $\mu^{(f)} = 0.21$

where v is volume fraction, while E and μ denote Young's modulus and Poisson ratio, respectively. The parameters of the damage evolution law are chosen as $\alpha^{(m)} = 8.2$, $\beta^{(m)} = 10.2$ and $\vartheta_0 = 0.05$ (MPa)$^{1/2}$. For simplicity, $\gamma^{(m)}$ is assumed to be constant, and set $\gamma^{(m)} = 4.5$ for low-cycle fatigue, and $\gamma^{(m)} = 15$ for high-cycle fatigue.

The static loading capacity of the plate is 103.6 N as shown in Figure 5.2. The cyclic loading is designed as a tension-to-zero loading with amplitude of 90 N. For low-cycle fatigue, the direct cycle-by-cycle simulation serves as a reference solution. Several allowable damage increments per cycle were selected to study the convergence of the method. The results summarized in Figures 5.3-5.5 demonstrate excellent convergence characteristics of the proposed fatigue model. For high-cycle fatigue problems the solution obtained by the forward Euler method with very small $\Delta\omega_a^{(m)}$ is used as a reference solution instead of the direct simulation which is computationally prohibitive. Similar observations can be made for high-cycle fatigue problem.

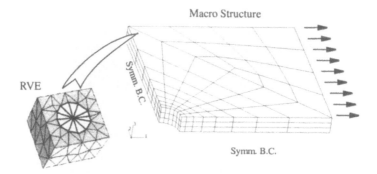

Figure 5.1 FE model of RVE and Macro Domain

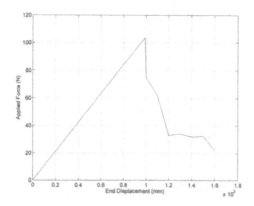

Figure 5.2 Static Loading Capacity

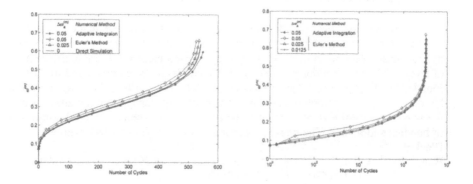

Figure 5.3 Fatigue Damage Cumulation for Low-Cycle and High-Cycle Fatigue

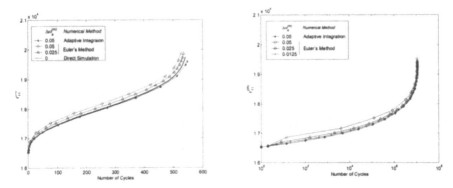

Figure 5.4 Strain Softening for Low-Cycle and High-Cycle Fatigue

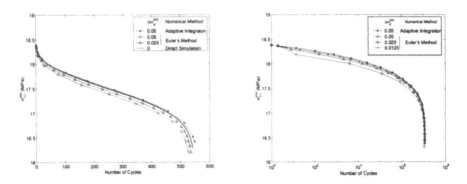

Figure 5.5 Local Stress Relaxation for Low-Cycle and High-Cycle Fatigue

5.2 Large Scale Fatigue Analysis For Woven Composites

In this section, we consider the tailcone exhaust structure made of Techni-weave T-Form Nextel312/Blackglas Composite System as shown in Figure 5.6. The matrix phase in the RVE has been removed to give a clear view of the architecture of fabrics. The fabric is designed using 600 denier bundles of Nextel 312 fibers surrounded by Blackglas 493C matrix material (Butler *et al.*, 1996). For simplicity, the bundles are assumed to be linear elastic (damage-free) throughout the analysis and the woven composite is considered to be a periodic two-phase material composed of bundles and matrix. The phase properties of RVE are summarized below:

Blackglas Matrix: $v^{(m)} = 0.565$, $E^{(m)} = 38.61\,\mathrm{GPa}$, $\mu^{(m)} = 0.26$

Bundle: $v^{(f)} = 0.435$, $E_A^{(f)} = 114.28\,\mathrm{GPa}$, $G_A^{(f)} = 45.19\,\mathrm{GPa}$, $\mu_A^{(f)} = 0.244$,
    $E_T^{(f)} = 112.10\,\mathrm{GPa}$, $G_T^{(f)} = 44.95\,\mathrm{GPa}$

where G denotes shear modulus; the subscripts A and T represent the axial and transverse directions for transversely isotropic material.

 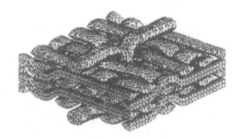

Figure 5.6 Geometric Model of the Tailcone and the Techniweave T-Form Woven
Microstructure

The compressive principal strains have been observed to have little effect on
the damage cumulation, so the constants in (3.6) are defined as $a_1 = 10^7$ and
$a_2 = 0$. The material constants $\alpha^{(m)}$, $\beta^{(m)}$ and $\bar{\vartheta}_0^{(m)}$ in (3.2) have been calibrated
based on the tensile test under uniform monotonic loading, which gave $\alpha^{(m)} = 7.6$,
$\beta^{(m)} = 10.9$ and $\bar{\vartheta}_0^{(m)} = 0.24$ $(\text{MPa})^{1/2}$. The endurance limit was taken as
$\bar{\vartheta}_{ini}^{(m)} = 0$. The predicted ultimate strength in weave plane was 105.8 MPa with
0.178% ultimate strain. In the direction normal to the weave plane, the ultimate
strength was 69.1 MPa and the ultimate strain 0.21%. Material constants were
selected so that numerical results at ultimate points were in good agreement with
the test data.

Similarly to fracture mechanics based fatigue models we assume that the
fatigue parameter $\gamma^{(m)}$ depends on the mean and the maximum values of principal
stress components as

$$\gamma^{(m)} = (\bar{\sigma}_{max}^{(m)})^{c_1}\{c_2 + c_3(\bar{\sigma}_{max}^{(m)} - \bar{\sigma}_{mean}^{(m)}) + c_4(\bar{\sigma}_{max}^{(m)} - \bar{\sigma}_{mean}^{(m)})^2\} \qquad (5.1)$$

where c_1-c_4 are material constants; $\bar{\sigma}_{max}^{(m)}$ and $\bar{\sigma}_{mean}^{(m)}$ are defined following (3.8) in
the matrix phase at a given Gauss point; $\sigma_u^{(m)}$ is the ultimate strength of the matrix
phase. The calibration of the material constants in (5.1) is not trivial, since $\gamma^{(m)}$ is a
function of the nonlocal phase stresses, whereas the only experimental data available is the number of cycles to failure, or fatigue life N_{cr}. Thus to calibrate material
constants we set the following problem: Find the material constants, c_i, so that
$\left\|N_{cr}^*(c_i) - N_{cr}\right\|_2$ is minimal, where $N_{cr} = [N_{cr}^1, ..., N_{cr}^k]^T$ is a set of experimental
fatigue life predictions obtained with various cyclic loading conditions; $N_{cr}^*(c_i)$ is
a set of computed fatigue life predictions. The Jacobian matrix for the least square
analysis is evaluated using the finite difference method. Calibration is performed
for the uniform tension-to-tension cyclic loading test in the weave plane. The minimal tensile loading was ten percent of the maximum value.

Figure 5.7 compares the computed fatigue life with material constants
$c_1 = 0.5$, $c_2 = 0.554$, $c_3 = 1.35$, $c_4 = -2.68$ and test data. The ultimate

strength of the matrix phase is $\sigma_u^{(m)} = 69.1\,\text{MPa}$. Figure 5.8 illustrates the evolution of nonlocal matrix damage parameter and the nonlocal equivalent matrix stress. Once the fatigue damage model has been calibrated, we turn to the evaluation of fatigue life of the tailcone structure. The finite element mesh of one-eighth of the tailcone model in the neighborhood of the attachment hole is shown in Figure 5.9. It consists of 3,154 nodes, 3,242 thin shell elements and 385 spring elements totaling 17,766 degrees of freedom. It can be seen that the damage initiates at the location of the attachment and then quickly spreads around the supporting ring causing the overall structural failure. The critical state is reached after 1.12 million cycles. No experiments have been conducted up to failure to verify this result.

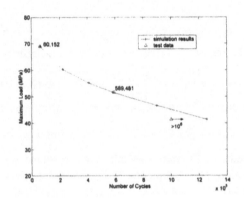

Figure 5.7 Predicted Fatigue Life and Test Data

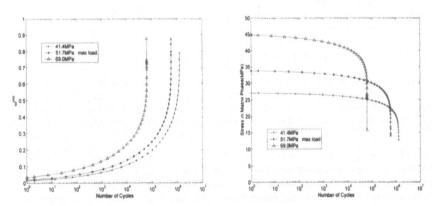

Figure 5.8 Damage Evolution and Stress Relaxation under Uniaxial tension-tension Cyclic Loading

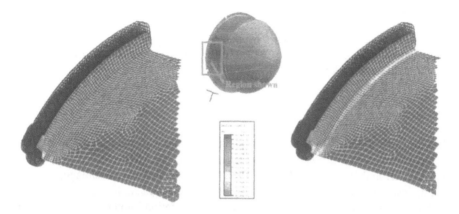

Figure 5.9 Damage Distribution in Critical Region of the Tailcone

6.0 Conclusions and Future Research Directions

Traditionally, life predictions for macrocrack initiation are carried out using S-N or ε-N curves in conjunction with some parameters designated to take into account the differences between actual components and test specimens such as geometry, fabrication, environmental conditions, etc. Due to the coupled nature of the present multiscale CDM based fatigue model, mechanical and fatigue damage cumulation analyses are carried out simultaneously without relying on S-N or ε-N curves. The accelerating technique for the integration of the fatigue damage cumulation makes it possible to simulate the damage evolution by fully coupled analysis for large scale structures. Thus large amount of specimen tests can be avoided, while complex geometrical features, material imperfections, multiaxial loading conditions, and material data can be readily incorporated into the computational model.

For macrocrack propagation problem, the present model has certain limitations. As observed in (Geers, 1997; Geers *et al.*, 1998), the nonlocal damage theory may lead to the spurious damage zone widening phenomenon, especially when the crack opening is accompanied with large strains. Physically, with the evolution of macrocracks ($\omega^s \to 1$), the nonlocal domain in the vicinity of the macrocrack should be finally collapsed into the localized discontinuity, i.e. crack line. As noted in Section 3, however, the size of characteristic volume is assumed to be constant and the value of the damage variable in our model is not allowed to reach one to ensure the regularity of the solution. As a result, the localized discontinuity can never be formed and the widening of the damaged zone is unavoidable. In the future work we will explore a transient-gradient damage model (Geers, 1997), in which the characteristic length is assumed to be history dependent. As an alternative, we will also consider a possibility of switching to the fracture mechanics approach (Newman, 1997). Finally, a unified methodology linking the nonlocal damage theory and fracture mechanics will be also explored (Mazars and Pijaudier-Cabot, 1996; van Vroonhoven and deBorst, 1999).

Acknowledgment

This work was supported in part by the Allison Engines, the National Science Foundation under grant number CMS-9712227, and Sandia National Laboratories under grant number AX-8516.

References

Bazant, Z.P., 1991, Why continuum damage is nonlocal: micromechanical arguments. *Journal of Engineering Mechanics,* **117**(5), pp. 1070-1087.

Bazant Z.P. and Pijaudier-Cabot G., 1988, Nonlocal continuum damage, localization instability and convergence,. *Journal of Applied Mechanics,* **55**, pp. 287-293.

Belytschko T. and Lasry D., 1989, A study of localization limiters for strain softening in statics and dynamics. *Computers and Structures,* **33**, pp. 707-715.

Bhattacharya, B. and Ellingwood, B., 1999, A new CDM-based approach to structural deterioration. *International Journal of Solids and Structures,* **36**, pp. 1757-1779.

Butler, E.P., Danforth, S.C., Cannon, W.R. and Ganczy, S.T., 1996, *Technical Report for the ARPA LC³ program,* ARPA Agreement No. MDA 972-93-0007.

Chaboche, J.L., 1988, Continuum damage mechanics I: General concepts & II: Damage growth, crack initiation and crack growth. *Journal of Applied Mechanics,* **55**, pp. 59-72.

Chaboche, J.L. and Lesne, P.M., 1988, A nonlinear continuous fatigue damage model. *Fatigue and Fracture of Engineering Materials and Structures,* **11**(1), pp. 1-17.

Chow, C.L. and Wei, Y., 1994, A fatigue damage model for crack propagation, In: Advances in Fatigue Life time Predictive Techniques. In *ASTM STP 1292,* edited by Mitchell, M.R. and Landgraf, R.W., (Philadelphia: American Society of Testing and Materials), pp 86-99.

Dauskardt, R.H. and Ritchie, R.O., 1991, Cyclic fatigue of ceramics. In *Fatigue of Adavanced Materials,* edited by Ritchie, R.O., Dauskardt, R.H. and Cox, B.N., (Birmingham: Materials and Component Engineering Publications Ltd.), pp. 133-151.

Fatemi, A. and Yang, L., 1998, Cumulative fatigue damage and life prediction theories: a survey of the start of the art for homogeneous materials. *International Journal of Fatigue,* **20**(1), pp. 9-34.

Fish, J., Yu, Q., and Shek, K., 1999, Computational damage mechanics for composite materials based on mathematical homogenization. *International Journal For Numerical Methods In Engineering,* **45**, pp. 1657-1679.

Geers, M.G.D., 1997, *Experimental and Computational Modeling of Damage and Fracture.* Ph.D Thesis, Technische Universiteit Eindhoven, The Netherlands

Geers, M.G.D., Peerlings, R.H.J., deBorst, R. and Brekelmans, W.A.M., 1998, Higher-order damage models for the analysis of fracture in quasi-brittle materials. In *Material Instabilities in Solids,* edited by deBorst R. and van der Giessen,

E., (Chichester: John Wiley & Sons), pp. 405-423.

Marigo, J.J., 1985, Modelling of brittle and fatigue damage for elastic material by growth of microvoids. *Engineering Fracture Mechanics*, **21**(4), pp. 861-874.

Mazars, J. and Pijaudier-Cabot, G., 1996, From damage to fracture mechanics and conversely: A combined approach. *International Journal of Solids and Structures* **33**(20-22), pp. 3327-3342.

Newman, J.C, 1997, The merging of fatigue and fracture mechanics concept: a historical respective. In *Fatigue and Fracture Mechanics: 28th Volume, ASTM ASTP 1321*, edited by Underwood, J.H., MacDonald, B.D. and Ritchell, M.R., (Philadelphia: American Society of Testing and Materials), pp. 1-35

Odqvist, F.K.G, 1974, *Mathematical theory of creep and creep rupture*, (Oxford, Clarendon Press).

Papa, E., 1993, A damage model for concrete subjected to fatigue loading. *European Journal of Mechanics: A/Solids*, **12**(3), pp. 429~440.

Pass, M.H.J.W., Schreurs, P.J.G., and Brekelmans, W.A.M., 1993, A continuum approach to brittle and fatigue damage: Theory and numerical procedures. *International Journal of Solids and Structures*, **30**(4), pp. 579-599.

Sanchez-Palencia, E., 1980, *Non-Homogeneous Media and Vibration Theory*, (Berlin: Springer-Verlag).

Simo, J.C. and Ju, J.W., 1987, Strain- and stress-based continuum damage models - I. Formulation. *International Journal of Solids and Structures*, **23**(7), pp. 821-840.

Stoer, J. and Bulirsch, R., 1992, *Introduction to Numerical Analysis: Second Edition*, (New York: Springer-Verlag New York Inc.).

Suresh, S., *Fatigue of Materials*, 1991, (Cambridge: Cambridge University Press).

van Vroonhoven, J.C.W. and deBorst, R., 1999, Combination of fracture and damage mechanics for numerical failure analysis. *International Journal of Solids and Structures*, **36**, 1169-1191.

Part II

Thermal and Fluid Sciences

Life-Cycle and Durability Predictions of Elastomeric Components

Alan Wineman[1], Alan Jones[1] and John Shaw[2]

1. INTRODUCTION

Research into the predictions of durability and service life for rubber (or elastomeric) products has been actively pursued over the past decade because elastomers are finding increasingly abundant applications in the automotive, aerospace and medical industries. Elastomers are used in many critical service applications that require long trouble free life such as gaskets, seals, bridge pad bearings, tires, medical components and rocket fuel binders. For example, elastomeric seals are used in many applications where they are relatively inaccessible and the cost of replacement is high. Thus, a fifty-year service life for seals is now a clearly stated objective for the energy industry (Hertz, 1997). The life expectancy of elastomeric components on vehicles has changed dramatically over the past few years (Pett, 1997). Previously, many rubber components were viewed as items that would be replaced once or more during the life of a vehicle. Now, elastomeric components are expected to last the life of the vehicle, which is currently defined as 10 years/150,000 miles for both passenger cars and light trucks.

Service life predictions of elastomeric components become an increasingly important part of the engineering design process. Elastomeric materials are frequently used under severe thermal, chemical and mechanical stress conditions. The wide range in conditions produces changes in properties large enough to cause failure. There is clearly a need for a robust model for use in the service life prediction of rubber components. Indeed, the Akron Rubber Development Laboratory (ARDL) has presented symposia on the topic of elastomer service life prediction in 1997 and 1998.

There are many physical factors which affect the durability of an elastomeric component, such as large deformation, conversion of mechanical energy to heat due to dissipative effects, heat transfer within the component, and changes in material properties due to changes in microstructure. An important goal, therefore, is the development of a thermo-mechanics model incorporating these factors. Through theoretical and experimental studies typical rubber components under simple states of thermo-mechanical loading, the important first steps are being taken towards understanding the essential elements that will lead to the development of realistic and accurate life and durability prediction models for rubber components.

The current focus is on the role of heating on the durability of an elastomeric component. There are two sources of heating of elastomeric

[1] Department of Mechanical Engineering, University of Michigan
[2] Department of Aerospace Engineering, University of Michigan

components. First, components are often used in hot environments, such as in the engine compartment of an automobile. Thus, an engine mount might operate in an environment of over 100°C. Second, internal mechanical heating, such as occurs during cyclic loading, can lead to a substantial temperature rise within the component. This is illustrated by an experiment conducted by us of an automobile suspension bushing at the Tenneco Automotive Corporation. As shown in Figure 1, the outer sleeve of the bushing was supported in a fixed housing and its center was connected to a movable post. The bushing design includes an air gap, which affects its compliance under displacements in the horizontal direction in the figure. The post was given a 1000 N vertical pre-load and then subjected to sinusoidally oscillating displacements in the vertical direction with respect to the figure. The displacement amplitude was 2 mm and the frequency was 10 Hertz. The initial bushing temperature was about 70° F. The temperature rise in the bushing was measured with a thermal imaging camera. As can be seen, after 2.5 minutes the temperature of the lower part of the bushing increased to 86° F. Since rubber is a poor heat conductor and bushings normally undergo oscillations for long periods of time, significant temperature rises within the bushing can occur in operation. Regardless of the mechanism, there can be significant degradation of the elastomer when the temperature becomes sufficiently high. Therefore, there is a need for a constitutive model for the thermo-mechanical response of an elastomer as it degrades at high temperatures. This model, along with the equations of thermo-mechanics, could be used in a computational code to determine the stresses, deformation, temperatures and material property changes in an elastomeric component in typical applications. In this paper, we present several issues underlying the development of such a model.

Section 2 contains a discussion of experimental results from the literature that characterize the process of degradation of rubber at elevated temperatures and a simple thermo-mechanical model which has been proposed to model this process. These results apply to fixed uniaxial extension and constant temperature. A three dimensional generalization of this model is introduced in Section 3, and an extension to time-temperature varying histories is given in Section 4. It provides a set of hypotheses, from which the basis of an experimental program. Experimental results are then presented in Section 5.

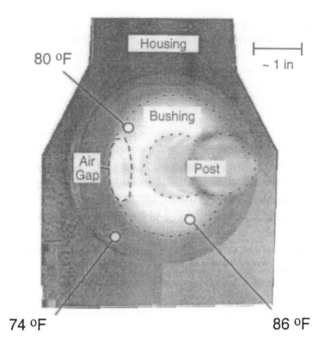

Figure 1. Temperature contours after 2.5 min in a bushing
subjected to 10 Hz vertical displacement of the post.

2. EXPERIMENTAL BACKGROUND

In the experiments conducted by Tobolsky (1960), a rubber strip at room
temperature was subjected to a fixed uniaxial stretch and then held at a higher
fixed temperature for some time interval. At temperatures above T_{cr}, (say
$100°$ C), called the chemorheological temperature, the stress decreased with time.
At the end of a specified time interval, the stress was removed and the specimen
was returned to its original temperature. The specimen was also observed to have
a permanent stretch. Tests were carried out for different applied stretches,
temperatures and time intervals. The decrease in tensile stress with time and the
permanent stretch were measured. The data were analyzed assuming neo-
Hookean behavior, for which the relation between tensile (Cauchy) stress $\sigma(t)$
and uniaxial stretch ratio λ is

$$\sigma(t) = 2n(t)kT(\lambda^2 - \frac{1}{\lambda}),$$
(1)

where T is the absolute temperature, k is the Boltzmann constant and $n(t)$ is the
current cross link density. The decrease in $\sigma(t)$ was attributed to scission of
molecular network cross links. The permanent stretch was attributed to a new

network which formed in the stretched state (healing). The stress–stretch relation for the system consisting of the two networks was assumed to be

$$\sigma(t) = 2n_1 kT(\lambda^2 - \frac{1}{\lambda}) + 2n_2 kT\left(\left(\frac{\lambda}{\hat{\lambda}}\right)^2 - \left(\frac{\hat{\lambda}}{\lambda}\right)\right), \tag{2}$$

where $\hat{\lambda}$ is the stretch ratio of the original network held at the high temperature, n_1 is the cross link density of the original network at the end of the test and n_2 is the cross link density of the new network. Equation (2) expresses the assumptions that (i) the stresses in each network are additive, (ii) each network acts as an incompressible isotropic neo-Hookean elastic material and (iii) the new network is formed stress free at the stretch ratio $\hat{\lambda}$ of the original network. Tobolsky's data suggested that n_1 and n_2 are independent of the stretch ratio $\hat{\lambda}$ for $\hat{\lambda} < 3$ to 4. This was supported by the results of Scanlan and Watson (1958). It was also assumed (Tobolsky, 1960, Tobolsky *et al*, 1944) that all broken cross-links re-formed to produce a new network in a stress free state. In other words, there was conservation of cross-links, $n_1 + n_2 = n(0)$, which is referred to hereafter as complete healing.

Neubert and Saunders (1958) carried out tests similar to those of Tobolsky, but for a pure shear deformation. They measured permanent biaxial stretch upon removal of stress and reduction of the temperature to its original value, and found that predictions based on the assumption of neo-Hookean response led to inaccurate predictions of permanent set. They modified Tobolsky's assumption (ii) by assuming the rubber could be described as a Mooney-Rivlin material, and showed that this model led to better agreement with measured permanent biaxial stretch. Fong and Zapas (1976) have since proposed using the Rivlin-Saunders model to determine the permanent biaxial stretch.

Several important results can be deduced from Equation (2). The permanent stretch on removal of stress is

$$\lambda_R^3 = \frac{n_1 + n_2\hat{\lambda}}{n_1 + \frac{n_2}{\hat{\lambda}^2}} > 1 \quad, \tag{3}$$

from which follows that $1 < \lambda_R < \hat{\lambda}$. Next, introduce the relative stretch ratio $\mu = \lambda/\lambda_R$. The response relative to the permanent stretch state, (Berry *et al*, 1956) can be obtained by writing Equation (2) in the form

$$\sigma = \chi\left(\mu^2 - \frac{1}{\mu}\right) \quad, \tag{4}$$

where

$$\chi = 2kTn_1\left(1+\frac{n_2}{n_1}\frac{1}{\hat{\lambda}^2}\right)^{1/3}\left(1+\frac{n_2}{n_1}\hat{\lambda}\right)^{2/3} . \tag{5}$$

According to Equation (5), the modulus of the original neo-Hookean network $2n(0)kT$ has reduced to χ since $n_1 < n(0)$. It is seen that χ depends on the cross-link densities of the networks and the stretch ratio $\hat{\lambda}$.

This simple model shows that permanent set and modified mechanical responses are consequences of scission and re-cross linking. These, in turn, depend on the temperature and time held at that temperature. Although Equations (3), (4) and (5) apply when both networks are neo-Hookean, it is to be expected that there will be permanent set and modified mechanical response for other assumed network models. However, the mathematical expressions corresponding to Equations (3), (4) and (5) will be more complicated. Moreover, the material will exhibit induced anisotropy relative to the permanent set. (Greene *et al*, 1965).

These changes in the response of an elastomeric component can have important implications on its operational life. A component can fail when its properties have changed to such an extent that it no longer meets the design specifications. Clearly, there is a need to develop a model which can be used to predict how long a component will meet these specifications, and which can be used in numerical simulations.

3. A CONSTITUTIVE EQUATION

The experiments of Tobolsky were carried out for uniaxial tests at a fixed stretch and constant temperature and led to Equation (2). This equation is now used as a foundation for a proposed constitutive framework for the three dimensional response of a rubber subjected to arbitrary homogeneous deformation and temperature histories. For a detailed discussion of the constitutive equation, see Wineman and Min (2001).

Consider a rubbery material in a stress free reference configuration at a temperature T_0. It is assumed that there is a range of deformations and temperatures in which the material response can be regarded as incompressible, isotropic and nonlinearly elastic. If $\mathbf{x}$ is the position at current time t of a particle located at $\mathbf{X}$ in the reference configuration, the deformation gradient is $\mathbf{F} = \partial\mathbf{x}/\partial\mathbf{X}$. The left Cauchy-Green tensor is defined as $\mathbf{B} = \mathbf{FF}^T$. The Cauchy stress σ is given by

$$\sigma = -p^o\mathbf{I} + \sigma^o(\mathbf{B}, T) = -p^o\mathbf{I} + 2\frac{\partial W^o}{\partial I_1}\mathbf{B} - 2\frac{\partial W^o}{\partial I_2}\mathbf{B}^{-1} , \tag{6}$$

where p^o, an arbitrary hydrostatic pressure, arises from the constraint that deformations are isochoric, I_1, I_2 are invariants of $\mathbf{B}$ and $W^o(I_1, I_2, T)$ is the strain energy density associated with the original material. In Equation (6), $\sigma, \mathbf{B}, T$ depend on the current time t, which is omitted from the notation for brevity.

For temperatures $T < T_{cr}$, no microstructural changes occurs. All of the material is in its original state and the total stress is given by Equation (6). For temperatures $T \geq T_{cr}$, scission of the original microstructural network is assumed to occur continuously in time. A scalar-valued function $a(t) > 0$ is introduced which represents the rate at which volume fraction of new network is formed at time t. Thus, $a(t)dt$ is interpreted as the volume fraction of original material that has reformed as time increases from t to $t + dt$. The volume fraction of original network remaining at time t is denoted as $b(t)$. $b(t) \in [0,1]$ and is a monotonically decreasing function of t. For the sake of simplicity, and consistent with Tobolsky's experimental results, it is assumed that $a(t)$ and $b(t)$ do not depend on the deformation. It is assumed further that the rate of formation of new networks is given by

$$a(t) = -\eta \frac{db(t)}{dt}, \tag{7}$$

where $\eta \in [0,1]$, called the healing parameter, a scalar parameter that depends on the particular rubber system being considered (Wineman and Shaw, 2001). Tobolsky's assumption of network conservation corresponds to complete healing, or $\eta = 1$. Complete scission, with no new network formation can be modeled with $\eta = 0$.

Now consider an intermediate time $\hat{t} \in [0,t]$ and the corresponding deformed configuration of the original material. Due to the formation of new cross links, a network is formed in the interval from $\hat{t}$ to $\hat{t} + d\hat{t}$ whose reference configuration is the configuration of the original material at time $\hat{t}$. As suggested by Tobolsky (1960) and Tobolsky *et al* (1944), this is assumed to be an unstressed configuration for the newly formed network. Under subsequent deformation, the motion of the newly formed network coincides with the motion of the original network. Stress arises in this newly formed network due to its deformation relative to its unstressed configuration at time $\hat{t}$. At the later time t, the material formed at earlier time $\hat{t}$ has the relative deformation gradient $\hat{\mathbf{F}} = \partial x/\partial \hat{x}$, where $\hat{x}$ is the position of the particle in the configuration corresponding to time $\hat{t}$ and x is its position at time t.

For simplicity in presentation, the new network is assumed to respond as an incompressible isotropic nonlinear elastic material with temperature dependence. The left Cauchy-Green tensor $\hat{\mathbf{B}} = \hat{\mathbf{F}}\hat{\mathbf{F}}^T$ is introduced for deformations of this network. The Cauchy stress $\hat{\sigma}$ at time t in the network formed at time $\hat{t}$ is then given by

$$\hat{\sigma} = -\hat{p}\mathbf{I} + \hat{\sigma}(\hat{\mathbf{B}}, T) = -\hat{p}\mathbf{I} + 2\frac{\partial \hat{W}}{\partial \hat{I}_1}\hat{\mathbf{B}} - 2\frac{\partial \hat{W}}{\partial \hat{I}_2}\hat{\mathbf{B}}^{-1}, \tag{8}$$

where $\hat{p}$ arises from the constraint that deformations are isochoric and $\hat{I}_1, \hat{I}_2$ are invariants of $\hat{\mathbf{B}}$. $\hat{W}(\hat{I}_1, \hat{I}_2, T)$ is the strain energy density associated with the newly formed network and can differ from that associated with the original network.

The total current stress in the material is taken as the superposition of the contribution from the remaining original network and stress in the new networks. Thus,

$$\sigma = -p\mathbf{I} + b\left[2\frac{\partial W^o}{\partial I_1}\mathbf{B} - 2\frac{\partial W^o}{\partial I_2}\mathbf{B}^{-1}\right] + \int_0^t a(\hat{t})\left[2\frac{\partial \hat{W}}{\partial I_1}\hat{\mathbf{B}} - 2\frac{\partial \hat{W}}{\partial I_2}\hat{\mathbf{B}}^{-1}\right]d\hat{t}, \qquad (9)$$

where $b, \mathbf{B}, T, \sigma$ are evaluated at the current time t. This generalizes Tobolsky's assumption (i). Since $\hat{\mathbf{B}} = \mathbf{I}$ in an undeformed newly formed network, it follows from Equation (8) that the stress has the form $q\mathbf{I}$. This can be absorbed into the term $-p\mathbf{I}$ in Equation (9), a superposed hydrostatic stress. The undeformed new network does not contribute to the stress and can be interpreted as stress free. This generalizes Tobolsky's assumption (iii). Although Tobolsky assumed the response of the original and newly formed networks to be neo-Hookean, Neubert and Saunders (1958) and Fong and Zapas (1976) considered other possibilities. Thus, $W^o(I_1, I_2, T)$ and $\hat{W}(\hat{I}_1, \hat{I}_2, T)$ are unspecified. This generalizes Tobolsky's assumption (ii).

4. A FORM FOR b(t)

Now we turn to a discussion of the form of b(t). Initially, $T < T_{cr}$ and $b = 1$. Since there is no scission unless the temperature exceeds T_{cr}, it is assumed that

$$\frac{db}{dt} = 0, \quad T < T_{cr}. \qquad (10)$$

Next, let the temperature vary so that $T(s) > T_{cr}, s \in [0, t]$, but let the deformation be fixed. Then the constitutive equation (9) reduces to

$$\sigma = -p\mathbf{I} + b\left[T(s)\big|_{s=0}^t\right]\left[2\frac{\partial W^o}{\partial I_1}\mathbf{B} - 2\frac{\partial W^o}{\partial I_2}\mathbf{B}^{-1}\right], \qquad (11)$$

where $b\left[T(s)\big|_{s=0}^t\right]$ denotes dependence on the temperature history. For a fixed uniaxial extension with stretch ratio λ, Equation (11) reduces to

$$\sigma(t) = b\left[T(s)\big|_{s=0}^{t}\right]\left(\lambda^2 - \frac{1}{\lambda}\right)\left[2\frac{\partial W^o}{\partial I_1} + 2\frac{1}{\lambda}\frac{\partial W^o}{\partial I_2}\right].\tag{12}$$

Since $b(0) = 1$ and b is independent of the deformation, so is the ratio

$$\frac{\sigma(t)}{\sigma(0)} = b\left[T(s)\big|_{s=0}^{t}\right].\tag{13}$$

When scission occurs at a constant temperature T, i.e. $T(s) = T > T_{cr}, s \in [0,t]$, let $b\left[T(s)\big|_{s=0}^{t}\right] = \beta(T,t)$. Then Equation (13) is written as

$$\frac{\sigma(t)}{\sigma(0)} = \beta(T,t).\tag{14}$$

$\beta(T,t)$ can be obtained experimentally, (see Tobolsky (1960, Figure V.4). Tobolsky (1960, p. 226) suggested for a number of rubbers that $\beta(T,t)$, $T \geq T_{cr}$, can be represented in the form

$$\beta(T,t) = \phi(\alpha(T)t).\tag{15}$$

where Tobolsky provided a particular form of $\alpha(T)$ for his rubber.

An expression for $b\left[T(s)\big|_{s=0}^{t}\right]$ is developed from a graphical construction which is omitted here for the purpose of brevity. (details have been provided by Wineman and Min, 2001) The underlying assumption is that scission at each instant depends on the temperature only at that instant. The construction leads to the expression

$$b\left[T(s)\big|_{s=0}^{t}\right] = \phi(\xi(t)),\tag{16}$$

where

$$\xi(t) = \int_{t_0}^{t}\alpha(T(s))ds.\tag{17}$$

When the temperature is constant, Equations (16) and (17) reduce to (15). Combining Equations (7), (9), (16) and (17) for uniaxial extension gives

$$\sigma(t) = \phi(\xi(t)) \left(\lambda^2 - \frac{1}{\lambda} \right) \left(2 \frac{\partial W^o}{\partial I_1} + \frac{1}{\lambda} 2 \frac{\partial W^o}{\partial I_2} \right)$$

$$+ \int_0^t (-\eta) \frac{d\phi(\xi(\hat{t}))}{d\hat{t}} \left(\frac{\lambda^2}{\lambda(\hat{t})^2} - \frac{\lambda(\hat{t})}{\lambda} \right) \left(2 \frac{\partial \hat{W}}{\partial \hat{I}_1} + 2 \frac{\lambda(\hat{t})}{\lambda} \frac{\partial \hat{W}}{\partial \hat{I}_2} \right) d\hat{t} \qquad (18)$$

5. EXPERIMENTAL RESULTS

The constitutive equation presented in Sections 3 and 4 incorporate hypotheses which are being evaluated through an experimental program. Results to date of an experimental program are presented in this section. These results are to be used to evaluate the constitutive equation presented in the previous sections, and to suggest modifications.

Uniaxial extension experiments were performed on an EnduraTEC ER32 pneumatic tension-torsion machine having a capacity of 2800 lb in tension and 1350 in-lb in torsion. A custom-built thermal chamber enclosed the test section. It has a $\pm 0.5^\circ$ C spatial variation of temperature in the range 20° C $- 150^\circ$ C. Low thermal mass grips were also custom-built. Grip and specimen temperatures were measured using exposed junction K-type thermocouples. Stretch was measured directly from the specimen test section through a window in the thermal chamber using a laser extensometer with a 1 micron resolution (Electric Instrument Research, Irwin, PA). The extensometer produced a planar laser sheet, avoiding parallax problems through the chamber window. A programmable control system was developed which allowed stretch control, force control or mixed stretch-force control. Data was acquired using Labview software on a Macintosh G3. Uniaxial specimens were 0.5 in. by 2 in. strips cut from commercial grade 0.08 in. thick rubber sheets provided by Tenneco Automotive Corporation.

Uniaxial extension experiments were carried out at a number of fixed stretches and fixed temperatures. Since the tensile force $F(t)$ and Cauchy tensile stress $\sigma(t)$ on a specimen of cross-sectional area A_o are related by $F(t) = \sigma(t) A_o / \lambda$, it follows that $F(t)/F(0) = \sigma(t)/\sigma(0)$. Figure 2 shows plots of $F(t)/F(0)$ vs. time carried out at a fixed stretch of $\lambda = 2$ at various temperatures. In effect, Figure 2 shows plots of $\beta(T,t)$ introduced in Equation (14). At 95° C, $F(t)/F(0)$ relaxes to a value of 0.2 after 4500 min and to 0.1 after 9000 min. The time for this relaxation to occur is seen to decrease significantly as the temperature increases. Figure 3 shows plots of $F(t)/F(0)$ vs. time for numerous stretches carried out at 125° C. According to Equation (14), the assumption that $b(t)$ is independent of the stretch implies that $F(t)/F(0)$ vs. time should also be independent of the stretch, i.e. plots for different stretches should coincide. Figure 3 shows that there is some dependence on the stretch. Figure 4 shows a magnified view of the beginning of this plot, where the decrease in $F(t)/F(0)$ is fastest. A different view of these results is shown in Figure 5, which shows the values of engineering stress $\sigma_E = F/A = \sigma/\lambda$ vs. $\log(t)$.

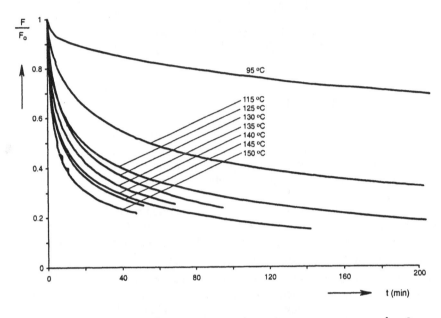

Figure 2. F(t)/F(0) relaxation for various temperatures at $\lambda = 2$.

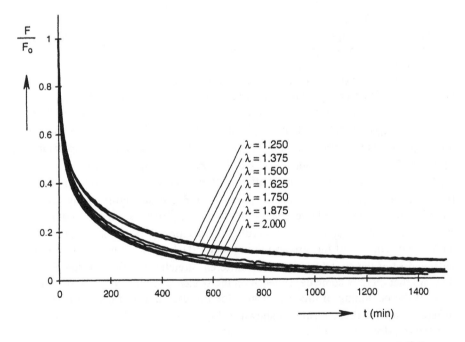

Figure 3. F(t)/F(0) relaxation for various stretches at $T = 125^{\circ}C$.

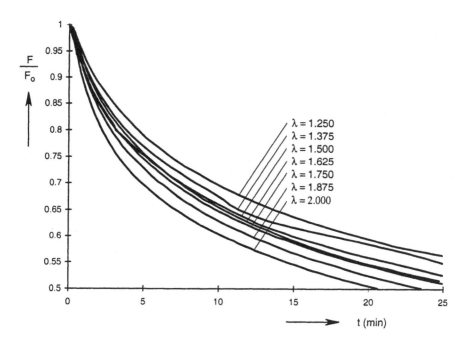

Figure 4. Magnified view of $F(t)/F(0)$ relaxation at $T = 125^{\circ}C$.

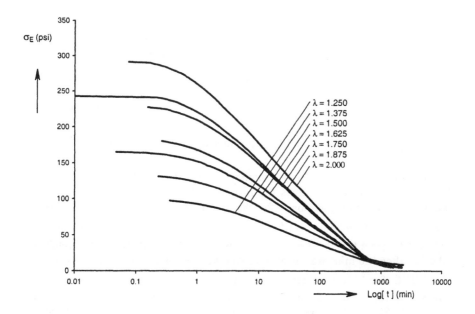

Figure 5. Engineering stress relaxation at $T = 125^{\circ}C$.

Equation (12) expresses the assumption that when a specimen is held at a fixed stretch, new networks formed by re-cross linking are stress free. This is true whether or not $b(t)$ is independent of stretch. Suppose $b(t)$ is independent of stretch, and $\partial W^o/\partial I_1$ and both $\partial W^o/\partial I_2$ are constants. A plot of $\sigma_E\big/\left(\lambda - \lambda^{-2}\right)$ vs. $1/\lambda$ at a fixed time (the Mooney-Rivlin plot) should be a straight line. As time increases, the decrease in $b(t)$ would cause the slopes and intercepts of the lines to decrease. Moreover, horizontal lines would indicate the response is neo-Hookean. Figure 6 shows that the plots are approximately straight lines with non-zero slopes. The slopes do not appear to vary with time. Thus, the response does not appear to be neo-Hookean, and $\partial W^o/\partial I_1$ and $\partial W^o/\partial I_2$ vary with time.

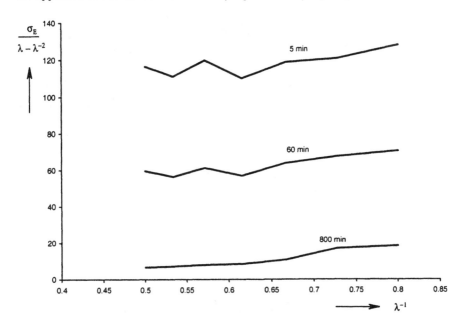

Figure 6. Mooney-Rivlin plots for three different times at $T = 125^\circ C$

Three room temperature stress–stretch plots are shown in Figure 7. The plot labeled 'virgin material' shows the response of the rubber before there is any thermally induced degradation. After the initial stress-stretch plot was carried out the specimen was held at $\lambda = 1.5$ and $T = 125^\circ C$ for 120 min. The specimen was returned to room temperature, the stress was removed and the stress-stretch plot labeled 'after 120 min' was measured. Another specimen was held at $\lambda = 1.5$ and $T = 125^\circ C$ for 1558 min, after which it was returned to room temperature, the stress was removed and the stress-stretch plot labeled 'after 1558 min' was measured. Figure 7 shows that permanent stretch increases and the material softens with the time held at $T = 125^\circ C$. This response demonstrates the softening suggested in Equations (4) and (5).

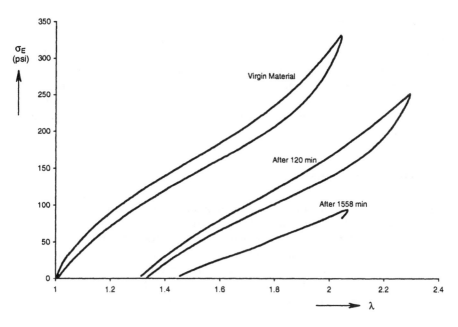

Figure 7. Stress-stretch plots at room temperature for virgin material and for two samples subsequently held at $T = 125^\circ C$ and $\lambda = 1.58$.

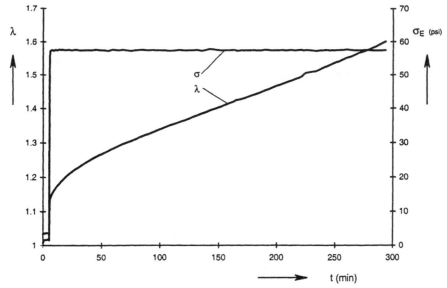

Figure 8. Creep under constant engineering stress at $T = 125^\circ C$.

The preceding results were obtained under stretch control conditions at constant stretch. Figure 8 shows results of a creep experiment under constant engineering stress at $T = 125^\circ C$. As the stress is rapidly increased to $\sigma_E = 58\,psi$, the stretch rises immediately to $\lambda = 1.14$. The stretch continues to increase to about $\lambda = 1.58$ after 300 minutes.

The results presented here illustrate the complexity of the behavior under uniaxial deformations. The development of a constitutive equation also requires verification under multi-axial deformation and more general time-temperature histories. Methods are being developed to provide such thermo-mechanical biaxial experimental data.

6. CLOSING COMMENTS

The experimental results of a typical commercially available natural vulcanized rubber presented herein show substantial stress relaxation at elevated temperatures, material softening, permanent set, and creep due to microstructural changes. Each can occur in an elastomeric component operating in a high temperature environment. The temperature environment can be caused by external influences or from internal mechanical heating. These effects contribute to a time dependent change in response characteristics that affect the performance and durability of an elastomeric component over time. Thus, it is important to develop a thermo-mechanical theory that accounts for these effects and can be used to simulate the operating conditions and thereby calculate a component's effective lifetime. The first steps at such a model have been presented and experimental verification is currently underway.

Acknowledgements:

The financial support of the National Science Foundation (NSF grant CTS-9908925) is acknowledged with thanks. The interaction and materials supplied by Tenneco Automotive Corp. are also greatly appreciated.

6. REFERENCES

Berry, J. P., Scanlan, J. and Watson, W. F., 1956, Cross-link formation in stretched rubber networks. *Transactions of the Faraday Society*, **52**, pp. 1137-1151.

Fong, J. T. and Zapas, L. J., 1976. Chemorheology and the mechanical behavior of vulcanized rubber. *Transactions of the Society of Rheology,* **20**, pp. 319-338.

Greene, A., Smith, K. J., Jr. and Ciferri, A., 1965, Elastic properties of networks formed from oriented chain molecules, part 2 - composite networks. *Transactions of the Faraday Society*, **61**, pp. 2772-2783.

Hertz, D., Jr., 1997, The aging process for elastomers, *Proceedings of the Elastomer- Service Life Prediction Forum '97*, organized by the Akron Rubber Development Laboratory, Akron, OH.

Neubert, D., Saunders, D. W., 1958. Some observations of the permanent set of cross-linked natural rubber samples after heating in a state of pure shear. *Rheological Acta*, **1**, pp. 151-157.

Pett, R., 1997, Predicting the life of automotive power steering hose materials, *Proceedings of the Elastomer-Service Life Prediction Forum '97*, organized by the Akron Rubber Development Laboratory, Akron, OH.

Scanlan, J. and Watson, W. F., 1958, The interpretation of stress-relaxation measurements made on rubber during aging, *Transactions of the Faraday Society*, **54**, pp. 740 –750.

Tobolsky, A. V., 1960. *Properties and Structures of Polymers*, New York: Wiley, Chapter V, pp. 223-265.

Tobolsky, A. V., Prettyman, I. B., Dillon, J. H., 1944. Stress relaxation of natural and synthetic rubber stocks, *Journal of Applied Physics*, **15**, pp. 380-395.

Wineman, A. and Min, J.-H., 2001,Time dependent scission and cross-linking in an elastomeric cylinder undergoing circular shear and heat conduction, *Journal of the Mechanics and Physics of Solids*, submitted.

Wineman, A. and Shaw, J. A., 2001, Scission and Healing in a Spinning Elastomeric Cylinder at Elevated Temperature, *Journal of Applied Mechanics*, submitted.

Radiative Transfer Modeling and Experiments Using Short Pulse Lasers

Sunil Kumar[1], Pei-feng Hsu[1], Kunal Mitra[1], Bruce Garetz[1], Zhixiong Guo[1] and Janice Aber[1]

1 INTRODUCTION

Thermal radiative models for utilization as tools in simulation based engineering of systems are not well established at present. Many issues remain unresolved, in particular the formulation and accuracy of different models, the relative merit of these methods vis-a-vis their respective computational resource requirements, and the development of models to capitalize on the computational power of teraflop-class parallel computers currently available or anticipated. Steps to further the development of accurate and viable models for simulation include computational studies of various methods with emphasis on irregular geometries and non-uniform properties, experimental validation of models to identify their respective accuracies, and the consideration of mathematical techniques such as parallelization to reduce computational time and resource requirements.

This chapter summarizes developments to date of an integrated theoretical and experimental analysis of the interaction of radiation with scattering-absorbing media, with a primary focus on identification of accurate models for analyzing radiative transfer, examination of model fidelity, validation of models via carefully designed experimentation, and the quantification of model uncertainty.

The specific case chosen for the model selection and validation aspect of the study is the interaction of short pulse radiation with participating media. Short pulse radiation is selected because it offers unique features that have not been previously exploited in the study of radiative transport and which present opportunities to obtain accurate high-precision data and evaluate different models.

The unique features of short pulses of radiation and the characteristics of their interaction with participating medium that make them a valuable research tool are identified here. Firstly, short pulse monochromatic radiation can be easily created in a laboratory via a short pulse laser. Accurate detection with high spatial, temporal, and signal resolution is easily feasible through modern instrumentation. This is in contrast with non-laser traditional radiation measurements that rely on emitting sources comprising of a wide variety of wavelengths. This ability to control the radiation source and precisely measure all aspects of the data when a short pulse laser is employed permits the collection of benchmark data that can be used to validate radiative transport models. Secondly, the scattered, reflected and transmitted signals measured when short pulse lasers interact with scattering-absorbing media possess a unique feature in addition to the usually measured magnitude and direction of the radiation signal. The distinct feature is the multiple

---

[1] Brooklyn Polytechnic

scattering induced temporal signature that persists for time periods greater than the duration of the source pulse and is a function of the source pulse width, the scattering and absorbing properties of the medium, and the location in the medium where the properties undergo changes. This variation of the radiative signal with time is an additional parameter to compare and contrast various models, either between themselves or with experimental data for validation.

Thus the transient modeling and the examination of predictions from such models facilitates the determination of the accuracy of models, because if a model matches experiments in the entire transient domain then it is expected to match at steady-state. This may be in contrast with steady-state modeling efforts where two models could predict very similar steady-state results, thereby creating a dilemma about which is the more accurate. The two could be differentiated through the transients where the temporal history of one may not match the experimental data.

Short pulse radiation is rapidly being deployed in many new applications such as bio-medical optical tomography, in-situ property evaluation, evaluation of particle size distribution, remote sensing of oceans and atmosphere, and others, where it is becoming imperative that accurate radiative models be developed to analyze, design, and optimize these applications. Therefore, in addition to the general identification of the most accurate radiative transfer models as discussed in the previous paragraphs, the study has also a direct impact on new applications.

2 MODEL DEVELOPMENT

In recent years, aided by continuing improvements in computing hardware and new or improved solution methods, radiative transfer problems in participating media have been solved by various researchers. Due to the integro-differential nature of the radiation transport, many algorithms have been possible and were developed in the past to solve the radiation transfer equation. Review papers (Menguc and Viskanta, 1983; Viskanta, 1984; Viskanta and Menguc, 1987, Howell, 1988) outline the different approaches and the vast body of literature is therefore not cited here.

Because of the complexity of the radiative transfer equation and the paucity of precise data available for benchmarking or validation, error estimation of the computations is usually unavailable, and it is not uncommon that large differences can be found in the results for the same problem using different methods. This became apparent at the First Symposium on Radiative Heat Transfer Solution Methods held at the 28th National Heat Transfer Conference in San Diego, August 1992, where extensive discussions were held regarding the modeling and validation of radiative transfer and many papers were presented on this topic (see report by Tong and Skocypec, 1992). Subsequent symposia, namely the Second Symposium at the 6th Joint Thermophysics and Heat Transfer Conference, Colorado Springs, August 1994, the Third Symposium at the 31st National Heat Transfer Conference, Houston, August 1996, and the NSF Workshop on Use of High Performance Computing for Radiative Transfer at Sandia, Albuquerque, March 1994, also addressed these issues. The results of the discussions, summarized in a Sandia Report (Gritzo *et al.*, 1995) and a review article (Howell, 1995), indicate that there is no uniform consensus in the research community about which models are, or will be, the most appropriate for simulations, especially of the unprecedented magnitude in size, scope, resolution, complexity, and comprehensiveness that are envisioned for simulation-based engineering of systems.

The radiative transfer equation to describe laser radiation transport in scattering, absorbing and emitting turbid media can be written as (Siegel and Howell, 1992; Brewster, 1992; Modest, 1993; Kumar and Mitra, 1999):

$$\frac{1}{c}\frac{\partial I(\mathbf{r},\hat{s},t)}{\partial t}+\hat{s}\cdot\nabla I\ (\mathbf{r},\hat{s},t)+(\sigma_a+\sigma_s)\,I(\mathbf{r},\hat{s},t)=$$
$$\sigma_a I_b(\mathbf{r},\hat{s},T,t)+\frac{\sigma_s}{4\pi}\int_{4\pi}I(\mathbf{r},\hat{s}',t)\,\Phi(\hat{s}'\to\hat{s})\,d\omega+S(\mathbf{r},\hat{s},t)\qquad(1)$$

where I is the radiation intensity, ∇ the gradient operator, t the time, $\mathbf{r}$ the spatial location vector, $\hat{s}$ the unit vector in the direction of intensity, S the source term, σ_a the absorption coefficient, σ_s the scattering coefficient, ω the spatial solid angle, and $\Phi(\hat{s}'\to\hat{s})$ the scattering phase function.

2.1 P-1 Approximation

The P_1 method (Menguc and Viskanta, 1983) gives accurate transmission values for the case of optically thick media. Even though this method has limitations regarding its usage, it is computationally much less expensive and is easy to implement. Under the P_1 approximation the intensity is considered to be a linear function of direction cosine for the case of one-dimensional geometry (see Mitra *et al.*, 1997, for two-dimensional formulations). The resulting equation in one dimensional system is given as (Kumar *et al.*, 1996)

$$\frac{3}{c^2}\frac{\partial^2 u}{\partial t^2}-\frac{\partial^2 u}{\partial x^2}+\frac{3}{c}[\sigma_a+\sigma_e-\sigma_s\overline{p}]\frac{\partial u}{\partial t}+3[\sigma_e-\sigma_s\overline{p}]\sigma_a u=$$
$$[\sigma_e-\sigma_s\overline{p}]\frac{3}{2}\int_{-1}^{1}Sd\mu-\frac{3}{2}\int_{-1}^{1}\frac{\partial S}{\partial x}\mu d\mu+\frac{3}{2c}\int_{-1}^{1}\frac{\partial S}{\partial t}d\mu\qquad(2)$$

where $\overline{p}$ is an integrated phase function, u is the average intensity over all angles, $\mu=\cos\theta$ and θ is the polar angle measured from the primary propagation direction.

2.2 Two-Flux Approximation

The two-flux method (Brewster and Tien, 1982) yields relatively accurate results in certain cases if one is interested in looking at the back-scattered flux, such as in remote sensing of oceans and atmospheres and remote fisheries management (Mitra and Churnside, 1999). This method is computationally less expensive and implementation is very simple. The method considers the intensity to be constant over the forward (I^+) and backward facing hemispheres (I^-). The resulting coupled equations are given by the following for a one-dimensional medium, where B is an integrated phase function (Mitra and Kumar, 1999),

$$\frac{1}{c}\frac{\partial I^+}{\partial t}+\frac{1}{2}\frac{\partial I^+}{\partial x}+(\sigma_a+\sigma_s B)I^+-\sigma_s BI^-=\int_0^1 Sd\mu\ ,$$

$$\frac{1}{c}\frac{\partial I^-}{\partial t}+\frac{1}{2}\frac{\partial I^-}{\partial x}+(\sigma_a+\sigma_s B)I^--\sigma_s BI^+=\int_{-1}^{0}Sd\mu \ . \tag{3}$$

2.3 Discrete Ordinates Method

The discrete ordinates method (Fiveland, 1988) has been one of the most widely applied methods in steady state radiative analyses since it requires a single formulation to invoke higher order approximations of discrete ordinate quadrature sets, integrates easily into control volume transport codes, and is applicable to the complete anisotropic scattering phase function and inhomogeneous media. Recently, Guo and Kumar (2001a) and Guo (2001) have developed the discrete ordinates method for transient radiative transfer in two- and three-dimensional problems. In a three-dimensional rectangular enclosure, the transient radiative transfer equations in discrete ordinates form can be formulated as

$$\frac{1}{c}\frac{\partial I^l}{\partial t}+\xi^l\frac{\partial I^l}{\partial x}+\eta^l\frac{\partial I^l}{\partial y}+\mu^l\frac{\partial I^l}{\partial z}+\sigma_e I^l=\sigma_e S^l, \qquad l=1,2,..,n \ , \tag{4}$$

where ξ^l,η^l and μ^l are the three direction cosines in a discrete ordinate direction $\hat{s}^l$, I^l is the radiation intensity in the discrete direction and is a function of position (x, y, z) and time instant t. S^l is the radiative source term and can be expressed as:

$$S^l=(1-\omega)I_b+\frac{\omega}{4\pi}\sum_{i=1}^{n}w^i\,\Phi^{il}\,I^i+S_c^l, \qquad l=1,2,...,n \ , \tag{5}$$

where Φ^{il} represents scattering phase function $\Phi(\hat{s}^i\rightarrow\hat{s}^l)$, and S_c^l is the source contribution of collimated irradiation such as laser incidence. A quadrature set of n discrete ordinates with the appropriate angular weight w^l ($l=1,2,...,n$) is used in the "S_N-approximation".

To solve the discrete transfer equation, Equation (4), the enclosure is divided into a number of small control volumes in an orthogonal grid. The discretized transfer equation in three-dimensional geometry can be expressed as

$$\frac{V}{c\,\Delta t}(I^l-I_P^{l0})_P+\left|\xi^l\right|(A_{xu}I_{xu}^l-A_{xd}I_{xd}^l)+\left|\eta^l\right|(A_{yu}I_{yu}^l-A_{yd}I_{yd}^l)$$
$$+\left|\mu^l\right|(A_{zu}I_{zu}^l-A_{zd}I_{zd}^l)=\sigma_e PV(-I_P^l+S_P^l) \tag{6}$$

where, $A_{xu}=A_{xd}=\Delta y\Delta z$, $A_{yu}=A_{yd}=\Delta x\Delta z$, $A_{zu}=A_{zd}=\Delta x\Delta y$, and $V=\Delta x\Delta y\Delta z$. I_{xd}^l, I_{yd}^l, and I_{zd}^l are the radiation intensities on the downstream surfaces in the direction $\hat{s}^l$. I_{xu}^l, I_{yu}^l, and I_{zu}^l are those on the upstream surfaces in the direction. Subscript P denotes the nodal of the control volume. I_P^{l0} is the radiation intensity in the previous time step.

A zero initial intensity field is assumed in the present study. In order to solve Equation (6), the weighted finite differencing scheme is usually used to relate the radiation intensities on the cell surfaces with those at the cell centers:

$$I_P^l = \gamma_x^l I_{xd}^l + (1-\gamma_x^l)I_{xu}^l = \gamma_y^l I_{yd}^l + (1-\gamma_y^l)I_{yu}^l = \gamma_z^l I_{zd}^l + (1-\gamma_z^l)I_{zu}^l \qquad (7)$$

The final discretization equation for the cell intensity in a generalized form becomes

$$I_P^l = \frac{\dfrac{1}{c\,\Delta t}I_P^{l0} + \sigma_e S_P^l + \dfrac{|\xi^l|}{\gamma_x^l \Delta x}I_{xu}^l + \dfrac{|\eta^l|}{\gamma_y^l \Delta y}I_{yu}^l + \dfrac{|\mu^l|}{\gamma_z^l \Delta z}I_{zu}^l}{\dfrac{1}{c\,\Delta t} + \sigma_e + \dfrac{|\xi^l|}{\gamma_x^l \Delta x} + \dfrac{|\eta^l|}{\gamma_y^l \Delta y} + \dfrac{|\mu^l|}{\gamma_z^l \Delta z}}. \qquad (8)$$

2.4 Radiation Element Method

The radiation element method with ray tracing model was initially developed to deal with surface radiation problems with arbitrary shapes (Maruyama, 1993; Guo et al., 1997). It is further extended to analyze radiative heat transfer in participating media including nongray and inhomogeneous properties and good accuracy for example three-dimensional problems was found (Guo and Maruyama, 2000). In this method, the arbitrary configuration is considered by various radiation elements consisting of numerous polygons and polyhedrons such as triangles, quadrilaterals, wedges, and hexahedrons. New concepts of absorption view factor and diffusely scattering view factor are introduced, and their values are obtained via ray tracing method. An effective area for each radiation element including volume element is calculated, and it is identical to the surface area when a surface element is concerned. The analysis of radiative transfer is based on the net radiation method commonly used in surface radiation transfer. Radiative heat flux at the boundaries and radiative flux divergence inside the medium can be directly obtained.

In transient radiation element method, the rate of radiation energy leaving the ith element can be expressed in a generalized form as (Guo and Kumar, 2001b; Guo, 2001)

$$Q_{J,i}(t) \equiv A_i^R [\varepsilon_i E_{b,i}(t) + \Omega_i G_i(t)] , \qquad (9)$$

where $\varepsilon_i = 1-\Omega_i$, $E_{b,i} = \pi I_{b,i}, G_i = \pi I_i^D$, and an effective radiation area A_i^R is defined as

$$A_i^R = \frac{1}{\pi}\int_{A_i(\hat{s})} [1-\exp(-\int_0^S \sigma_{ei}dS')]d\omega\,dA \qquad (10)$$
$$\approx \frac{1}{\pi}\int_{4\pi} A_i(\hat{s})[1-\exp(-\int_0^{\bar{S}} \sigma_{ei}dS')]d\omega$$

The net rate of radiative heat generation can be derived from the radiation heat balance on the radiation element as

$$Q_{X,i}(t) = A_i^R \varepsilon_i \big[E_{b,i}(t) - G_i(t)\big] , \qquad (11)$$

The expression of emissive power $Q_{T,i}$ of the radiation element is

$$Q_{T,i}(t) = A_i^R \varepsilon_i E_{b,i}(t) .\tag{12}$$

The expression of irradiation energy $Q_{G,i}$ is

$$Q_{G,i} = A_i^R G_i .\tag{13}$$

If the system is consisted of N volume and surface elements, the above can be rewritten as

$$\left.\begin{aligned}Q_{J,i}(t) &= Q_{T,i}(t) + \sum_{j=1}^{N} F_{j,i}^D Q_{J,j}(t_{ij}) \\ Q_{X,i}(t) &= Q_{T,i}(t) - \sum_{j=1}^{N} F_{j,i}^A Q_{J,j}(t_{ij})\end{aligned}\right\}, \qquad i = 1, 2, \ldots\ldots ,N \tag{14}$$

in which, the absorption view factor $F_{i,j}^A$ and diffuse scattering view factor $F_{i,j}^D$ are introduced. If the radiative properties of the media are not functions of media temperature, the view factors are not functions of time and are identical to steady state definitions (Guo and Maruyama, 2000).

Special attention should be paid to t_{ij} in Equation (14). It is the time instant t subtracted by the averaged flight time of rays between elements i and j:

$$t_{ij} = t - S_{ij}/c ,\tag{15}$$

where S_{ij} is the distance between elements i and j.

2.5 Integral Formulation

The integral model predicts the time dependence and wave behavior of radiation propagation inside the participating medium, which can't be obtained accurately with the existing approximate models. It also avoids the numerical dispersion and dissipation errors usually associated with the integro-differential formulation. The integral equation solutions have good agreement with the steady state finite volume and quadrature methods' solutions at large time step. The comparisons with the transient Monte Carlo solutions under different boundary conditions also show excellent agreement (Hsu, 2001). These efforts verified good accuracy of the integral formulation.

It is found that the Lagrangian viewpoint can simplify the analysis of the time-of-flight of photon and allow the deduction of the domain of influence (Tan and Hsu, 2000). For the case of isotropic scattering, $\Phi(\hat{s}' \to \hat{s}) = 1$, and without emission, the integrated intensity can be obtained as

$$G(x, y, z, t) = \int_{4\pi} I(x_w, y_w, 0, \hat{s}, t - s/c)\exp(-\kappa s)d\omega$$
$$+ \frac{1}{4\pi} \int_{4\pi} \left[\int_0^s \sigma(x', y', z')G(x', y', z', t')\exp(-\kappa(s-s'))ds'\right]d\omega$$

$$= \iint_{A(t)} \frac{1}{s^2} I(x_w, y_w, 0, \hat{s}, t - s/c) \exp(-\kappa s) \cos \theta dA$$

$$+ \frac{1}{4\pi} \iiint_{V(t)} \frac{1}{(s-s')^2} \sigma(x', y', z') G(x', y', z', t') e^{-\kappa(s-s')} dV \qquad (16)$$

This is a Volterra integral equation of the second kind. An essential part for a complete solution is the specification of the time-dependent domain of integration or domain of influence, $A(t)$ and $V(t)$. The exact forms depend on the boundary condition and the physical geometry of the participating medium. Two different integration schemes have been used: the discrete rectangular volume (Tan and Hsu, 2001) and YIX methods (Hsu *et al.*, 1993).

Many practical problems have multi-dimensional geometry with nonhomogeneous radiative property distribution. It is essential to develop a transient radiation model that can treat such situation. The model must also allow realistic simulations without requiring significant computing resources. It is noteworthy that the moment of intensity, i.e., G is solved directly instead of the intensity in the integro-differential treatments. After solving the G, the intensity and radiative flux can be readily obtained by direct integraions. Therefore, computationally the integral formulation has much smaller memory requirement than the integro-differential formulations, even lower than the Monte Carlo method in the transient transfer analysis (Hsu, 2000). If the memory requirement are $O(N_I)$ and $O(N_{ID})$ for integral formulation and the integro-differential formulation, respectively, then typically $O(N_{ID})/O(N_I) \sim N_w$. N_I or N_{ID} is the number of unknowns to be determined in each formulation and N_w is the number of angular intensities to be solved in a volume node. The N_w can be in the range of 10^2 to 10^4 depending on the accuracy requirement. This becomes critical when dealing with complex multi-dimensional geometries in many applications where memory requirement of $O(N_{ID})$ will be very high. On the other hand, the computational effort is $O(N_I^2)$ for the integral formulation and $O(N_{ID})$ for the traditional integro-differential treatments. Many numerical schemes developed for the Fredholm and Volterra integral equations can be used to solve G and q functions.

2.6 Monte Carlo Method

In the Monte Carlo approach, radiative transfer is simulated using the calculated movements of a statistically large number of radiative energy bundles. As each bundle proceeds from its initial location and initial emission time through the participating medium, it experiences scattering, reflection, and absorption. Upon completion of the simulation, predictions of heat transfer are determined based on the averaged behavior of the set of individual energy bundles. The difficulties inherent in the solution of the time-dependent integro-differential radiative transfer equation and the complexity in dealing with the Fresnel reflection and anisotropic scattering can be avoided. The advantage of the Monte Carlo method is that even the most complicated problem can be solved with relative ease. For a trivial problem, setting up the appropriate photon sampling techniques alone may require more effort than finding the analytical solution. As the complexity of the problem increases, the complexity of formulation and the solution effort increase much more rapidly for conventional numerical techniques. For problems beyond a certain complexity, the Monte Carlo solution will be preferable. The method suffers from two drawbacks: the statistical nature of the results it produces, and the large computation time sometimes necessary to obtain the precision and accuracy required for engineering applications. Monte Carlo methods have been used extensively in the simulations of radiative heat transfer in the steady state. General reviews by Siegel and Howell (1992), Modest (1993), Yang *et al.* (1995), and Farmer and Howell (1996) provided the details for implementation of the method.

In transient analyses, the treatment of the flight time of photons is the only addition as compared to steady state. Further, Fresnel reflection is usually encountered in ultrafast laser transport in turbid media, such as living tissues. The following steps form the algorithm to track bundles which is used to simulate the short-pulse laser radiation in the scattering-absorbing turbid media:

1. Choose initial emission location of a photon bundle with unity energy at the incident surface and determine the initial time.
2. Choose the direction of initial propagation of the photon bundle.
3. Choose the path length to the next interaction between the photon and the medium.
4. Calculate the new position of the photon bundle. If the position is out of the medium, adjust the path length and let the new position locate at the boundary. Determine the reflection direction.
5. Calculate the flight time of the photon between the two successive interactions. Accumulate the flight time of a photon from its initial time.
6. Calculate the remaining energy of the photon. If the interaction is inside the medium, the absorption fraction of the energy is determined using the scattering albedo. If the interaction occurred at a boundary, the absorption fraction of the energy is determined using Beer's absorption law, and the remaining part is further determined according to Fresnel's reflection equations or the absorption and reflection properties of the surface.
7. Trace the photon bundle until it is fully absorbed by the medium, or out of the medium, or the remaining energy has been reduced to a very small pre-assigned value.
8. Every time the photon bundle strikes a detector, the absorbed energy fraction and the flight time of the photon are recorded.

Repeat steps 1 to 8 for many bundles. Upon the completion of ray tracing, the detected temporal signals at all specified locations are obtained. They are finally normalized by the incident radiation intensity. Detailed description for transient Monte Carlo method can be seen from Guo *et al.* (2000) and Guo and Kumar (2000).

3 EXPERIMENTAL APPROACH

The validation of models developed is performed via comparison with experiments designed to provide well characterized data for various configurations. The experimental approach is described below.

3.1 Experimental Setup

The experimental setup is shown in Figure 1. The output from a Coherent Antares, a mode-locked, frequency doubled, Nd^{+3}:YAG laser (532 nm, 60 ps FWHM, 76 MHz, 2W average power maximum) is split in two optical paths, in a 10/90 intensity ratio. The 10% path is designed to generate an undistorted temporal marker pulse before the beginning of the scattered signal, to as unambiguously as possible set zero time for the scattered signal, and minimize the effect of timing jitter and drift. The 90% path is incident on the scattering sample. A retroreflector on an optical rail gives a variable time delay, and is set such that the scattering signal is separated from and later in time than the marker pulse. In both paths, a halfwave plate / polarizer combination is included to control the intensity incident on the sample. A Coherent LabMaster power meter with a LM-2 photodiode head is used as a power monitor, and scattering signals are corrected for drift in incident average power.

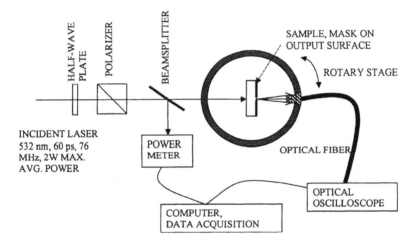

Figure 1 Schematic of the experimental setup (10% marker pulse path not shown for clarity).

The signal is collected by an optical fiber mounted on a rotary stage. The optical fiber, typically 100 μm diameter, numerical aperture 0.290, full acceptance angle 33.7°, is multimode, and focusing optics are not necessary to collect sufficient light for the sampling optical oscilloscope, Hamamatsu OOS-01. The OOS-01 specifies < 10 ps time resolution (equivalent to > 30 GHz bandwidth), <20 ps jitter, 2MHz maximum sampling frequency, 2.5 ps minimum sampling interval, and an effective wavelength range of 350-850 nm. Attempts had also been made to use ultrafast photodiodes, but their sensitivity was inadequate for all but the optically thinnest samples.

The mount is designed to give discrete adjustable distances from the fiber optic input to the sample face, typically 3 cm and 6 cm. Alternatively, the fiber is mounted on an arm on an x-y-z micrometer driven stage and can be independently adjusted with respect to the incident beam and any inclusion in the sample. This configuration is used when the distance from the fiber optic input to the sample face is nominally zero. When angular measurements are made, the fiber (input to detector) is off the sample face so a mask is placed on the output face to interrogate only a discrete spatial region. The mask has an aperture of typically 1 mm or 3 mm diameter. A micrometer-driven stage is used to align the mask coaxially with the laser beam prior to mounting the sample. By making rotary and screw adjustments on the sample mount, the reflection from the sample surface is aligned with the aperture before the sample, to assure the sample surface is perpendicular to the incident beam.

The power meter and optical oscilloscope are interfaced with a Pentium-based PC via GPIB, and data collected using Labview. In the data specifications, the actual incident average power is given if it was not altered in the experiment. Often the incident power has to be adjusted when data is taken at various positions and/or angles so as not to saturate the A/D converter. In these cases, the data is scaled to round number, typically 1 W or 1.5 W.

The oscilloscope used is a scanning optical oscilloscope, and averages the signal in two different ways. The "accumulate" specifies how many scanning sweeps at a fixed data point are taken and accumulated before moving on to repeat the process at the next data point. The "scan" is the number of fully scanned signal curves that are averaged together.

The competing effects are as follows. At high scan, high accumulate the signal to noise ratio is improved on short time scales (the scans do not have as much high frequency noise), but the data collection runs take much more time and the longer-term laser timing jitter and drift show up as large, lower-frequency ripples in the data.

3.2 Sample Preparation

To model short pulse propagation in scattering media, samples consisting of cast solid plastics with well-dispersed glass or silica microspheres were identified as desirable models. Ideally, a sample used in modeling calculations would have the refractive indices and absorption coefficients adjustable and well characterized for both the matrix and for the scattering centers. The scattering centers should have a well-characterized shape and size distribution, and the matrix should be homogeneous and the surface should be of good optical quality.

Samples are made of polystyrene with silica spheres acting as the scatterers with $n(matrix)$ = 1.59 and $n(particle)$ = 1.417 (unless otherwise noted), respectively where n is the refractive index at 532 nm. Polystyrene softens at 240 °C, so melting is not feasible, since a sonicator is used to disperse the spheres. Solvent casting was investigated and discarded for polystyrene due to excessive shrinkage. Tests were run on the polystyrene / Monospher 1000 (1000 nm diameter silica spheres +/- 10%) system. Various initiator/styrene monomer ratios were tested at 80 °C, to see if there were problems with polymerization rate, bubbling (since free radical polymerization releases gas), and discoloration. A w/w ratio of 0.8 +/- .2 benzoyl peroxide/styrene was satisfactory. The general range of sample loadings (scatterer/matrix) are 0.2-1.6% w/w, or about 0.1-0.85%

v/v (density of microspheres 1.96 g/ml; density of polystyrene 1.05 g/ml), with sample thickness ranges of 0.5-1.5 cm. In all cases, the 1000 nm diameter silica spheres are dispersed by repetitively sonicating and stirring at approximately 60 °C for 1-4 hours, depending on the loading. Polymerization of the matrix starts under these conditions, and the solution would gradually become slightly viscous. This is particularly necessary at higher loadings. Tests showed settling in 1% w/w (about 0.54% v/v) loadings, when relatively short dispersion times were used. Using these longer dispersion times at elevated temperatures, samples with 0.8 % v/v loading could be made routinely. The molds are preheated to the initial reaction temperature, 80°C. Small samples were reacted overnight at 80°C, typically 18 hours, and larger samples are further reacted at 120°C, typically 4 hours more, to reduce crazing.

4 SELECTED RESULTS

For the sake of brevity, only selected results are presented in this chapter. Further details are available from publications by the research team that are listed in the reference section. First, the results of models are compared to each other, followed by validation by comparison with experimental data. The emphasis is on the four new methods developed, namely transient discrete ordinates, transient radiation element, transient integral formulation, and the Monte Carlo method.

To compare different models a simple case of a turning an intensity signal "on" at one boundary at zero time is considered. Three methods are considered in Figure 2. The optical path length is unity and the medium is a cube. It is seen that the transient discrete ordinates method indicates that transmitted light will be observed at a time that is less than the actual physical time that it would take the light to travel the distance. The transient radiation element and the transient discrete ordinates methods do not exhibit such behavior.

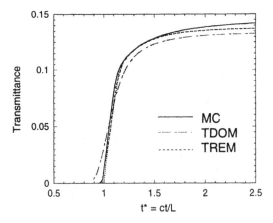

Figure 2 Comparison of absolute transmittance for Monte Carlo (MC), transient discrete ordinates (TDOM), and transient radiation element (TREM) methods for a slab with an "on" condition at one boundary.

In Figure 3 the results from a Monte Carlo simulation are compared with the experimental data for laser transmission through a slab. The match is good, thereby instilling confidence in the data and in the simulations. The details of the sample are indicated in the figure. The data is collected by setting the scan values to 10,000 and accumulation to 10. The laser beam width FWHM is 1.3 mm with an average power of 1.5W. The Monte Carlo simulation is done with 3 million bundles per run and 30 runs were averaged to obtain the results shown.

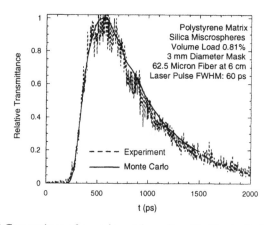

Figure 3 Comparison of experimental results with Monte Carlo simulation for a 19.95 mm thick slab.

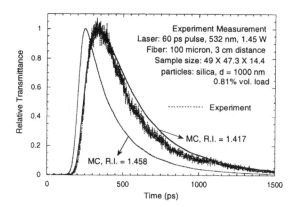

Figure 4 Sensitivity of particle refractive index on transmittance: Monte Carlo (MC) simulation and experimental data.

The sensitivity of the transmitted signal to the properties of the medium, are highlighted in the results of Figure 4. A less than 3% variation in the refractive index of the scatterers shows a significant change in the transmitted profile. This demonstrates the potential of the method to non-invasively probe the properties of a scattering medium. The two values of refractive index are chosen to compare the data for bulk silica to the measured refractive index for the spheres used (Firbank *et al.*, 1995).

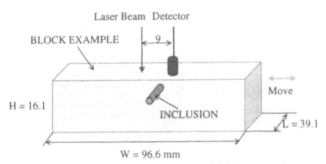

Figure 5a Schematic of the sample with inhomogeniety.

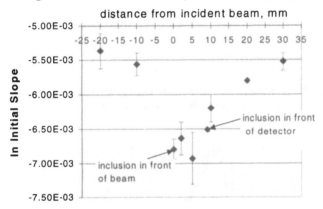

Figure 5b Experimentally obtained log of slopes of the reflectance signals at early time after peak (150 ps after peak).

The utilization of reflectance signal to detect inhomogenieties or inclusions in an otherwise uniform homogenous scattering medium is shown in Figure 5. By examining the initial slope of the reflectance after the peak, the presence of the inclusion is detected.

5 SUMMARY AND CONCLUSIONS

Study of the transport of short pulse lasers through scattering-absorbing media provides valuable insights about the characteristics of various models. By analyzing the entire temporal range of the reflected and transmitted signals, the differences between models are highlighted. It is found that for simulation based life cycle engineering the most useful model is the Monte Carlo model since it can accurately model the geometry of the boundaries, the variation of properties in the media, reflections and refractions at interfaces, and the variations associated with the radiation source. It has excellent accuracy for all optical thicknesses, from optically thin to thick. However, due to statistical errors and the need for computational resources (especially for large optical thicknesses), other methods should be considered whenever possible. These are the radiation element model, the discrete ordinates model, and the integral formulation.

6 REFERENCES

Brewster, M.Q., and Tien, C.L., 1982, Examination of the two-flux model for radiative transfer in participating systems, *International Journal of Heat and Mass Transfer*, **25**, pp. 1905-1907.

Brewster, M.Q., 1992, *Thermal Radiative Transfer and Properties*, Wiley.

Farmer, J.T., and Howell, J.R., 1996, Comparison of Monte Carlo strategies for radiative transfer in participating media, *Advances in Heat Transfer*, **31**, pp. 333-429.

Firbank, M., Oda, M., Delpy D.T., 1995, An improved design for a stable and reproducible phantom material use in near infra red spectroscopy and imaging, *Physics in Medicine and Biology*, **40**, pp. 955-961.

Fiveland, W.A., 1988, Three-dimensional radiative heat transfer by the discrete-ordinates method, *Journal of Thermophysics and Heat Transfer*, **2**, pp. 309-316.

Gritzo, L.A., Skocypec, R.D., and Tong, T.W., 1995, The use of high-performance computing to solve participating media radiative heat transfer problems - Results of an NSF workshop, *Sandia National Laboratory, Albuquerque, New Mexico*, Report SAND95-0225.

Guo, Z., Maruyama, S. and Tsukada, T., 1997, Radiative heat transfer in curved surfaces in Czochralski furnace, *Numerical Heat Transfer A*, **32**, pp. 595-611.

Guo, Z. and Kumar, S., 2000, Equivalent isotropic scattering formulation for transient short-pulse radiative transfer in anisotropic scattering planar media, *Applied Optics*, **39**, pp. 4411-4417.

Guo, Z. and Maruyama, S., 2000, Radiative heat transfer in inhomogeneous, nongray, and anisotropically scattering media, *International Journal of Heat and Mass Transfer*, **43**, pp. 2325-2336.

Guo, Z., Kumar, S., and San, K. C., 2000, Multi-dimensional Monte Carlo simulation of short pulse laser radiation transport in scattering media, *Journal of Thermophysics and Heat Transfer*, **14**, pp.504-511.

Guo, Z., 2001, *Ultrashort Laser Transport in Turbid Media*, PhD Dissertation, Polytechnic University.

Guo, Z. and Kumar, S., 2001a, Discrete-ordinates solution of short-pulsed laser transport in two-dimensional turbid media, *Applied Optics*, **40**, pp. 3156-3163.

Guo, Z. and Kumar, S., 2001b, Radiation element method for transient hyperbolic radiative transfer in plane-parallel inhomogeneous media, *Numerical Heat Transfer, Part B: Fundamentals*, **39**, pp. 371-387.

Howell, J.R., 1988, Thermal radiation in participating media: The past, the present, and some possible futures, *ASME Journal of Heat Transfer*, **110**, pp.1220-1229.

Howell, J.R., 1995, Radiative heat transfer: Opportunities and challenges, *ASME / JSME Thermal Engineering Joint Conference, Maui, Hawaii,* Vol. 3, pp.23-33.

Hsu, P.-F., and Farmer, J.T., 1992, Benchmark solutions of radiative heat transfer within nonhomogeneous participating media using the Monte Carlo and YIX methods, *ASME Journal of Heat Transfer,* **119**, pp.185-188.

Hsu, P.-F., 2000, Effects of multiple scattering and reflective boundary on the transient radiative transfer process, *National Heat Transfer Conference Proceedings* paper NHTC2000-12078, Pittsburgh, PA, August.

Kumar, S., Mitra, K., and Yamada, Y., 1996, Hyperbolic damped-wave models for transient light-pulse propagation in scattering media, *Applied Optics,* **35**, pp. 3372-3378.

Kumar, S., and Mitra, K., 1999, Microscale aspects of thermal radiation transport and laser applications, *Advances in Heat Transfer,* **33**, pp.187-294.

Maruyama, S., 1993, Radiation heat transfer between arbitrary three-dimensional bodies with specular and diffuse surfaces, *Numerical Heat Transfer,* Part A, **24**, pp. 181-196.

Menguc, M.P., and Viskanta, R., 1983, Comparison of radiative transfer approximations for highly forward scattering planar medium, *Journal of Quantitative Spectroscopy and Radiative Transfer,* **29**, pp. 381-394.

Mitra, K., and Kumar, S., 1999, Development and comparison of models for light-pulse transport in scattering absorbing media, *Applied Optics,* **38**, pp.188-196.

Mitra, K. and Churnside, J. H., 1999, Transient radiative transfer equation applied to oceanographic lidar, *Applied Optics,* **38**, pp. 889-895.

Mitra, K., Lai, M.-S., Kumar, S., 1997, Transient radiation transport in participating media within a rectangular enclosure," *AIAA Journal of Thermophysics and Heat Transfer,* **11**, No. 3, pp. 409-414.

Modest, M.F., 1993, *Radiative Heat Transfer,* McGraw Hill.

Siegel, R., and Howell, J.R., 1992, *Thermal Radiation Heat Transfer,* 3rd ed., McGraw-Hill, New York.

Tan, Z.-M., and Hsu, P.-F., 2000, An integral formulation of transient radiative transfer: Theoretical investigation, *National Heat Transfer Conference Proceedings* paper NHTC2000-12077, Pittsburgh, PA, August.

Tan, Z.-M., and Hsu, P.-F., 2001, Transient radiative transfer in a three-dimensional participating medium, *International Symposium on Radiative Transfer,* Antalya, Turkey, June.

Tong, T. W., and Skocypec, R.D., 1992, Summary on comparison of radiative heat transfer solutions for a specified problem, *National Heat Transfer Conference, San Diego, California,* HTD-Vol. 203, pp.253-264.

Viskanta, R., and Menguc, M.P., 1987, Radiation heat transfer in combustion systems, *Progress in Energy and Combustion Science,* **13**, pp. 97-160.

Viskanta, R., 1984, Radiative heat transfer, *Progress in Chemical Engineering (Fortschritte der Verfahrenstechnik)* Section A, **22**, pp. 51-81.

Yang, W. J., Taniguchi, H., and Kudo, K., 1995, Radiative heat transfer by the Monte Carlo method, *Advances in Heat Transfer,* **27**, pp. 1-215.

Acknowledgement: This ongoing work is partially supported by Sandia National Laboratories via contract AW-9963. Dr. Shawn Burns serves as the project manager for the contract.

Aerothermodynamic Predictions for Hypersonic Reentry Vehicles

Steven P. Schneider[1] and Craig Skoch[1]

1 Abstract

Laminar-turbulent transition is the largest source of uncertainty in the prediction of aerodynamic heating for ballistic reentry vehicles. Quiet-flow wind tunnels with low operating costs are needed to develop and validate mechanism-based prediction methods with improved accuracy. Purdue University continues to develop a 9.5-inch Mach-6 Ludwieg tube for quiet-flow operation to high Reynolds number. Preliminary measurements of the tunnel performance are reported here. A fast transducer was used to make pitot-pressure measurements of the mean flow and fluctuations on the nozzle centerline. The run time is about 10 seconds. The mean Mach number decreases from about 6.0 at 1 atm. stagnation pressure to about 5.5 at 10 atm. The pitot-pressure fluctuations decrease from about 4% at a stagnation pressure of 1 atm. to about 1% at 10 atm. Problems with the bleed slot flow are suspected to be the cause of the lack of quiet flow; the pressure fluctuations at the geometrical minimum of the bleed slot range from 0.4% at 1 atm. to 0.2% at 10 atm.

2 Introduction

2.1 Hypersonic Laminar-Turbulent Transition

Laminar-turbulent transition in hypersonic boundary layers is important for prediction and control of heat transfer, skin friction, and other boundary layer properties. However, the mechanisms leading to transition are still poorly understood, even in low-noise environments. Applications hindered by this lack of understanding include reusable launch vehicles such as the X-33 (Berry, Horvath, Hollis, Thompson, & Hamilton II, 1999), high-speed interceptor missiles (Korejwo & Holden, 1992), hypersonic cruise vehicles (AGARD, 1997), and ballistic reentry vehicles (Lin, Grabowsky, & Yelmgren, 1984).

A closely relevant example is given in Fig. 8 of Hamilton II, Millman, and Greendyke (1992), which presents computations and measurements of the surface heat transfer during the Reentry-F test of a ballistic reentry vehicle. The vehicle was a 13-ft. beryllium cone that reentered at a peak Mach number of about 20 and a total enthalpy of about 18 MJ/kg. The cone half-angle was 5-degrees, the angle of attack was near zero, and the graphite nosetip

---

[1] School of Aeronautics and Astronautics, Purdue University, West Lafayette

had an initial radius of 0.1 inches (Schneider, 1999). Heating rises by a factor of 3 during transition. Hamilton's computations were carried out using a variable-entropy boundary-layer code that includes equilibrium chemistry effects. Agreement is good for both the laminar and turbulent regions, once the transition location is known. Harris Hamilton from NASA Langley says that typical accuracies are 20-25% for the turbulent boundary layer, and 15-20% for the laminar layer. The largest uncertainty, a factor of 2-3, is due to transition, which causes a large increase in the heat transfer rates (private communication, March 1999). The dominant uncertainty sources are the transition onset location and the extent of transitional flow.

Many transition experiments have been carried out in conventional ground-testing facilities over the past 50 years. However, these experiments are contaminated by the high levels of noise that radiate from the turbulent boundary layers normally present on the wind tunnel walls (Beckwith & Miller III, 1990). These noise levels, typically 0.5-1% of the mean, are an order of magnitude larger than those observed in flight (Schneider, 2001, 1999). These high noise levels can cause transition to occur an order of magnitude earlier than in flight (Beckwith & Miller III, 1990; Schneider, 1999). In addition, the mechanisms of transition operational in small-disturbance environments can be changed or bypassed altogether in high-noise environments; these changes in the mechanisms change the parametric trends in transition (Schneider, 2001).

For example, linear instability theory suggests that the transition Reynolds number on a 5 degree half-angle cone should be 0.7 of that on a flat plate, but noisy tunnel data showed that the cone transition Reynolds number was about twice the flat plate result. Only when quiet tunnel results were obtained was the theory verified (Chen, Malik, & Beckwith, 1989). This is critical, since design usually involves consideration of the trend in transition when a parameter is varied. Clearly, transition measurements in conventional ground-test facilities are generally not reliable predictors of flight performance.

2.2 Development of Quiet-Flow Wind Tunnels

Only in the last two decades have low-noise supersonic wind tunnels been developed (Beckwith & Miller III, 1990; Wilkinson, Anders, & Chen, 1994). This development has been difficult, since the test-section wall boundary-layers must be kept laminar in order to avoid high levels of eddy-Mach-wave acoustic radiation from the normally-present turbulent boundary layers. Quiet tunnels typically have fluctuation levels of 0.1% or less; these are sometimes measured as pitot-pressure fluctuations, using fast pressure transducers, and sometimes as static pressure or massflow fluctuations, using hot wires (e.g., Schneider and Haven (1995)).

A Mach 3.5 tunnel was the first to be successfully developed at NASA Langley (Beckwith, Creel, Chen, & Kendall, 1983). Langley then developed a Mach 6 quiet nozzle, which was used as a starting point for the new Purdue nozzle (Blanchard, Lachowicz, & Wilkinson, 1997). Unfortunately, this nozzle was removed from service due to a space conflict. Langley also attempted to develop a Mach 8 quiet tunnel (Wilkinson et al., 1994); however, the high temperatures required to reach Mach 8 made this a very difficult and expensive effort, which was ultimately unsuccessful. The old Langley Mach-6 quiet nozzle is now to be put back in service (Steve Wilkinson, private communication, April 2001). Since this will take some time, the new Purdue Mach-6 quiet flow Ludwieg tube may become the only operational hypersonic quiet tunnel in the world.

2.3 Background of the Boeing/AFOSR Mach-6 Quiet Tunnel

A Mach-4 Ludwieg tube was constructed at Purdue in 1992, using a 4-inch nozzle of conventional design that was obtained surplus from NASA Langley. By early 1994, quiet-flow operation was demonstrated at the low Reynolds number of about 400,000 (Schneider & Haven, 1995). Since then, this facility has been used for development of instrumentation and for measurements of instability waves under quiet-flow conditions (e.g., Schneider et al. (1996), Schmisseur, Collicott, and Schneider (2000), Salyer, Collicott, and Schneider (2000)). However, the low quiet Reynolds number imposes severe limitations; for example, the growth of instability waves under controlled conditions on a cone at angle of attack was only about a factor of 2 (Ladoon & Schneider, 1998). This is far smaller than the factor of $e^9 - e^{11}$ typically observed prior to transition, and small enough to make quantitative comparisons to computations very difficult.

A facility that remains quiet to higher Reynolds numbers was therefore needed. The low operating costs of the Mach-4 tunnel had to be maintained. However, hypersonic operation was needed in order to provide experiments relevant to the hypersonic transition problems described above. Operation at Mach 6 was selected, since this is high enough for the hypersonic 2nd-mode instability to be dominant under cold-wall conditions, and high enough to observe hypersonic roughness-insensitivity effects, yet low enough that the required stagnation temperatures do not add dramatically to cost and difficulty of operation. Schneider (1998b) describes the overall design of the facility, and the detailed aerodynamic design of the quiet-flow nozzle, carried out using the e^N method.

Attached flow should be maintained in the contraction of the nozzle, since the separation bubbles sometimes observed in wind tunnel contractions are generally unsteady, and would transmit noise downstream into the Mach-6 nozzle. Preliminary analyses have suggested that the low-frequency fluctuations present in the Langley Mach-6 quiet nozzle (Blanchard et al., 1997) may be caused by such a separation. Therefore, a detailed aerodynamic design of the contraction was also carried out (Schneider, 1998a). Schneider (1998a) also supplies a preliminary report on the detailed mechanical design of the nozzle and contraction, which was carried out during 1997-98. This mechanical design is not trivial, since quiet uniform flow requires very tight tolerances on surface contour and finish.

After some initial tests of fabrication procedures, a purchase order for fabrication of the nozzle and contraction was awarded in January 1999. Schneider (2000a) reported on design and testing of some of the component parts, including the driver-tube heating, the as-measured contraction contour, the throat-region mandrel fabrication and polishing experience, and so on. Schneider (2000a) also reports the results of numerous measurements of the surface roughness of the mandrel and various test pieces. Finally, Schneider (2000a) reports temperature measurements carried out on the heated driver tube. Schneider (2000b) reports on the design and fabrication of the support structure, diffuser, and second-throat section (which also serves as the sting support). It also reports experience with final contraction fabrication, and with operation of the vacuum system. Schneider (2000b) also reports on the contour measurements on the third attempt at throat-mandrel fabrication. Schneider, Rufer, Randall, and Skoch (2001) reports on (1) the completion of the upstream portion of the nozzle, including the wall-contour measurements through section 5, (2) the contraction-region heating apparatus, (3) the burst-diaphragm tests, and (4) the bleed-slot suction system.

Schneider and Skoch (2001) reports the remainder of the nozzle-wall contour measurements, along with initial measurements of the mean flow and freestream noise. This reference is summarized here. During Spring 2001, the tunnel was officially named the Boeing/AFOSR Mach-6 Quiet Tunnel.

2.4   The Boeing/AFOSR Mach-6 Quiet Tunnel

Quiet facilities require low levels of noise in the inviscid flow entering the nozzle through the throat, and laminar boundary layers on the nozzle walls. These features make the noise level in quiet facilities an order of magnitude lower than in conventional facilities. To reach these low noise levels, conventional blow-down facilities must be extensively modified. Requirements include a 1 micron particle filter, a highly polished nozzle with bleed slots for the contraction-wall boundary layer, and a large settling chamber with screens and sintered-mesh plates for noise-reduction (Beckwith & Miller III, 1990). To reach these low noise levels in an affordable way, the Purdue facility has been designed as a Ludwieg tube (Schneider & Haven, 1995). A Ludwieg tube is a long pipe with a converging-diverging nozzle on the end, from which flow exits into the nozzle, test section, and second throat (Figure 1).   A diaphragm is placed downstream of the test section. When the diaphragm bursts, an expansion wave travels upstream through the test section into the driver tube. Since the flow remains quiet after the wave reflects from the contraction, sufficient vacuum can extend the useful runtime to many cycles of expansion-wave reflection, during which the pressure drops quasi-statically.

Figure 2 shows a portion of the nozzle of the new facility.   The region of useful quiet flow lies between the characteristics marking the onset of uniform

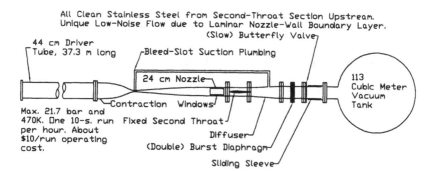

Figure 1: Schematic of Boeing/AFOSR Mach-6 Quiet Tunnel

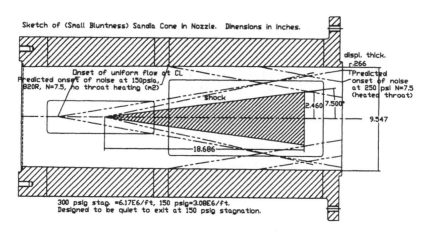

Figure 2: Schematic of Mach-6 Quiet Nozzle with Model

flow, and the characteristics marking the upstream boundary of acoustic radiation from the onset of turbulence in the nozzle-wall boundary layer. The cone is drawn at the largest size for which it is likely to start (Bountin, Shiplyuk, & Sidorenko, 2000).

3   Initial Performance Tests

The first run of the new tunnel was carried out on 19 April 2001. Operations have been fairly smooth, and initial experience with cleanliness is good. However, the tunnel has been running with noise levels well above the quiet level. Initial measurements have focused on preliminary determination of facility performance, and on determining the cause of the lack of quiet flow. Problems with the bleed-slot flow are the suspected cause of the lack of quiet flow. Since changing the bleed-slot flow requires irreversible machining of small contraction pieces, substantial characterization of the tunnel flow is first to be carried out.

3.1   Initial Operating Procedures

Procedures for tunnel operation are now being developed; the following describes those used for most of the runs to date. The driver tube and contraction are heated ahead of time, as these take at least 24 hours to reach equilibrium. Typically, the driver tube and the two downstream sections of the contraction were heated to $170°C$, while the upstream section of the contraction was heated to $160°C$. The upstream section was heated slightly less, because the silicone O-ring in this region showed signs of excessive temperature, suggesting that the actual contraction temperature might have been substantially higher. The thermocouples on the contraction are clamped to the outside of the uninsulated vessel, so they may be reading an artificially low value.

The vacuum pump is turned on and the vacuum tank is pumped down with the gate valve closed (Fig. 1). The diaphragms are inserted and the gate valve is then opened. The driver tube is then filled with air that has been heated by the circulation heater. The circulation heater is turned on to preheat, just before the air begins to flow into the driver tube; the temperature of the circulation heater and the preheat time are additional variables that need to be determined. Typically, the circulation heater was set to $170°C$. For the first run of the day, it was turned on 90 sec. before flow was started; for later runs, it was turned on 30-45 sec. before flow started. This air is allowed to equilibrate for approximately 30 minutes before the run is made (as suggested by the results of Munro (1996)). The electrically-controlled pressure regulator that supplies air to the driver tube is then turned down to zero; little air flows backwards to the regulator from the tube due to a check valve. In all of the runs the initial pressure in the vacuum tank has been below 4 torr. The double diaphragms are burst by bleeding the gap air to vacuum, starting the run.

After the run, the gate valve is closed to allow the vacuum pump to continue pumping down for the next run. After a run the pressure in the driver tube is below atmospheric. It has been observed that the pressure regulator tries to relieve this condition by venting air in from the room, if it is not commanded to put air in from the compressed air tanks. This may be a problem because the room air has not been dried, raising the dewpoint in the driver tube. To prevent this, it is necessary to fill the driver tube with dry air as soon as the gate valve is closed. This concern was not noticed until after most of the runs were done, and this part of the procedure varied. This causes uncertainty regarding the dewpoint of the driver-tube air.

3.2 Instrumentation

The mean flow and fluctuations are measured in the nozzle using fast pressure transducers (for details, see Schneider and Skoch (2001)). The driver-tube temperature is monitored using thermocouples on the outside of the tube, and also using a thermocouple in the air at the upstream end. No cold-wire measurements of the actual gas stagnation temperature have yet been made in the nozzle. The driver-tube pressure before the run starts is measured with a used Heise gauge that was recalibrated to $\pm 0.1\%$ by a NIST-traceable lab. The initial pressure drop due to the expansion wave in the driver tube is very small, due to the very low driver-tube Mach number of about 0.003; these effects are therefore neglected here.

3.3 Preliminary Tunnel Characterizations

Preliminary measurements of the pitot pressure were obtained for various initial driver tube pressures. For these runs, the pitot probe is again located on the centerline at $z = 84.31$ in., where z is an axial coordinate with $z = 0$ at the nozzle throat. The mean pitot pressure is normally obtained from the portion of the record between 0.2 and 0.3 seconds after the run was triggered. The traces were examined visually, and in some cases the average was taken between 0.25 and 0.35 sec., to avoid a longer startup period.

Fig. 3 shows the resulting Mach numbers, computed using the Rayleigh pitot formula, using the initial driver tube pressure for the stagnation pressure. The initial driver tube pressure is obtained from the pre-run data for the bleed-slot transducer, which agrees with the data recorded from the Heise gauge. There are four points that don't fit within the trend at the high stagnation pressures. These points were all taken on the same day, early in the test series, with run conditions similar to all of the other data. These high pitot pressures have not been repeated since these points, and might be the result of a problem with moisture causing a condensation shock. Neglecting these runs, the Mach number decreases as stagnation pressure increases up to

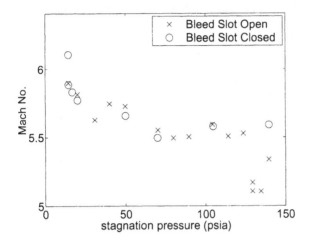

Figure 3: Mean Mach Number vs. Stagnation Pressure

about 50 psia, and then is nearly constant to 140 psia. The results are nearly independent of whether the bleed valves are open or closed. In earlier Langley measurements in the Mach-6 quiet tunnel, carried out by Frank Chen (Steve Wilkinson, private communication, 15 May 2001), the mean Mach number decreased about 3% when the bleed slots were closed. The change here is much smaller, suggesting that the boundary-layer is turbulent in both cases, and that the spillage of flow over the bleed lip with the slot closed does not greatly change the massflow through the nozzle throat.

The mean Mach number decreases with increasing pressure. This is puzzling, for one would normally expect thinner boundary layers on the nozzle walls at higher pressures, and so a higher Mach number, due to the larger area ratio from the throat. The variation is substantial, and remains to be explained. Transitional effects in the nozzle-wall boundary-layer are one possible cause.

The mean pressure in the geometric minimum of the bleed slot is shown in Fig. 4. Here, $p_{th,mean}$ is the mean pressure in the bleed-slot throat, and p_0 is the initial driver-tube pressure. It is apparent that the pressure is significantly above the 0.528 value for sonic flow. It thus appears that the transducer is consistently upstream of the sonic line.

The rms pitot fluctuations are shown in Fig. 5. Here, $p_{t2,rms}$ is the rms pitot pressure, and $p_{t2,mean}$ is the mean pitot pressure. The rms is determined by postprocessing the digitized data over the interval from 0.25 to 1.75 sec. after the trigger. The rms pitot pressure decreased from about 4% at 1 atm total pressure to about 1% at 140 psia. Since the nominal maximum for quiet flow is about 0.1%, these are not quiet levels. Spectral analysis is presently

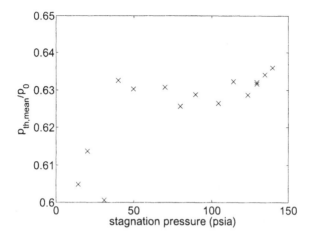

Figure 4: Mean Pressure in Bleed-Slot Throat

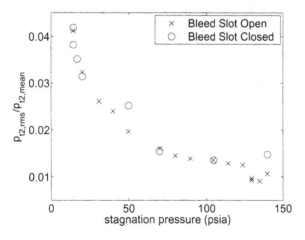

Figure 5: Pitot Pressure Fluctuations vs. Stag. Pressure

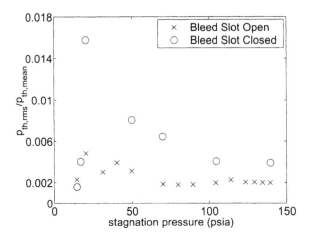

Figure 6: Fluctuations in Bleed-Slot Throat Pressure

being carried out.

The rms fluctuations in the bleed-slot throat pressure $(P_{th,rms})$ are shown in Fig. 6. With the valve open, so that suction air is passing through the slot, the fluctuations decrease about 0.4% at 1 atm. total pressure to about 0.2% at 10 atm. For a sample run at 69.9 psia, the rms throat noise is 0.19% and the electrical noise in the prerun portion of the signal is 0.03%, so the signal/noise ratio is still sufficient to resolve these levels. The noise levels are larger than expected, considering that the mean pressure is above the sonic value (Fig. 4), so that the transducer is located upstream of the sonic point in the slot flow. It should then see only the same level of fluctuation as in the contraction, which should be very low. This suggests that some unsteadiness or separation may be occuring in the bleed slot flow.

Fig. 6 shows an increased noise level in the throat Kulite when the bleed slot is closed, as might be expected, except for the three lowest pressure runs. These low pressure runs produced lower fluctuation levels, for unknown reasons. The lowest pressure run was performed twice at the same conditions with nearly identical results, so that only two low-noise low-pressure points are visible. It should be noted that the lowest pressure run had longer than the usual 30 minutes to reach equilibrium, since it was performed at atmospheric pressure, and the waiting time for the run was only dependent on when the vacuum pressure was sufficiently low. The two next higher pressures, including the run with the highest noise level, only had about 10 minutes to reach equilibrium, due to the difficulty of holding the pressure with the 0.010-inch acetate diaphragms required at those pressures.

Preliminary measurements of the mean flow uniformity have also been performed, on the tunnel centerline. Fig. 7 shows the Mach number distribu-

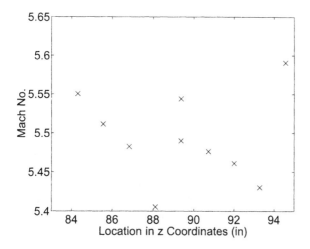

Figure 7: Mach Number Distribution on Tunnel Centerline

tion down the centerline, for a driver pressure of 69.75 ± 0.3 psia and a driver temperature of $170°C$. The nominal onset of uniform flow is at $z = 75.13$ in., and the end of the contoured nozzle is at $z = 101.98$ in. The scatter is about $\pm 2\%$; the cause of this scatter is unknown. It could be related to dewpoint or calibration problems.

4   Summary and Future Plans

Laminar-turbulent transition is the dominant source of uncertainty in predicting the heat-transfer to ballistic reentry vehicles. To help develop prediction methods based on the transition mechanisms, a Mach-6 quiet Ludwieg tube has been developed. The facility has now been operational for about 3 months. The mean Mach number on the centerline decreases from about 6.0 at $Re \simeq 3 \times 10^5$/ft. to about 5.5 at $Re \simeq 3 \times 10^6$/ft. The pitot fluctuations decrease from about 4% at the lower Reynolds numbers to about 1% at the higher ones, so the tunnel is not yet running quiet. Problems with moisture, condensation, and the bleed-slot flow are suspected.

Systematic studies of the driver tube temperature, the contraction temperature, and the temperature and preheat-time of the circulation heater are to be carried out. Cold-wire measurements of the stagnation temperature are to be made in the uniform-flow region. A plexiglas window is to be installed (Kwon & Schneider, 2002), and blockage tests are to be performed with round-cone and slab-delta models. The massflow through the bleed slot is to be varied by machining a throat-ring that forms the geometrical minimum of the slot. The amount of massflow captured by the bleed can also be varied, by machining the last section of the contraction. The temperature of the nozzle throat can be raised to a maximum of about 350F, using equipment already in place. These and other measurements are now to be carried out (Schneider, Skoch, Rufer, & Matsumura, 2002).

5   Acknowledgements

The research is funded by AFOSR under grant F49620-00-1-0016, monitored by Steve Walker, by Sandia National Laboratory, under contract BG-7114, and by NASA Langley, under Grant NAG1-01-027. Graduate student Shann Rufer assisted with various aspects of the tunnel shakedown and operation. Fabrication of the tunnel was supported primarily by a gift from the Boeing Company and two grants from the Defense University Research Instrumentation Program (F49620-98-1-0284 and F49620-99-1-0278). The first grant was funded by AFOSR, and the second was funded jointly by AFOSR and BMDO. Tunnel fabrication has also been supported by Sandia National Laboratory, and by a gift in memory of K.H. Hobbie. The tunnel design and fabrication has been carried out in cooperation with Dynamic Engineering Inc., of Newport News, Virginia. The generous cooperation of Dr. Steve Wilkinson and the rest of the NASA Langley quiet tunnel group has been critical to our progress.

References

AGARD (Ed.). (1997, April). *Sustained hypersonic flight.* (CP-600, vol. 3.)

Beckwith, I., Creel, T., Chen, F., & Kendall, J. (1983). *Freestream noise and transition measurements on a cone in a Mach-3.5 pilot low-disturbance tunnel* (Technical Paper No. 2180). NASA.

Beckwith, I., & Miller III, C. (1990). Aerothermodynamics and transition in high-speed wind tunnels at NASA Langley. *Annual Review of Fluid Mechanics, 22*, 419-439.

Berry, S. A., Horvath, T. J., Hollis, B. R., Thompson, R. A., & Hamilton II, H. H. (1999). *X-33 hypersonic boundary layer transition* (Paper No. 99-3560). AIAA.

Blanchard, A. E., Lachowicz, J. T., & Wilkinson, S. P. (1997). NASA Langley Mach 6 quiet wind-tunnel performance. *AIAA Journal, 35*(1), 23-28.

Bountin, D., Shiplyuk, A., & Sidorenko, A. (2000). Experimental investigations of disturbance development in the hypersonic boundary layer on a conical models. In H. Fasel & W. Saric (Eds.), *Laminar-turbulent transition. Proceedings of the IUTAM symposium, Sedona, 1999* (p. 475-480). Berlin: Springer-Verlag.

Chen, F.-J., Malik, M., & Beckwith, I. (1989). Boundary-layer transition on a cone and flat plate at Mach 3.5. *AIAA Journal, 27*(6), 687-693.

Hamilton II, H., Millman, D., & Greendyke, R. (1992). *Finite-difference solution for laminar or turbulent boundary layer flow over axisymmetric bodies with ideal gas, CF4, or equilibrium air chemistry* (Technical Paper No. 3271). NASA.

Korejwo, H., & Holden, M. (1992). *Ground test facilities for aerothermal and aero-optical evaluation of hypersonic interceptors* (Paper No. 92-1074). AIAA.

Kwon, S., & Schneider, S. P. (2002). *Stress analysis for the window of the Purdue Mach-6 quiet-flow Ludwieg tube* (Paper No. 2002-XXXX). AIAA. (Submitted to the Jan. 2002 AIAA Aerospace Sciences Meeting.)

Ladoon, D. W., & Schneider, S. P. (1998). *Measurements of controlled wave packets at Mach 4 on a cone at angle of attack* (Paper No. 98-0436). AIAA.

Lin, T. C., Grabowsky, W. R., & Yelmgren, K. E. (1984). The search for optimum configurations for re-entry vehicles. *J. of Spacecraft and Rockets, 21*(2), 142-149.

Munro, S. E. (1996). *Effects of elevated driver-tube temperature on the extent of quiet flow in the Purdue Ludwieg tube.* Unpublished master's thesis, School of Aeronautics and Astronautics, Purdue University. (Available from the Defense Technical Information Center as AD-A315654.)

Salyer, T. R., Collicott, S. H., & Schneider, S. P. (2000). *Feedback stabilized laser differential interferometry for supersonic blunt body receptivity experiments* (Paper No. 2000-0416). AIAA.

Schmisseur, J., Collicott, S. H., & Schneider, S. P. (2000). Laser-generated localized freestream perturbations in supersonic and hypersonic flows. *AIAA Journal, 38*(4), 666-671.

Schneider, S. P. (1998a). *Design and fabrication of a 9-inch Mach-6 quiet-flow Ludwieg tube* (Paper No. 98-2511). AIAA.

Schneider, S. P. (1998b). *Design of a Mach-6 quiet-flow wind-tunnel nozzle using the e**N method for transition estimation* (Paper No. 98-0547). AIAA.

Schneider, S. P. (1999). Flight data for boundary-layer transition at hypersonic and supersonic speeds. *Journal of Spacecraft and Rockets, 36*(1), 8-20.

Schneider, S. P. (2000a). *Fabrication and testing of the Purdue Mach-6 quiet-flow Ludwieg tube* (Paper No. 2000-0295). AIAA.

Schneider, S. P. (2000b). *Initial shakedown of the Purdue Mach-6 quiet-flow Ludwieg tube* (Paper No. 2000-2592). AIAA.

Schneider, S. P. (2001). Effects of high-speed tunnel noise on laminar-turbulent transition. *Journal of Spacecraft and Rockets, 38*(3), 323-333.

Schneider, S. P., Collicott, S. H., Schmisseur, J., Ladoon, D., Randall, L. A., Munro, S. E., & Salyer, T. (1996). *Laminar-turbulent transition research in the Purdue Mach-4 quiet-flow Ludwieg tube* (Paper No. 96-2191). AIAA.

Schneider, S. P., & Haven, C. E. (1995). Quiet-flow Ludwieg tube for high-speed transition research. *AIAA Journal, 33*(4), 688-693.

Schneider, S. P., Rufer, S., Randall, L., & Skoch, C. (2001). *Shakedown of the Purdue Mach-6 quiet-flow Ludwieg tube* (Paper No. 2001-0457). AIAA.

Schneider, S. P., & Skoch, C. (2001). *Mean flow and noise measurements in the Purdue Mach-6 quiet-flow Ludwieg tube* (Paper No. 2001-2778). AIAA.

Schneider, S. P., Skoch, C., Rufer, S., & Matsumura, S. (2002). *Transition research in the Boeing/AFOSR Mach-6 quiet tunnel* (Paper No. 2002-XXXX). AIAA. (Submitted to the Jan. 2002 AIAA Aerospace Sciences Meeting.)

Wilkinson, S. P., Anders, S. G., & Chen, F.-J. (1994). *Status of Langley quiet flow facility developments* (Paper No. 94-2498). AIAA.

Numerical Study of Turbulence-Radiation Interactions in Reactive Flow

Michael F. Modest[1], Lance R. Collins[2], Genong Li[1] and T. Vaithianathan[2]

1 INTRODUCTION

In order to predict radiative heat transfer accurately in turbulent reacting flows, it is necessary to couple the radiation calculations with turbulence calculations. This requires the solution of many coupled partial differential equations and, in practice, these equations need to be Reynolds or Favre averaged. The difficulty arises from the averaging process: many unclosed terms appear as a result of turbulence-radiation interactions (TRI) that need to be modeled. Traditional moment methods fail to obtain closure for these terms because too many additional partial differential equations need to be modeled and solved simultaneously, which generally exceeds the power of current computers (Mazumder and Modest 1998). As a consequence, traditional modeling of radiating reactive flows has generally ignored turbulence-radiation interactions, i.e., radiation calculations have been based on mean temperature and concentration fields (Viskanta 1998) even though experimental work has suggested that mean radiative quantities may differ significantly from those based on mean scalar values (G. M. Faeth and Jeng 1989).

Some efforts have been made to couple radiation and turbulence calculations in a more rigorous way. Song and Viskanta (1987) have investigated a turbulent premixed flame inside a two-dimensional furnace. While turbulence-radiation interactions were considered, in order to obtain closure for their governing equations, correlation functions for gaseous properties had to be assumed. Gore et al. (1992) and Hartick et al. (1996) applied this approach to study diffusion flames, in which extended k-ϵ-g models were used. To consider turbulence-radiation interactions, the shape of the probability density function (PDF) of mixture fraction and total enthalpy had to be prescribed. Mazumder and Modest (1998) used the velocity-composition PDF method to investigate turbulence-radiation interactions. In their method, velocities, species concentrations and temperature were treated as random variables and the probability density function of these variables was considered. The uniqueness of this method is that it can resolve various interactions between turbulence, chemical reactions and radiative transfer in a rigorous way. Using this method, Mazumder and Modest were able to calculate the important unclosed terms as a result of turbulence-radiation interactions exactly. However, the

---

[1] Department of Mechanical & Nuclear Engineering, Penn State University
[2] Department of Chemical Engineering, Penn State University

velocity-composition PDF method, although powerful in potential, is still at an early stage of development, making it difficult to apply to general 2D and 3D problems.

In this study, turbulence-radiation interactions were investigated by a composition PDF approach, in which only species concentrations and temperature/enthalpy (called the composition variables, since they determine the composition of the mixture) are treated as random variables. The composition PDF method has been developed and successfully used in numerical simulations of nonradiating reactive flows (Pope 1985). We have demonstrated that the method is able to treat turbulence-radiation interactions exactly in the context of the optically thin-eddy approximation (Li and Modest 2001). PDF methods have the advantage over conventional moment methods: they represent arbitrary-order nonlinear terms (such as chemical reactions and thermal radiative emission) exactly. One disadvantage of the PDF framework is that scalar mixing, a two-point statistic, must be modeled. We developed a new closure for the mixing term that explicitly accounts for the broad range of length and time scales in turbulent flows. The new closure is based on a well-established theory in the literature (the eddy damped quasi-normal Markovian theory, Orszag 1970) that we extended to the present circumstance (Ulitsky and Collins 2000). A second drawback of the PDF method is that it has too many dimensions to be solved efficiently by finite volume/finite element methods. Instead, a Monte Carlo approach has been used, in which the probability density function is represented by a large number of particles that evolve according to a system of stochastic differential equations. Although significant progress has been made in the development of the PDF/Monte Carlo method, several important weaknesses in the method's numerical implementation still remain. Conventional particle-tracing schemes do not permit the use of cell systems with large variations in size in an economical way, even though this kind of cell system is often required to capture sharp gradients in the flow. In this study, first a new particle tracing scheme was developed, which allows the PDF/Monte Carlo method to be used efficiently with any cell system; with that effective scheme implemented in the code, turbulence-radiation interactions were exhaustively investigated.

2 EFFICIENT PARTICLE TRACING SCHEME

Conventional PDF/Monte Carlo methods populate the computational mesh with particles of equal mass. For meshes with cells of varying size, this can result in too many particles in large cells and too few particles in small cells. This imbalance of particle populations makes particle tracing inefficient because tracing more particles than necessary in a cell wastes computer time, while tracing too few particles leads to large statistical errors. Another weakness of conventional PDF/Monte Carlo methods is the fact that they customarily use a constant time step for the integration of particle evolution equations, which again tends to waste computer time because the time step is usually constrained by small cells and is unnecessarily small for particles in large cells.

A new particle tracing scheme was developed to overcome the shortcomings of the conventional schemes, in which we introduced the concepts of variable time steps and variable particle mass. Using locally-adaptive time steps in the integration of particle equations, the particles' parameters are updated more frequently in regions of strong gradients than in those of flat gradients, which greatly improves the time-efficiency of particle tracing. To reduce statistical errors, a particle splitting and combination procedure was devised. With this procedure the number of particles in each cell is maintained at a prescribed level. Consequently, statistical errors of mean quantities are more uniformly distributed over the computational domain. The new scheme allows the PDF/Monte Carlo code to use any grid that is constructed by a finite volume code, making it possible to couple PDF/Monte Carlo methods to all popular commercial CFD codes, thereby extending the capabilities of existing CFD codes to simulate radiation transport in turbulent reactive flows.

The numerical performance of the new particle tracing scheme and the solution procedure were illustrated by considering a turbulent heat transfer problem in a parallel channel and a turbulent jet diffusion flame in a cylindrical combustor (Li and Modest 2000a; Li and Modest 2000b). The first problem was solved on a structured rectangular mesh. The results showed that the time step splitting procedure decreased the CPU time by as much as a factor of 5, for a given number of particles, and the particle splitting and combination procedure reduced the statistical error significantly, allowing the use of 10 times less particles in the simulation for the same level of statistical convergence. In the first test problem, the variation of cell sizes was relatively mild ($V_{max}/V_{min} = 13$) to make the problem amenable to the conventional method. In the second test problem, the variation of cell sizes and the disparities of time scales over cells were so large ($V_{max}/V_{min} = 5 \times 10^5$) that the conventional method could not be used, while the new method handled it with ease. In addition, an unstructured mesh was used to demonstrate the particle tracing procedure in a general mesh system.

3 PDF MIXING MODEL

Although the PDF approach yields the chemical source and TRI terms in closed form, turbulent and molecular mixing must be modeled. The errors associated with modeling the mixing terms can be significant, often yielding PDF's that are not even in qualitative agreement with experiments or DNS. The problem is due in part to the standard assumption that turbulence has a single length and time scale (usually taken to be the integral length scale and large eddy turnover time respectively). In fact, turbulence is composed of a full spectrum of length and time scales that interact through nonlinear couplings arising from the convective terms and the chemical source terms. These interactions do not naturally arise in the PDF framework, since the PDF is fundamentally a single-point quantity.

Consequently, mixing in most PDF models is based on the turbulence integral time scale, defined as k/ϵ in most k–ϵ models. This time scale is a reasonable one for mixing of the bulk species (e.g., fuel and oxidizer), which generally are broken into smaller fragments by the actions of turbulence at a rate that is determined by the large scales (Corrsin 1951). However, inhomogeneities that arise at other length scales are not well represented by this mixing paradigm. An example would be species that are produced by the reactions in the flame, which may scale with the flame thickness rather than the integral length scale. Furthermore, effects such as differential diffusion (Bilger and Dibble 1982; Yeung and Pope 1993; Kerstein et al. 1995), whereby species migrate due to differences in the molecular diffusivities, are completely ignored.

The goal of this part of the study is to introduce a multiple-scale model of mixing that accounts for the flux of scalar from large to small scales (transfer) and allows mixing at different scales to proceed at independent rates. To accomplish this, we will associate several length scales with each Monte Carlo particle. That is, for a given Monte Carlo particle, there will be a concentration for each of the scalar species at *each of the designated length scales*. Exchanges of scalar between different "scales" corresponds to transfer, while exchanges of scalar between particles is mixing. The rates of transfer and mixing will be determined based on the eddy damped quasi-normal Markovian theory (or EDQNM for short). EDQNM is a relatively rigorous theory, originally developed by Orszag (1970), that has been shown to predict energy (Andre and Lesieur 1977; Lesieur 1987) and scalar (Herring et al. 1982; Lesieur and Herring 1985) spectra well. We have adapted the theory to describe the mixing of multiple scalars.

Rigorous application of EDQNM requires that the system be statistically homogeneous. There have been a few attempts to extend the theory beyond this idealized case; however, the approach has been limited to predicting two-point correlations or spectra. Here, our goal is to predict the single-point PDF by modeling the mixing using the EDQNM closure. The present application forces us to relax the rigor with which the EDQNM equations were derived. In its place we propose a *multi-scale* model, wherein each "scale" in the model corresponds roughly to a wavenumber in the EDQNM theory. The correspondence will be exploited to obtain the transfer functions in closed form from the EDQNM theory in such a way as to ensure that the model relaxes to the EDQNM theory under the circumstance that the turbulence is homogeneous and isotropic. For more general flows, the model can be thought of as treating the turbulence as locally isotropic.

The model is still under development; however, the basis for the modeling is described in the next two sections.

4 SPECTRAL MODEL

Scalar mixing involves the interaction of scalar inhomogeneities, at a multitude of length scales, with the energy spectrum. The classical theory of mixing (Kolmogorov 1941; Obukhov 1949; Corrsin 1951) suggests that the scalar spectrum takes on the following form in the inertial-convective range

$$E_\phi(k) = \beta \epsilon^{-1/3} \chi k^{-5/3} , \qquad (1)$$

where ϵ and χ are the turbulent energy and scalar dissipation rates respectively, k is the wavenumber (reciprocal of the length) and β is a universal constant of order unity. Equation (1) is a reasonable approximation for a scalar spectrum that has relaxed into equilibrium, but it can be quantitatively wrong at short times or when the energy spectrum is dynamically changing. Moreover, the above spectrum does not describe multiple scalars that may correlate or decorrelate due to differences in their molecular diffusivities (Bilger and Dibble 1982; Yeung and Pope 1993; Kerstein et al. 1995; Smith et al. 1995; Saylor and Sreenivasan 1998).

Earlier work has shown that the dynamics of the scalar spectrum are well represented by the EDQNM theory (Herr et al. 1996); however, previous studies were restricted to a single scalar. We extended the EDQNM model to account for multiple scalars (Ulitsky and Collins 2000) and encountered a problem with the "realizability" of the covariance spectrum. For a binary system, we define the scalar variance and covariance as

$$\int_0^\infty E_\phi^{\alpha\alpha}(k)dk \equiv \overline{\phi_\alpha'^2} ,$$

$$\int_0^\infty E_\phi^{\alpha\beta}(k)dk \equiv \overline{\phi_\alpha'\phi_\beta'}$$

where $\phi_\alpha' \equiv \phi_\alpha - \overline{\phi_\alpha}$ is the fluctuation in the concentration of species α and $\phi_\beta' \equiv \phi_\beta - \overline{\phi_\beta}$ is the same for species β. Following Yeung and Pope (1993), we define the correlation coefficient and coherency spectrum respectively as

$$\tilde{\rho} \equiv \frac{\overline{\phi_\alpha'\phi_\beta'}}{\sqrt{\overline{\phi_\alpha'^2}\ \overline{\phi_\beta'^2}}} , \qquad \rho(k) \equiv \frac{E_\phi^{\alpha\beta}(k)}{\sqrt{E_\phi^{\alpha\alpha}(k)\ E_\phi^{\beta\beta}(k)}} .$$

The Cauchy-Schwartz inequality requires that: $-1 \leq \tilde{\rho} \leq +1$ and $-1 \leq \rho(k) \leq +1$ is satisfied at each wavenumber. Figure 1(left) shows the results from the standard EDQNM model. Notice that $\rho(k)$ exceeds unity over a range of wavenumbers, in clear violation of the constraint.

Ulitsky and Collins (2000) showed that the problem in the conventional EDQNM model arises due to the Markovianization approximation. Ironically, this approximation renders the single scalar equation *realizable* (i.e., positive definite), but causes the covariance spectrum in a multiple scalar system to violate the Cauchy-Schwartz inequality. Through an analysis based on

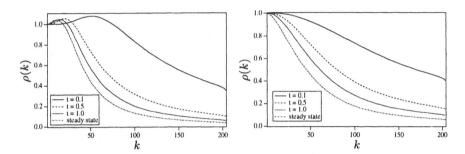

Figure 1: Coherency spectrum from the standard EDQNM model (left) and the revised EDQNM model (right) for two forced isotropic scalars with Schmidt numbers of 1.0 and 0.25 respectively. Notice that $\rho(k)$ exceeds unity over a range of wavenumbers in the former graph but is below unity at all wavenumbers in the latter.

a stochastic representation of the EDQNM model, Ulitsky and Collins (2000) developed a simple fix that involved changing the definition of the inverse time scales in the eddy damping factors. Figure 1(right) shows the modified, realizable model.

A subsequent study (Ulitsky et al. 2001) made extensive comparisons of the realizable model to direct numerical simulations. Virtually all comparisons were in good agreement with the simulations. Figure 2 shows a comparison of the correlation coefficient, $\tilde{\rho}$, as a function of time for the DNS and the EDQNM model.

4.1 Bimolecular Reactions

The EDQNM theory has been extended to the case of binary scalars undergoing a bimolecular reaction with a constant reaction rate constant. The scalar fields obey the following equations

$$\frac{\partial \phi_\alpha}{\partial t} + \nabla \cdot (\mathbf{u}\phi_\alpha) = \mathcal{D}_\alpha \nabla^2 \phi_\alpha - k_R \phi_\alpha \phi_\beta ,$$

$$\frac{\partial \phi_\beta}{\partial t} + \nabla \cdot (\mathbf{u}\phi_\beta) = \mathcal{D}_\beta \nabla^2 \phi_\beta - k_R \phi_\alpha \phi_\beta ,$$

where k_R is constant. This simplified reaction involves only a quadratic nonlinearity that is amenable to the EDQNM analysis. This problem is therefore an interesting testbed for our new mixing paradigm, as it will be possible to compare the complete EDQNM model with the PDF/EDQNM hybrid and assess the strength and limitations of the latter. Figure 3 shows some preliminary calculations of the mean scalar, $\overline{\phi_\alpha}$, and the scalar variance, $\overline{\phi_\alpha'}^2$, as a function of time. Notice the good agreement between the model and DNS.

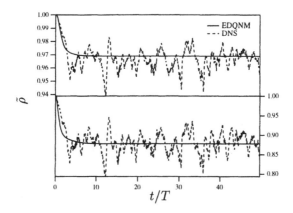

Figure 2: Comparison of the EDQNM prediction for the correlation coefficient, $\tilde{\rho}$, as a function of time with DNS for a Schmidt number ratio of 4 (top) and 16 (bottom), taken from Ulitsky et al. (2001).

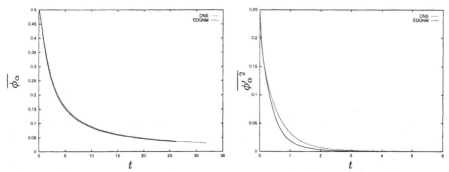

Figure 3: Average scalar and scalar variance for scalars undergoing bimolecular reactions. The solid line is the DNS and the dashed line is the EDQNM model.

5 LANGEVIN EQUATION

The EDQNM model shows promise for describing scalar mixing; however, its direct implementation into a PDF calculation is not feasible without incorporating it into the Monte Carlo procedure that is used to update the PDF. To facilitate this, it is useful to consider the Langevin equation that was developed by Ulitsky and Collins (2000). They showed that for a binary system, the following Langevin equation for scalar concentrations precisely recovers the EDQNM statistics when appropriate averages are taken

$$\frac{\partial \Phi(\mathbf{k}, t)}{\partial t} + \mathbf{A}(k, t) \cdot \Phi(\mathbf{k}, t) = \mathbf{r}(\mathbf{k}, t) \ ,$$

where $\Phi(\mathbf{k}, t) \equiv (\hat{\phi}_\alpha(\mathbf{k}, t), \hat{\phi}_\beta(\mathbf{k}, t))$ is the Fourier transform of the scalar con-

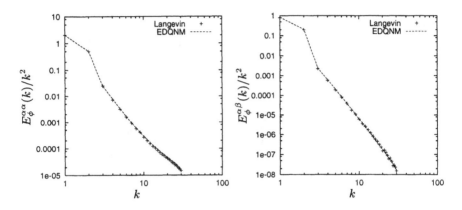

Figure 4: Comparison of 5,000 realizations of the Langevin equation (Eq. 2) prediction for the variance- (left) and covariance- (right) spectra with the EDQNM model.

centration vector, $\mathbf{A}(k,t)$ is a second-order deterministic matrix and $\mathbf{r}(\mathbf{k},t)$ is a random forcing vector. Although this formulation faithfully reproduces the EDQNM statistics, its form is combersome for implementing as a mixing model in a PDF calculation because it requires that the scalar be defined at a representative number of points throughout the *wavevector* volume, which would be prohibitive in a three- or even two-dimensional calculation.

Vaithianathan and Collins (2001) derived a simplified Langevin equation that is only a function of the *wavenumber*. The resulting equation is still consistent with the EDQNM model (highlighting the non-uniqueness of the stochastic differential equation), but only requires that the wavenumber space be sampled

$$\frac{\partial \tilde{\Phi}(k,t)}{\partial t} + \mathbf{A}(k,t) \cdot \tilde{\Phi}(k,t) = \tilde{\mathbf{r}}(k,t) \;, \tag{2}$$

where $\tilde{\Phi}(k,t) \equiv (\phi'_\alpha(k,t), \phi'_\beta(k,t))$ and $\tilde{\mathbf{r}}(k,t)$ depend only on the wavenumber $k \equiv |\mathbf{k}|$ and time. Figure 4 shows a comparison between the Langevin model (Eq. 2 averaged over 5,000 realizations) and the EDQNM model. The agreement confirms the validity of Eq. (2).

6 RADIATION MODELS

In the investigation of turbulence-radiation interactions, a solution technique for the radiative transfer equation (RTE) and a model for the spectral nature of molecular gas radiation are needed. The radiative transfer equation is a spatially and directionally dependent integro-differential equation, which is extremely difficult to solve even on a spectral basis. Several approaches are available to reduce this equation to a simpler form, such as the method of spherical harmonics (P_N), the discrete ordinates method (S_N), and the zonal method among others. Among many solution techniques available, one of the simplest, yet very powerful methods is the P_1 approximation, in which the general radiative transfer equation is transformed to a Helmholtz equation, which is relatively easy to solve. For the vast majority of important engineering problems (i.e., in the absence of extreme anisotropy in the intensity field), this method provides high accuracy at very reasonable computational cost. Furthermore, the P_1 approximation can be easily combined with sophisticated spectral models. This solution approach has been used in our study.

Another challenge in radiation calculation comes from the strong spectral variation of radiation properties. Molecular gases such as water vapor and carbon dioxide exhibit so called 'vibration-rotation bands,' which are formed as a result of many thousands of overlapping and broadened spectral 'lines.' These lines are the result of vibrational and rotational transitions between the different quantum mechanical states of the gas molecule. The absorption coefficient of a gas is far from constant across the spectrum, and therefore the common assumption of a gray medium almost always fails in media where radiative heat transfer is dominated by emitting and absorbing molecular gases. The most accurate approach is to integrate over the spectrum, calculating radiative transfer on a "line-by-line" basis. However, line-by-line calculations are too time-consuming for any practical combustion system. Recently, Modest and Zhang (2000) developed a new global model, the so-called "Full-Spectrum Correlated-k Distribution Method" (FSCK), which is nearly as accurate as the line-by-line calculations, but is much more efficient.

In the FSCK method the absorption coefficient, which oscillates rapidly in spectral space, is reordered as a function of an equivalent fractional Planck function, g, (which is the cumulative distribution function of the absorption coefficient calculated over the whole spectrum and weighted by the Planck function). As a result, the integration over spectral space can be transformed into the integration over g space. For example, the radiative source term in the energy equation can be rewritten as

$$S_{\text{radiation}} = - \int_0^\infty \kappa_\eta (4\pi I_{b\eta} - G_\eta) d\eta = - \int_0^1 k_g u (4\pi a_g I_b - G_g) dg \qquad (3)$$

where κ_η is the spectral absorption coefficient of the radiating gas, $I_{b\eta}$ is the spectral blackbody intensity (or Planck function), and G_η is the spectral incident radiation; the only approximation in this method is the scaling assumption, requiring that spectral and spatial dependencies of the absorption coefficient can be separated, i.e., $\kappa_\eta(\eta, \phi) = k(\eta)u(\phi)$, where k and u are the spectral and spatial dependence of the absorption coefficient, respectively and a in the equation is an FSCK weight factor.

The advantage of the reordering is that the absorption coefficient is much better behaved in reordered coordinates, e.g., it increases monotonically and is free of discontinuities. As a result, far fewer numerical quadrature points

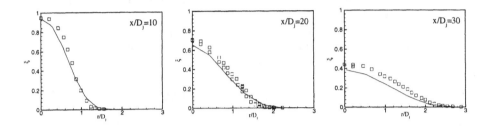

Figure 5: Profiles of mixture fraction at different cross-sections for Flame L from the Masri and Bilger database (see Masri 1997).

are needed—say 10—to capture the spectral dependence of radiative properties in the calculation of Eq. (3) as opposed to almost one million in a typical line-by-line calculation. Another advantage of the FSCK model is that it can be incorporated into any arbitrary RTE solution technique, including the P_1 approximation used here. This novel spectral model has been implemented into our PDF code.

7 CODE VALIDATION

A number of experiments have been designed to study the structure of non-premixed flames. Among them is a series of piloted jet methane flames investigated by Masri and Bilger at Sydney University (Masri and Bilger 1988) and by Barlow and Frank at Sandia National Laboratories (Barlow and Frank 1998). The data on these flames is well documented and available on the *World Wide Web* (see Masri 1997; Barlow 2000). Data include detailed profiles of composition variables at different flame locations, and the data from the second flame also contain some radiative quantities, offering an opportunity to further check our radiation submodels. The hybrid FV/PDF Monte Carlo method was used to simulate these two flames. The velocity field was obtained from the standard k-ϵ model in the finite volume code, while composition variables were computed in the PDF/particle code using a traditional mixing model (Dopazo 1973). Figure 5 shows a typical comparison—computed and measured mixture fraction at different locations for Flame L of Masri and Bilger's database. Considering the complexity of the problem and the fact that no model parameter has been fine tuned to this particular problem, the agreement is reasonably good.

Table 1: Computed total radiative heat loss from the combustor and computed total radiative heat fluxes through the boundaries for various treatments of the TRI terms. Angle brackets are ensemble averages (see Eq. 4) while the overbar refers to the value obtained by substituting the mean scalar values.

different scenarios	$k\langle u\rangle$	$k\langle ua I_b\rangle$	$\iiint \nabla \cdot \underline{q}^R dV$ (KW)	$\iint \hat{n} \cdot \underline{q}^R dS$ (KW)		
				inlet	wall	exit
no TRI	$k\bar{u}$	$k\bar{u}\bar{a}\bar{I}_b$	17.7	71.9	29.8	-84.0
partial TRI-1	$k\langle u\rangle$	$k\bar{u}\bar{a}\bar{I}_b$	14.8	71.6	27.2	-84.0
partial TRI-2	$k\langle u\rangle$	$k\langle u\rangle\bar{a}\bar{I}_b$	17.6	71.8	29.7	-83.9
partial TRI-3	$k\bar{u}$	$k\bar{u}\bar{a}\langle I_b\rangle$	23.6	75.5	33.2	-85.1
partial TRI-4	$k\bar{u}$	$k\bar{u}\langle a I_b\rangle$	19.8	74.3	31.3	-85.8
partial TRI-5	$k\bar{u}$	$k\langle ua I_b\rangle$	32.3	75.0	42.8	-85.5
full TRI	$k\langle u\rangle$	$k\langle ua I_b\rangle$	29.1	74.9	39.7	-85.5

8 INVESTIGATION OF TRI

To couple radiation with flow calculations, the radiative source term in the energy equation, Eq. (3), and the P_1 equation need to be Reynolds averaged. The time averaging process requires two correlations, $\langle ua I_b\rangle$ and $\langle uG\rangle$, which are unclosed in conventional moment methods. For most applications, the optically thin eddy approximation is used, which assumes that the radiative intensity (governed by the entire flow field) and the local absorption coefficient are only weakly correlated, and as a result the second correlation can be further simplified to $\langle u\rangle\langle G\rangle$. This leaves the moments $\langle ua I_b\rangle$ and $\langle u\rangle$ to be determined. These quantities are calculated from the composition PDF as follows

$$\langle u\rangle = \int u f(\underline{\psi})d\underline{\psi}, \qquad \langle ua I_b\rangle = \int ua I_b f(\underline{\psi})d\underline{\psi}, \qquad (4)$$

where $f(\underline{\psi})$ is the PDF of the composition variables that is determined by the Monte Carlo method described above.

The method was first demonstrated by considering a simplified methane/air diffusion flame formed in a cylindrical combustor(Li and Modest 2001). To investigate the significance of TRI, numerical experiments were performed with and without TRI. In the latter case, radiative fluxes were calculated based on the mean temperature and mean species concentrations. Comparing the two calculations, we observed that TRI enhances the flame emission and, as a result, the flame was cooler than the flame calculation that ignored TRI. TRI were found to increase the radiative heat loss by more than 30%.

To dissect the effects of TRI further, we show in Table 1 the radiative flux computed from a single realization in the combustor at one instant of time using various degrees of treatment of the TRI terms. Besides the no-TRI and full-TRI, we show five additional intermediate scenarios in which different levels of approximation of the TRI have been used. In Table 1, the angle bracket refers to the ensemble average (see Eq. 4) while the overbar refers to the value obtained by substituting the mean scalar quantities. Firstly, neglecting turbulence-radiation interactions in turbulent radiating flow calculations is generally unacceptable. The total radiative heat loss would be significantly underpredicted (40% in this case) when TRI are ignored. Secondly, it appears that all contributions to TRI are important and cannot be neglected. For example, neglecting $\langle ua I_b \rangle$ in the TRI-1 case or $\langle u \rangle$ in the TRI-5 case either underpredicts or overpredicts the radiative heat loss substantially. Thirdly, by considering the intermediate scenarios, it is clear that some correlations cancel the contribution of others. Apparently, both emission and absorption are enhanced by TRI. Finally, the consideration of temperature self-correlation (which is embodied in $\langle I_b \rangle$ for gray media and $\langle a I_b \rangle$ for non-gray media) alone is not enough, even though the non-linearity of the Planck function is commonly believed to be the most significant contributor to TRI. Comparing the results of the TRI-3 case and the TRI-4 case, it is also interesting to note that the consideration of $\langle a I_b \rangle$ diminishes the total radiative heat loss. This indicates that TRI effects may be less important in strongly non-gray media than in gray media.

9 SUMMARY

The composition PDF method can treat turbulence-radiation interactions in a rigorous way. To implement this method, first an efficient particle tracing scheme was developed, which allowed the PDF/Monte Carlo method to be incorporated into any finite volume CFD code. The effects of turbulence-radiation interactions were demonstrated by considering a simple methane/air diffusion flame. The results clearly showed the importance of TRI to the overall energy transfer.

A new mixing model based on a multi-scale representation of the scalar has been developed. The nonlinear couplings between the different scales was modeled by appealing to the eddy damped quasi-normal Markovian theory. EDQNM was extended to multiple scalars and a Langevin equation was derived. The EDQNM theory was also developed for a binary system undergoing a bimolecular reaction. This will be used to test the PDF/EDQNM hybrid model currently under development.

10 FUTURE WORK

As a first attempt to capture turbulence-radiation interactions, the fast chemistry approximation has been invoked in our calculations. While the fast-chemistry assumption has been used with great success in combustion modeling (e.g., in Bilger's flamelet model), it is recognized that finite-rate effects may be very important in many practical applications, especially those involving flame ignition and/or extinction. In those circumstances, consideration of chemical kinetics is essential and radiation and turbulence are even more closely coupled: TRI increases emission, lowering the local flame temperature; this in turn dramatically reduces the chemical reaction rate, further suppressing the local flame structure. As a result, turbulence-radiation interactions are expected to be even stronger for situations with "slow" reactions. A two-step reduced chemical mechanism (Dryer and Glassman 1972) is being implemented; the results will be compared to the fast-chemistry approximation.

We are also incorporating the new EDQNM mixing model into the existing PDF/Monte Carlo formulation. This involves defining several "length" scales for each Monte Carlo particle and allowing exchanges of scalar between the different scales (transfer) and between the different particles (mixing). Simultaneously, we are developing numerical procedures to speed up the more costly mixing model so that it will be competitive with current mixing models.

ACKNOWLEDGMENTS

The authors gratefully acknowledge their collaborators: Drs. Mark Ulitsky (Los Alamos National Laboratory), Louis Gritzo and Shawn Burns (Sandia National Laboratory). This work was supported by the National Science Foundation, Grant CTS-9732223.

References

Andre, J. C. and Lesieur, M., 1977, Influence of helicity on the evolution of isotropic turbulence at high Reynolds number. *Journal of Fluid Mechanics*, **81**, pp. 187.

Barlow, R. S., 2000, Web site: `http://www.ca.sandia.gov/tdf/Workshop.html`.

Barlow, R. S. and Frank, J. H., 1998, Effects of turbulence on species mass fractions in methane/air jet flames. *International Symposium on Combustion*, **27**, pp. 1087–1095.

Bilger, R. W. and Dibble, R. W., 1982, Differential molecular diffusion effects in turbulent mixing. *Combustion Science and Technology*, **28**, pp. 161.

Corrsin, S., 1951, On the spectrum of isotropic temperature fluctuations in isotropic turbulence. *Journal of Applied Physics*, **22**, pp. 469.

Dopazo, C., 1973, *Non-isothermal turbulent reactive flows: stochastic approaches*. Ph. D. thesis, State University of New York, StonyBrook.

Dryer, F. L. and Glassman, I., 1972, High-temperature oxidation of CO and CH$_4$. *International Symposium on Combustion*, **14**, pp. 987.

Faeth, G. M., Gore, J. P., Shuech, S. G. and Jeng, S. M., 1989, Radiation from turbulent diffusion flames. *Annual Review of Numerical Fluid Mechanics and Heat Transfer*, **2**, pp. 1–38.

Gore, J. P., Ip, U. S. and Sivathanu, Y. R., 1992, Coupled structure and radiation analysis of acetylene/air flames. *Journal of Heat Transfer*, **114**, pp. 487–493.

Hartick, J. W., Tacke, M., Fruchtel, G., Hassel, E. P. and Janicka, J., 1996, Interaction of turbulence and radiation in confined diffusion flames. *International Symposium on Combustion*, **26**, pp. 75–82.

Herr, S., Wang, L.-P. and Collins, L. R., 1996, EDQNM model of a passive scalar with a uniform mean gradient. *Physics of Fluids*, **8**, pp. 1588–1608.

Herring, J. R., Schertzer, D., Lesieur, M., Newman, G. R., Chollet, J. P. and Larcheveque, M., 1982, A comparitive assessment of spectral closures as applied to passive scalar diffusion. *Journal of Fluid Mechanics*, **124**, pp. 411.

Kerstein, A. R., Cremer, M. A. and McMurtry, P. A., 1995, Scaling properties of differential molecular effects in turbulence. *Physics of Fluids*, **7**, pp. 1999.

Kolmogorov, A. N., 1941, The local structure of turbulence in an incompressible viscous fluid for very large Reynolds numbers. *Dokl. Akad. Nauk. SSSR*, **30**, pp. 299–303.

Lesieur, M., 1987, *Turbulence in fluids, stochastic and numerical modeling*. (Boston: M. Nijhoff).

Lesieur, M. and Herring, J., 1985, Diffusion of a passive scalar in two-dimensional turbulence. *Journal of Fluid Mechanics*, **161**, pp. 77.

Li, G. and Modest, M. F., 2000a, An effective particle tracing scheme in PDF/Monte Carlo methods. In *Proceedings of the 34th National Heat Transfer Conference (NHTC2000)*.

Li, G. and Modest, M. F., 2000b, A hybrid finite volume/PDF Monte Carlo method to capture sharp gradients in unstructured grids. In *Proceedings of the 2000 IMECE, ASME*.

Li, G. and Modest, M. F., 2001, Investigation of turbulence-radiation interactions in reacting flows using a hybrid FV/PDF method. In *Proceedings of the ICHMT 3rd International Symposium on Ratiative Transfer*.

Masri, A. R., 1997, Web site: http://www.mech.eng.usyd.edu.au/research/energy/resources.shtml.

Masri, A. R. and Bilger, R. W., 1988, Turbulent nonpremixed flames of methane near extinction: Mean structure from ramam measurements. *Combustion and Flame*, **71**, pp. 245–266.

Mazumder, S. and Modest, M. F., 1998, A PDF approach to modeling turbulence-radiation interactions in nonluminous flames. *International Journal of Heat and Mass Transfer*, **42**, pp. 971–991.

Modest, M. F. and Zhang, H., 2000, The full-spectrum correlated-k distribution and its relationship to the weighted-sum-of-gray-gases method. In *Proceedings of the 2000 IMECE, ASME.*

Obukhov, A. M., 1949, Structure of the temperature field in turbulent flows. *Izv. Akad. Nauk SSSR, Geogmagn and Geophys. Ser.*, **13**, pp. 58.

Orszag, S. A., 1970, Analytical theories of turbulence. *Journal of Fluid Mechanics*, **41**, pp. 363.

Pope, S. B., 1985, PDF methods for turbulent reactive flows. *Progress in Energy and Combustion Science*, **11**, pp. 119–192.

Saylor, J. R. and Sreenivasan, K. R., 1998, Differential diffusion in low Reynolds number water jets. *Physics of Fluids*, **10**, pp. 1135–1146.

Smith, L. L., Dibble, R. W., Talbot, L., Barlow, R. S. and Carter, C. D., 1995, Laser raman scattering measurements of differential molecular diffusion in turbulent nonpremixed jet flames of H_2/CO_2 fuel. *Combustion and Flame*, **100**, pp. 153–160.

Song, T. H. and Viskanta, R., 1987, Interaction of radiation with turbulence: Application to a combustion system. *Journal of Thermophysics*, **1**, pp. 56–62.

Ulitsky, M. and Collins, L. R., 2000, On constructing realizable, conservative mixed scalar equations using the eddy damped quasi-normal Markovian theory. *Journal of Fluid Mechanics*, **412**, pp. 303–329.

Ulitsky, M., Vaithianathan, T. and Collins, L. R., 2001, A spectral study of differential diffusion of passive scalars in isotropic turbulence. *Journal of Fluid Mechanics*, in review.

Vaithianathan, T. and Collins, L. R., 2001, An efficient Langevin formulation for scalar mixing. *Journal of Fluid Mechanics*, in preparation.

Viskanta, R., 1998, Overview of convection and radiation in high temperature gas flows. *International Journal of Engineering Science*, **36**, pp. 1677–1699.

Yeung, P. K. and Pope, S. B., 1993, Differential diffusion of passive scalars in isotropic turbulence. *Physics of Fluids A*, **5**, pp. 2467.

Material Deterioration in High Temperature Gradients and Transients

Ashwin Hattiangadi[1] and Thomas Siegmund[1]

1 INTRODUCTION

Engineering components considered in high temperature applications have to be designed to withstand expected and hypothetical events such as thermal shock and/or high heat flux. Only if realistic failure criteria are implemented into fully coupled thermo-mechanical simulations the required detailed prediction of life cycle limiting events can be achieved.

It is well established that cracks and delaminations are discontinuities that impede heat flow and subsequent redistribution temperature. This results in local thermal stresses that can induce and promote crack growth. While this process is fundamentally obvious, the actual failure analysis – including crack initiation, growth and arrest - of engineering components under consideration of these aspects is complex. In the past all approaches to this problem have been concerned with detailed stress analysis based on input from a separate heat transfer analysis. Fully coupled thermo-mechanical analyses for non-uniform and non-steady heat flow conditions with arbitrarily growing cracks have not been performed. A methodology that simultaneously accounts for both material failure and the heat transfer problem with moving boundaries due to crack growth is currently not available.

Past investigations of the structural integrity of components subjected to high heat flux and transients were based on applications of the framework of linear-elastic fracture mechanics. Early investigations by Florence and Goodier (1959, 1963) and Sih (1962), were concerned with conditions of steady state and remote uniform heat flux, and assumed full isolated cracks in infinite homogeneous isotropic bodies. The energy release rate, G, for the mode II crack tip fields present in such a case is given by:

$$G = \frac{(1+\nu)\pi Ea^3}{16(1-\nu)}\left(\alpha\frac{\partial\theta}{\partial y}\right)^2 \qquad (1)$$

where E is Young's modulus, ν Poisson's ratio, α the coefficient of thermal expansion and $(\partial\theta/\partial y)$ the temperature gradient normal to the crack. This result,

[1] School of Mechanical Engineering, Purdue University, IN 47907

and subsequent others derived by Barber (1979) and Chen and Huang (1992), demonstrated that the energy release rate associated with thermal effects could indeed be substantial. Subsequently, other investigators extended the linear elastic fracture mechanics analyses to cracks at bi-material interfaces, Lee and Shul (1991), Budde and Gao (1992), Itou (1993), Yan and Ting (1993), Chao and Chang (1994), Lee and Park (1995), Lee and Erdogan (1998), Qian et al. (1997, 98), Miller and Chona (1998), as well as to arbitrary cracks geometries, Hasebe et al. (1986). More importantly in the context of the present analysis, crack heat flow was included into the analyses by Martin-Moran et al. (1983), Barber and Comninou (1983), Kuo (1990), Lee and Park (1995), Hutchinson and Lu (1995), Miller and Chona (1998) and Qian et al. (1998). The results from this group of investigations clearly demonstrate a very strong dependence of the energy release rate on the crack conductance. For example, Hutchinson and Lu (1995) analyzed plate of thickness, H, which contains a long crack, and is loaded by an applied temperature difference (θ_{top}-θ_{bot}) between the top and bottom of the plate. They predict the maximum value of the energy release rate to occur for a crack at a distance 0.211H from the hot surface. With the thermal conductivity of the plate material, k_s, the energy release rate for this situation is given by:

$$G = \frac{EH\left[\alpha\left(\theta_{top} - \theta_{bot}\right)\right]^2}{32\left(1+Bi\right)^2} \tag{2}$$

where Bi is the Biot number defined as Bi=(H h_{cr})/k_s, with h_{cr} the conductivity across the crack. In contrast to the results from Equation (1), Hutchinson and Lu (1995) predict cracks to be under mixed mode loading, $45^{\circ}<\psi<90^{\circ}$, as long as the crack remains less than 0.5 H from the hot surface. These results clearly demonstrate a strong dependence of the structural response on the crack conductance, and thus the necessity to incorporate a correct description of h_{cr} into failure predictions of structures under high heat flux loading. As in Equation (2), crack conductance, however, has only been considered in simplified ways in the past. It was generally assumed to be constant, not only along the entire crack but also independent of crack opening, and rarely based on physically motivated input. Since variable crack conductance were not considered, nor was crack growth, sequential solutions of the temperature and displacement problems were sufficient. If, however, integrated multi-physics life cycle predictions of critical components are the goal, these simplifying assumptions have to be overcome.

The methods currently available for the analysis of fracture under thermal loading are in a significant contrast to the state of knowledge in fracture analysis under pure mechanical loading. There, the development of physically mechanism based material separation models for use in continuum mechanics crack growth simulations has greatly enhanced the capabilities of failure predictions for a broad range of materials. One of the promising approaches developed within these efforts is the so-called cohesive zone model. Its success stems from the fact that the cohesive zone model approach allows one to incorporate a broad range of different physical processes of material separation within a single numerical method. Based on early ideas of Barenblatt (1962) and Dugdale (1960), cohesive zone models were used by several investigators for studies of fracture under pure mechanical loading, e.g., Needleman (1987, 90), Tvergaard and Hutchinson (1992, 94), Xu and Needleman (1994), Camacho and Ortiz (1996), Siegmund and Needleman (1997), Geubelle and Baylor (1998), Chaboche et al. (1997), Siegmund and Brocks (1999), Rahul-Kumar et al. (2000), and Pandolfi et al. (2000). The softening constitutive equations of the cohesive zone are phrased in terms of a traction-separation law. The main cohesive constitutive parameters in this law, cohesive strength and separation energy per unit area, are characteristic of the underlying material separation processes. A length-scale, i.e. separation energy per unit area divided by cohesive strength, is directly part of the constitutive equation. This cohesive length is the material separation distance after which the stress carrying capacity of the material falls to zero. As a consequence of the presence of a length scale in the constitutive equations, convergent results with respect to the finite element mesh can in principle be obtained. All past applications of cohesive zone models as outlined above were directed towards the analysis of mechanical loading only. It is the purpose of this article to demonstrate that the cohesive zone model approach is indeed promising in cases where fracture is coupled to and influenced by other physical processes. To enable coupled thermomechanical analyses which account for the creation of new free surfaces due to crack growth, or break down of crack bridging, and the subsequent changing heat transfer conditions, a description of conductance across the crack, need to be defined and implemented. These cohesive zone conductance laws capture the energy transport across cracks or delaminations, and provide additional coupling between thermal and stress analysis. Variations in cohesive zone conductance of the order of several magnitudes occur depending on the stage of the processes associated with material failure. Material deterioration taking place in the cohesive zone leads to both large changes in stress carrying capability and conductance. As load transfer breaks down, the conductance in an open crack subsequently depends on the presence of a gas and/or radiative heat transfer. Crack face contact occurring at interfaces, under mixed mode loading and during unloading is to be accounted for.

While such a framework is potentially useful in series of engineering problems, the present article focuses on the application of the approach to studies of fiber reinforced ceramic matrix composites. Here, the influence of the fracture processes can be especially important due to extended crack bridging. Fiber reinforced ceramic matrix composites are promising candidates in high temperature applications for thermal protection systems for space vehicles, Freeman et al. (1997), Sumrall et al. (1999), subjected to heat flux of up to 100 W/cm^2, Papadopoulos et al. (1999); or actively cooled first wall liners in fusion reactors with expected peak heat fluxes of up to 15,000 W/cm^2, Tivey (1999). In addition to their thermal stability, fiber reinforced ceramic matrix composites also offer certain flexibility in microstructural design. With respect to the present investigation this is of interest since crack bridging - due to cross stitches, multidirectional microstructures or unbroken fibers in laminates - play an important role in both load, and heat transfer.

The present paper describes the general framework for coupled thermomechanical crack growth simulations within the cohesive zone model approach. The model is subsequently applied to a numerical study of the mechanics of bridging cracks in unidirectionally reinforced laminates.

2 FORMULATION

The finite element formulation and numerical method for coupled thermomechanical crack growth studies will extend the cohesive surface formulation established in the references given in the previous section. The mechanical equilibrium statement written as the principle of virtual work is given by:

$$\int_V \mathbf{s} : \delta\mathbf{F} \; dV - \int_{S_{int}} \mathbf{T}_{CZ} \cdot \delta\mathbf{\Delta} \; dS = \int_{S_{ext}} \mathbf{T}_e \cdot \delta\mathbf{u} \; dS \tag{3}$$

with the nominal stress tensor, $\mathbf{s} = \mathbf{F}^{-1} \det(\mathbf{F})\boldsymbol{\sigma}$, with $\boldsymbol{\sigma}$ the Cauchy stress, the displacement vector, $\mathbf{u}$, the deformation gradient, $\mathbf{F}$, as well as by the traction vector, $\mathbf{T}_e$, on the external surface of the body. Traction vectors are related to $\mathbf{s}$ by $\mathbf{T} = \mathbf{n} \, \mathbf{s}$, with $\mathbf{n}$ being the surface normal. The cohesive zone contributions over the internal surface, S_{int}, are given by the cohesive surface traction, $\mathbf{T}_{CZ}$, and the normal and tangential displacement jump across the cohesive surface, $\mathbf{\Delta} = (\Delta_n \mathbf{n} + \Delta_t \mathbf{t}) = \mathbf{u}^+ - \mathbf{u}^-$, with $\mathbf{u}^+$ and $\mathbf{u}^-$, the displacement discontinuity between two initially bonded points on the opposite crack surfaces.

In the unidirectionally fiber reinforced laminate of concern in the present study, the failure mechanism acting under normal loading of the bridging zone is that of fiber bending, peel from the matrix, and subsequent fracture at the fiber root due to bending stresses. In Kaute et al. (1993), these mechanisms were be characterized by a relationship between the force transmitted per fiber and the crack opening. At small values of normal separation jump across the cohesive zone, fiber bending leads to a linear relation between force per fiber and normal separation jump. Subsequently, peeling the fiber from the matrix leads to a steady-state plateau force. Under shear separation it is assumed that fiber pullout dominates, leading to a linear relation between force per fiber and the shear displacement jump. Failure is introduced via a Weibull function, which provides the number of fibers per area in dependence of the displacement jumps. Combining this statistical distribution with the force per fiber – displacement jump functions for the individual fibers the traction-separation law is obtained. The traction-separation law Needleman (1990) used here is based on a potential, ϕ:

$$\phi(\mathbf{\Delta}) = \frac{9}{16}\sigma_{max}\delta\left\{1 - \left[1 + z\left(\frac{\Delta_n}{\delta}\right) - \frac{1}{2}\alpha z^2\left(\frac{\Delta_t}{\delta}\right)\right]\exp\left[-z\left(\frac{\Delta_n}{\delta}\right)\right]\right\} \tag{4}$$

with $z=16e/9$ and $e=\exp(1)$. σ_{max} and δ denote cohesive strength and length, respectively. α is a ratio of the initial stiffness of the normal and shear separation functions, here $\alpha=1$ throughout the investigation. The normal and shear tractions are the derivatives of ϕ with respect to Δ_n and Δ_t, respectively. Within this framework, damage, i.e. the density of failed fibers, can be defined as

$$D = \frac{\phi[\max(\Delta_n), \max(\Delta_t)]}{\phi_{sep}} \tag{5}$$

where ϕ_{sep} is the separation energy, and D a damage variable. For the traction-separation law given in Equation (4):

$$\phi_{sep} = \frac{9}{16}\sigma_{max}\delta \tag{6}$$

Figure 1 depicts the normal traction-separation function together with the damage variable for the cases $\Delta_t=0.0$ and $\Delta_t=\Delta_n/2$.

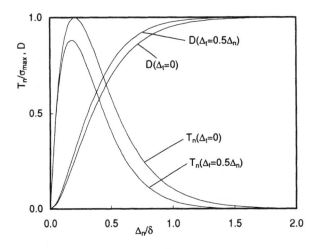

Figure 1 Dependence of normal traction and damage on normal separation.

In addition to the mechanical equilibrium equation the system must fulfill thermodynamic equilibrium, i.e. the first law of thermodynamics must be satisfied. In the variational form and using Fourier's law, this can be expressed by:

$$\int_V \rho c_p(\partial_t\delta\theta)\,\delta\theta\ dV - \int_{S_{int}} h_{cz}\,\Delta\theta\,\delta\Delta\theta\ dS + \int_V \frac{\partial\delta\theta}{\partial x}\cdot k\cdot\frac{\partial\delta\theta}{\partial x}dV = \int_V \delta\theta\,r\,dV + \int_{S_{ext}} \delta\theta\,q\,dS \tag{7}$$

Here, the standard parts of the equation (including the terms on V and S_{ext}) are described by the temperature field, θ, heat capacity, c_p, the conductivity matrix, **k**, the heat flux per unit area of the body flowing into the body, q, and the heat supplied externally into the body per unit volume, r. The cohesive contributions, representing the crack and the cohesive zone in front of the crack tip and in the bridged crack wake, are described by the integral over the internal surface, S_{int}, in terms of the cohesive zone conductance, h_{CZ}, and the temperature jump across the cohesive surface, $\Delta\theta$.

Coupling between stress and heat transfer part of the problem, Equations (3) and (7), respectively, occurs not only the thermal stresses but also via the cohesive zone conductance, h_{CZ}. This quantity, in general, is dependent on the cohesive zone traction, **T**, the displacement jump, Δ, and temperature jump across the cohesive surface, $\Delta\theta$:

$$h_{CZ} = f(\mathbf{T}, \mathbf{\Delta}, \Delta\theta) \tag{8}$$

In the present investigation only heat conductance mechanism, through gas in the crack and fibers, are assumed to contribute to h_{CZ}. The cohesive zone conductance model employed here is based on a thermal resistor model is used to account for the effect of fiber bridging on cohesive zone conductance given by:

$$h_{CZ} = V_f(1-D)\left(\frac{k_f \sin\gamma}{\Delta_n}\right) + [1 - V_f(1-D)]\left(\frac{k_g}{\Delta_n}\right) \tag{9}$$

where V_f is the fiber volume fraction, k_f and k_g the conductivity of fibers and gas, respectively. γ represents the angle of inclination of the bridging fibers relative to the crack surface. Equation (9) differs from the expression in the recent investigation of Lu and Hutchinson (1995), Donaldson et al. (1998), and McDonald et al. (2000) in that the relative contributions of the heat transfer through fibers and gas changes as material separation increases and material deterioration progresses. With this framework, the Biot number is no longer a constant but dependent on the mechanics of the bridging zone.

The heat transfer cohesive elements were implemented into the finite element code ABAQUS via the UEL capability. The implementation uses four noded elements with identical linear interpolation functions for the displacements and temperature. The continuum elements used were four node plane strain temperature-displacement elements. ABAQUS uses a backward-difference scheme to integrate temperatures, and the coupled system is solved using Newton's method, accounting for nonsymmetric stiffness matrix. Coupled temperature-displacement solutions in steady state were computed.

3 RESULTS AND DISCUSSION

The problem investigated is that of a plain strain crack in a fiber reinforced laminate of thickness H=2 mm. The crack possesses a length of 2a=15H, initially is fully bridged, and contains air.

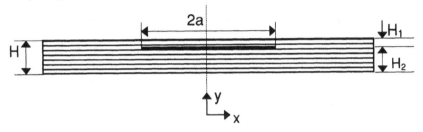

Figure 2 Specimen geometry.

The position of the crack is H_1=0.211H from the upper hot plate surface. Figure 2 depicts the specimen geometry. Uniform temperatures, θ_{top} and θ_{bot}, are applied to the top and bottom surfaces of the plate, respectively, with the maximum applied temperature difference 1200K. The material properties used are approximate values for an Al_2O_3 ceramic reinforced with SiC fibers. The properties of the solid were taken to be E=200 GPa, v=0.25, α=8E-6/K, k_s=30 W/(mK). In view of the parametric study, the material properties characterizing the traction separation behavior aim to provide a range of values possible for laminates with fiber bridging, σ_{max}=0.1, 0.4 or 1.0 MPa, δ=25, 50, 400 or 1000 μm. Kaute et al. (1993) found σ_{max}=0.4 MPa and δ=400 μm in their experimental study. The volume fraction of the bridging fibers was assumed to be V_f=0.3. The values for the conductivity of fiber and gas are k_f=1.6 W/(mK), and k_g=0.032 W/(mK), respectively. All properties were assumed independent of temperature. The fiber inclination angle γ was assumed to be constant and equal to 10°, McDonald et al. (2000).

Figure 3 depicts the predicted values of the energy release rate in dependence of the applied temperature difference for a cohesive strength of σ_{max}=0.1 MPa, and several values of the cohesive length. In addition to the predictions for the cracks bridged with fibers, Figure 3 shows the results obtained for a specimen with a crack containing air but without bridging, and those for a fully insulating crack. For the ideally insulating crack, a non-zero value of energy release rate is predicted, independent of the magnitude of the applied temperature difference. This type of behavior is consistent with the results given in Equation (2), which predicts the same type of behavior for any value of the Biot number. This is contrast to the present numerical results incorporating the cohesive zone conductance. If the cohesive zone conductance in dependence as of Equation (9) is accounted for, the crack stays closed and the energy release rate remains zero below a critical applied temperature difference. For the crack in which only air conducts heat across the

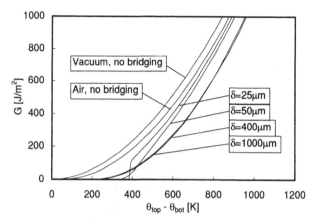

Figure 3 Predicted values of the energy release rate in dependence of the applied temperature difference for σ_{max}=0.1 MPa and various values of δ, a crack with air only, and a fully insulating crack.

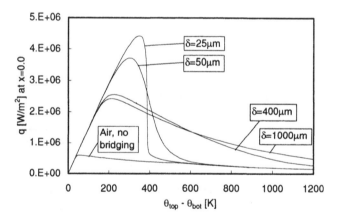

Figure 4 Predicted values of the specific heat flux at x=0.0 in dependence of the applied temperature difference for σ_{max}=0.1 MPa and various values of δ, a crack with air only, and a fully insulating crack.

crack, this critical temperature difference is small, approximately 100K. Similarly, the existence of a critical applied temperature difference was also predicted in Hutchinson and Lu (1995). However, in their analysis a correlation between crack opening and energy release rate based on a solution for a crack in an infinite medium was used. As a consequence, only small values of crack opening are predicted, and the critical temperature differences were considerably larger. For the bridged cracks the critical applied temperature difference is approximately two to four time larger than in the case of the crack containing air only. Nevertheless,

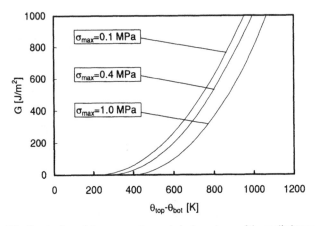

Figure 5 Predicted values of the energy release rate in dependence of the applied temperature difference for δ=400 μm and various values of σ_{max}.

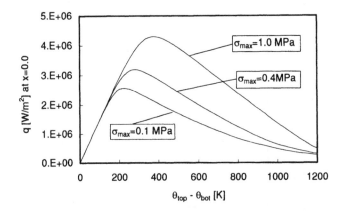

Figure 6 Predicted values of the specific heat flux at x=0.0 in dependence of the applied temperature difference for δ=400 μm and various values of σ_{max}.

even for the bridged cracks the energy release rate increases sharply after the critical temperature difference is passed. Significant levels of G are present, such that ceramic materials might in fact fail under these conditions. In the case of small δ, while the critical temperature difference was largest, the energy release rate rises quickly to values similar to those for the crack containing air only, and the beneficial effects of bridging fibers is lost. For the case with large cohesive length, the lowest values of the energy release rate are predicted. The desirable behavior of the specimen with δ=400 and 1000 μm can be contributed to two factors, the reduction of G due to continuous shielding in the presence of bridging tractions,

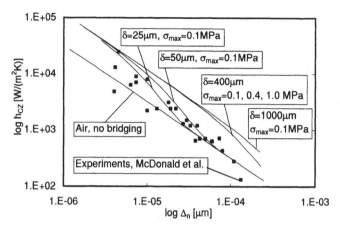

Figure 7 Predicted values of the cohesive zone conductance at x=0.0 in dependence of the normal separation. Comparison of predictions for the crack with air only, several bridging parameter combinations, and experimental data from McDonald et al. (2000).

and, more importantly, the contribution of the fibers to the cohesive zone conductance.

Figure 4 depicts the specific heat flux across the crack at x=0.0. For the insulating crack, obviously q=0.0. For the crack with air only, the specimen is fully conducting and obeys the Fourier law of the undisturbed solid during the initial stages of loading only. Subsequently, the heat flux breaks down and the crack becomes increasingly insulating as the crack opening increases with increased applied temperature difference. For the bridged cracks, ideal heat flux is sustained to higher applied temperature differences since the cracks are kept closed longer. Subsequently, a strong reduction of the heat flux occurs in the case of small values of δ as the crack opens, and damage accumulates. For small values of cohesive length, the heat flux drops quickly to that of a crack without any prior bridging. For the case of large δ the crack opens, too, however not enough to rapidly induce sufficient damage. Fibers continue to bridge the crack, and the heat flux remains at relatively large values up to large applied temperature differences. This, in terms of the results of Figure 3, causes a smaller temperature jump across the crack, and thus lowers values of energy release rates.

Figure 5 and 6 show the results corresponding to Figure 3 and 4, but for a fixed δ=400 μm, and several values of the cohesive strength. Figure 5 demonstrates that increasing the cohesive strength by a factor of ten increases the critical applied temperature difference by a factor of approximately 1.5 only. Obviously, the energy release rate is also reduced as σ_{max} is increased. It is, however, more interesting to observe here that crack opening, and the subsequent reduction in the cohesive zone conductance occur independent of the value of cohesive strength assumed, and the heat flux always reaches a maximum at the critical applied temperature difference and subsequently decreases, Figure 6.

The behavior of the bridged cracks can also be characterized in a plot of the cohesive zone conductance vs. normal separation. Figure 7 depicts such a plot for the location x=0.0, and for several conditions describing the cohesive zone. In addition, experimentally determined data from McDonald et al. (2000) are included for comparison. For small values of normal separation the bridged cracks possess a cohesive zone conductance of more than an order of magnitude larger than the crack containing with air only. For the cases δ=25 and 50 μm and low cohesive strength, damage via fiber breakage occurs as the crack opens, and the conductance is gradually reduced to that for the case of air only. Only if the cohesive length of the bridging law is large, the cohesive zone conductance remains at higher levels up to large normal separation. An increase in the cohesive strength had little, if any effect on the cohesive zone conductance, at least for the case of δ=400 μm. The data of McDonald et al. (2000) were obtained by loading a unidirectionally reinforced laminate in four-point bend to open the crack, and simultaneously measuring the heat transfer across the crack. This allows a comparison with the present predictions, since in temperature effects can be ignored. The experimental data are close to the conductance values for a crack with air only for large crack openings. For small crack opening large scatter in the data is present, nevertheless in principle the measurements show an increase of conductance across the crack similar to the present model.

4 CONCLUSIONS

A framework for investigations of fracture under thermomechanical loading has been outlined. Coupling between stress analysis and heat transfer analysis occurs not only due to thermal stresses and temperature dependent material properties, but also due to the changes in the conductance across cracks and delaminations. The coupling between load transfer across bridged crack wakes and across fracture process zones is described by the use of a cohesive zone conductance law. This relation accounts for both crack opening as well as the changing relative contributions of conductance mechanisms as the material deteriorates. The proposed framework is of interest especially in material where a large bridging zone exists.

In a first step in the exploration of the capabilities provided by this model, the mechanics of bridged cracks under thermal loading has been investigated for the case of a unidirectionally fiber reinforced ceramic matrix composite laminate. Crack bridging was found to influence the specimen behavior in several aspects. It not only provides mechanical shielding of the crack tip, but also holds the crack closed. This mechanism is beneficial in that no disturbance of the temperature gradient occurs, and the energy release rate remains zero. However, above a critical applied temperature difference, this mechanism can no longer be sustained. Due to the fact that the stiffness of the bridging mechanism is much smaller than the elastic stiffness of the solid, it is inevitable that the crack opens. Then, the heat flux across the crack breaks down, and the cohesive zone conductance tends to approach that of a crack without any bridging at all. Both crack opening itself as well as damage - depending on the mechanical response of the cohesive zone - contribute to this effect. Also, damage can be envisioned to result from either secondary mechanical loads, or previous events such as e.g. foreign-object impact or fatigue. For the laminate considered here, this transition from a conducting to an insulating crack occurs first at the center of the specimen and gradually proceeds towards the tip of the bridging zone. As the shielding of the crack tip is reduced and the applied temperature loading is increased, the conditions for crack propagation will thus be met, and further crack extension will set in. Overlying a second layer of cohesive zone elements along the assumed crack path, an aspect that has not been investigated yet, can then capture this stage of failure within the present modeling approach.

The effects of temperature on material properties and heat transfer mechanisms were not included in the present study. This is especially important with respect to the cohesive zone conductance law, Equation (9). With increasing temperature the conductivity of the solid as well as of the gas will change, and radiation is to be accounted for. The large number of possible interactions between the individual mechanical and thermal effects results warrant further investigations.

From a technology point of view the described framework for coupled thermomechanical investigations of material failure might potentially benefit many different applications under high heat flux conditions such as composite shell structures in gas turbines and rocket engines, high performance composite brakes, thermal barrier coatings to micro- and power electronic devices, as well as thermal shock loaded components such as ceramic superconductors. However, even bioengineering applications such as thermal shock of dental structures, or the shock freezing of tissue could benefit from the approach outlined here.

ACKNOWLEDGEMENTS

The financial support of Sandia National Laboratories through the NSF-Sandia LCE program is gratefully acknowledged.

5 REFERENCES

Barber, J.R., 1979, Steady-state thermal stresses caused by an imperfectly conducting penny shaped crack in an elastic solid. *Journal of Thermal Stresses*, **3**, pp. 77-83.

Barber, J.R. and Comninou, M., 1983, The penny-shaped interface crack with heat flow: 2. Imperfect contact. *Journal of Applied Mechanics*, **50**, pp. 770-776.

Barenblatt, G.I., 1962, The mathematical theory of equilibrium cracks in brittle fracture, *Advances in Applied Mechanics*, **7**, pp. 55-129.

Budde, L. and Gao, S.M., 1992, Dynamic thermal stress in metal-adhesive bonds. *Journal of Adhesion Science and Technology*, **6**, pp. 1189-1204.

Camacho, G.T. and Ortiz, M., 1996, Computational modeling of impact damage in brittle materials. *International Journal of Solids and Structures*, **33**, pp. 2899-2938.

Chao, C.K. and Chang, R.C., 1994, Analytical solutions and numerical examples for thermoelastic interface crack problems in dissimilar anisotropic media. *Journal of Thermal Stresses*, **17**, pp. 285-299.

Chaboche, J.L., Girad, R. and Schaff, A., 1997, Numerical analysis of composite systems by using interphase/interface models. *Computational Mechanics*, **20**, pp. 3-11.

Chen, W.H. and Huang, C.C., 1992, Three-dimensional thermal analysis of an infinite solid containing an elliptical surface crack. *International Journal of Fracture*, **54**, pp. 225-234.

Donaldson, K.Y., Trandell, B.D., Lu, Y., Hasselman, D.P.H., 1998, Effect of delamination on the transverse thermal conductivity of a SiC-fiber-reinforced SiC-matrix composite, *Journal of the American Ceramic Society*, **81**, pp. 1583-1588.

Dugdale, D.S., 1960, Yielding in steel sheets containing slits, *Journal of the Mechanics and Physics of Solids,* **8**, pp. 100-104.

Freeman, D.C., Talay, T.A., Austin, R.E., 1997, Reusable launch vehicle technology program, *Acta Astronautica*, **41**, pp. 777-790.

Florence, A.L. and Goodier, J.N., 1959, Thermal stresses at spherical cavities and circular holes in uniform heat flow. *Journal of Applied Mechanics*, **26**, pp. 293-294.

Florence, A.L. and Goodier, J.N., 1963, The linear thermoelastic problem of uniform heat flow disturbed by a penny-shape insulated crack. *International Journal of Engineering Sciences*, **1**, pp. 553-540.

Geubelle, P.H., Baylor, J.S., 1998, Impact-induced delamination of composites: a 2D simulation, *Composites*, **29B**, pp. 589-602.

Hasebe, N., Tamai, K. and Nakamura, T., 1986, Analysis of kinked crack under uniform heat-flow. *Journal of Engineering Mechanics*, **112**, pp. 31-42.

Hutchinson, J.W. and Lu, T.J., 1995, Laminate delamination due to thermal gradients. *Journal of Engineering Materials and Technology*, **117**, pp. 386-390.

Itou, S., 1993, Thermal stresses around a crack in an adhesive layer between two dissimilar elastic half-planes. *Journal of Thermal Stresses*, **16**, pp. 373-400.

Kaute, D.A.W., Shercliff, H. R. and Ashby, M.F., (1993), Delamination, fibre bridging and toughness of ceramic matrix composites. *Acta Metallurgica et Materialia*, 41(7), pp.1959-1970.

Kuo, A.Y., 1990, Effects of crack surface heat conductance on stress intensity factors. *Journal of Applied Mechanics*, **57**, pp. 354-358.

Lee, K.Y., and Shul, C.W., 1991, Determination of thermal stress intensity factors for an interface crack under vertical heat-flow. *Engineering Fracture Mechanics*, **40**, pp. 1067-1074.

Lee, K.Y., and Park, S.J., 1995, Thermal-stress intensity factors for partially insulated interface crack under uniform heat-flow. *Engineering Fracture Mechanics*, **50**, pp. 475-482.

Lee, Y.D. and Erdogan, F., 1998, Interface cracking of FGM coatings under steady-state heat flow. *Engineering Fracture Mechanics*, **59**, pp. 361-380.

Lu, T.J. and Hutchinson, J.W., 1995, Effect of matrix cracking on the overall thermal conductivity of fibre-reinforced composites. *Philosophical Transactions of the Royal Society of London A*, **351**, pp. 595-610.

Martin-Moran, C.J., Barber, J.R. and Comninou, M., 1983, The penny-shaped interface crack with heat flow: 1. Perfect contact. *Journal of Applied Mechanics*, **50**, pp. 29-36.

McDonald, K.R., Dryden, J.R., Majumdar, A. and Zok, F.W., 2000, Thermal conductance of delamination cracks in a fiber-reinforced ceramic composite. *Journal of the American Ceramic Society*, **83**, pp. 553-562.

Miller, T.C. and Chona, R., 1998, Finite element analysis of a thermally loaded interface crack in a ceramic coating. *Engineering Fracture Mechanics*, **59**, pp. 203-214.

Needleman, A., 1987, A continuum model for void nucleation by inclusion debonding. *Journal of Applied Mechanics*, **54**, pp. 525-531.

Needleman, A., 1990, An analysis of decohesion along an imperfect interface. *International Journal of Fracture*, **42**, pp. 21-40.

Pandolfi, A., Guduru, P.R., Ortiz, M. and Rosakis, A.J., 2000, Three dimensional cohesive-element analysis and experiments of dynamic fracture in C300 steel. *International Journal of Solids and Structures*, **37**, pp. 3733-3760.

Papadopoulos, P., Venkatapathy, E., Prabhu, D., Loomis, M.P. and Olynick, D., 1999, Current grid-generation strategies and future requirements in hypersonic vehicle design, analysis and testing. *Applied Mathematical Modeling* **23**, pp. 705-735.

Qian, G., Nakamura, T., Berndt, C.C. and Leigh, S.H., 1997, Tensile toughness test and high temperature fracture analysis of thermal barrier coatings. *Acta Materialia*, **45(4)**, pp. 1767-1784.

Qian, G., Nakamura, T. and Berndt, C.C., 1998, Effects of thermal gradient and residual stresses on thermal barrier coating fracture. *Mechanics of Materials*, **27(2)**, pp. 91-110.

Rahul-Kumar, P., Jagota, A., Bennison, S.J., Saigal, S., 2000, Interfacial failures in a compressive shear strength test of glass/polymer laminates, *International Journal of Solids and Structures*, **37**, pp. 7281-7305.

Rice, J.R.,1968, Mathematical analysis in the mechanics of fracture. Ed. Liebowitz,H. *Fracture*, **2**, Academic Press, New York, pp. 191-311.

Sih, G.C., 1962, On the singular character of thermal stresses near a crack tip. *Journal of Applied Mechanics*, **29**, pp. 587-589.

Siegmund, T. and Needleman, A., 1997, A numerical study of dynamic crack growth in elastic viscoplastic solids, *International Journal of Solids and Structures*, **34**, pp. 769-787.

Siegmund, T. and Brocks, W., 1999, Prediction of the work of separation and implications to the modeling. *International Journal of Fracture*, **99**, pp.97-116.

Sumrall, J., Lane, C. and Cusic, R., 1999, Venturestar[TM]: reaping the benefits of the X-33 program. *Acta Astronautica*, **44**, pp. 727-736.

Tivey, R., Ando, T., Antipenkov, A., Barabash, V., Chiocchio, S., Federici, G., Ibbott, C., Jakeman, R., Janeschitz, G., Raffray, R., Akiba, M., Mazul, I., Pacher, H., Ulrickson, M. and Vieider, G., 1999, ITER divertor, design issues and research and development. *Fusion Engineering Design*, **46**, pp. 207-220.

Tvergaard, V. and Hutchinson, J.W., 1992, The relation between crack growth resistance and fracture process parameters in elastic-plastic solids. *Journal of Mechanics and Physics of Solids*, **40**, pp. 1377-1397.

Tvergaard, V. and Hutchinson, J.W., 1994, Effect of T-stress on mode I crack growth resistance in a ductile solid. *International Journal of Solids and Structures*, **31**, pp. 823-833.

Yan, G. and Ting, T.C.T., 1993, The r(-1/2)ln(r) singularity at interface cracks in monoclinic and isotropic bimaterials due to heat-flow. *Journal of Applied Mechanics*, **60**, pp. 432-437.

Xu, X.P. and Needleman, A., 1994, Numerical simulations of fast crack growth in brittle solids. *Journal of Mechanics and Physics of Solids*, **42**, pp. 1397-1434.

Part III

Optimization and Uncertainty Quantification and Design

Protection Against Modeling and Simulation Uncertainties in Design Optimization

Hongman Kim[1], William H. Mason[1], Layne T. Watson[1], Bernard Grossman[1], Melih Papila[2] and Raphael T. Haftka[2]

ABSTRACT

Probabilistic models are applied to errors of optimal wing structural weight of a high-speed civil transport (HSCT). A Weibull model is successfully fit to the errors from structural optimization runs. The probabilistic model enables us to estimate average errors without performing very accurate optimization runs. In addition, modeling deficiencies of response surface models are analyzed. A study of the mean squared error yields an eigenvalue problem where the maximum eigenvalue at each point provides a bound for the bias error. A polynomial example and optimal wing structural weight of the HSCT are used to demonstrate the approach.

1. INTRODUCTION

Response surface (RS) models (*c.f.*, Myers and Montgomery, 1995) based on optimizations are now commonly used in engineering design. The availability of many optimization runs and the RS approximation provide an opportunity to estimate errors. The first part of this paper shows how we can use the runs to estimate the average optimization error, and the second part shows how we can estimate the error associated with fitting the data with low-order polynomials.

Numerical noise is inherent in many computational simulations. Computational simulation errors are deterministic in that the outputs are the same for the same input for repeated runs. However, simulation results can be very sensitive to small changes of input parameters, resulting in noisy and unpredictable output. Hence, a probabilistic model can be useful to characterize noise errors from computational simulations. We apply probabilistic models to the errors from a structural optimization procedure of a high-speed civil transport (HSCT).

Response surface approximations traditionally use low-order models such as quadratic polynomials, which may not always represent well the behavior of complex engineering systems. On the other hand, higher order models are generally impractical in high dimensional problems. Information concerning design regions where the low-order models incur large errors is therefore valuable. We derive a measure to determine such locations in the design space.

[1] Virginia Polytechnic Institute and State University, Blacksburg, Virginia
[2] University of Florida, Gainesville, Florida

2. HSCT OPTIMAL WING STRUCTURAL WEIGHT PROBLEM

The application problem in this paper is a 250-passenger HSCT design with a 5500 nautical miles range and cruise Mach number 2.4. A general HSCT model (Balabanov *et al.*, 1999) developed by the Multidisciplinary Analysis and Design (MAD) Center for Advanced Vehicles at Virginia Tech includes 29 configuration design variables. Of these, 26 describe the geometry, two the mission, and one the thrust. The takeoff gross weight (W_{TOGW}) was selected as the objective function to be minimized. We studied two simplified versions of the problem. First, a five-variable case following Knill *et al.* (1999), includes fuel weight, W_{fuel} and four wing shape parameters: root chord, c_{root}, tip chord, c_{tip}, inboard leading edge sweep angle, Λ_{ILE}, and the thickness to chord ratio for the airfoil, t/c. Second, a two-variable case includes c_{root} and Λ_{ILE} (Papila and Haftka, 2001). Figure 1 describes the five-variable HSCT design problem. The other variables such as fuselage, vertical tail, mission and thrust related parameters are kept unchanged at the baseline values. Table 1 shows the values and ranges of the design variables.

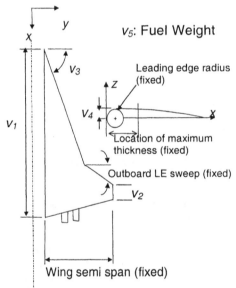

Figure 1: Configuration design variables for the five-variable HSCT problem.

W_{TOGW} is estimated using the weight equations from the Flight Optimization System (FLOPS) (McCullers, 1997) program. However, FLOPS weight equations were not accurate for the HSCT, particularly for the wing weight as a function of the wing planform shape, and structural optimization was adopted to obtain more accurate wing structural weight (W_s) (Balabanov *et al.*, 1999). GENESIS structural optimization software (VR&D, 1998) was used with a finite element (FE) model to minimize W_s. The structural optimization is a sub-optimization below the system level configuration design. The structural optimizations are performed *a priori* for many aircraft configurations and a response surface model is constructed as a function of the configuration design variables. The FE model uses 40 design variables, including 26 to control skin panel thickness, twelve to control spar cap areas, and two for the rib cap areas (Balabanov *et al.*, 1999). The HSCT codes calculate aerodynamic loads for each of four load cases used, and a mesh generator calculates the FE mesh and the applied load at the structural nodes, and creates the input for GENESIS.

Table 1: Simplified versions of the HSCT design problem.

HSCT configuration design variable (Total 29 variables)	Five-variable problem	Two-variable problem
Planform Variables		
Root chord, c_{root}	**150-190 ft**	**150-190 ft**
Tip chord, c_{tip}	**7-13 ft**	10 ft
Wing semi span, $b/2$	74 ft	74 ft
Length of inboard LE, s_{ILE}	132 ft	132 ft
Inboard LE sweep, Λ_{ILE}	**67°– 76°**	**67°– 76°**
Outboard le sweep, Λ_{OLE}	25°	25°
Length of inboard TE, s_{ITE}	Straight TE	Straight TE
Inboard TE sweep, Λ_{ITE}	Straight TE	Straight TE
Airfoil Variables		
Location of max. thickness, $(x/c)_{max-t}$	40%	40%
LE radius, R_{LE}	2.5	2.5
Thickness to chord ratio at root, $(t/c)_{root}$	**1.5-2.7 %**	2.1%
Thickness to chord ratio LE break, $(t/c)_{break}$	$(t/c)_{break} = (t/c)_{root}$	$(t/c)_{break} = (t/c)_{root}$
Thickness to chord ratio at tip, $(t/c)_{tip}$	$(t/c)_{tip} = (t/c)_{root}$	$(t/c)_{tip} = (t/c)_{root}$
Fuselage Variables		
Fuselage restraint 1 location, x_{fus1}	50 ft	50 ft
Fuselage restraint 1 radius, r_{fus1}	5.2 ft	5.2 ft
Fuselage restraint 2 location, x_{fus2}	100 ft	100 ft
Fuselage restraint 2 radius, r_{fus2}	5.7 ft	5.7 ft
Fuselage restraint 3 location, x_{fus3}	200 ft	200 ft
Fuselage restraint 3 radius, r_{fus3}	5.9 ft	5.9 ft
Fuselage restraint 4 location, x_{fus4}	250 ft	250 ft
Fuselage restraint 4 radius, r_{fus4}	5.5 ft	5.5 ft
Nacelle, Mission, and Empennage Variables		
Inboard nacelle location, $y_{nacelle}$	20 ft	20 ft
Distance between nacelles, $\Delta y_{nacelle}$	6 ft	6 ft
Fuel weight, W_{fuel}	**350000-450000 lb.**	315000 lb.
Starting cruise altitude	65000 ft	65000 ft
Cruise climb rate	100 ft/min	100 ft/min
Vertical tail area	548 ft²	548 ft²
Horizontal tail area	800 ft²	No horizontal tail
Engine thrust	39000 lb.	39000 lb.

3. ESTIMATING OPTIMIZATION ERROR

Optimization is typically an iterative procedure, and is rarely allowed to converge to high precision due to computational cost considerations. Consequently, optimization results are usually a noisy function of the parameters of the design problem. The structural optimization of the HSCT suffered from convergence difficulties and resulted in a noisy W_s response. It may be difficult to calculate the error of a single optimization. However, when many optimization runs are available such as in constructing response surface approximation of W_s, we can use statistical techniques to estimate the mean (average) error.

3.1 Effects of Convergence Criteria

In order to visualize the behavior of the error, Figure 2 shows W_s response for 21 HSCT designs spread uniformly along a line between two extreme points in the five-variable HSCT problem. Design 1 corresponds to (1, -1, 1, -1, 1) and design 21 corresponds to (-1, 1, -1, 1, -1) in a coded form of the HSCT configuration variables. Case A2, using the default convergence criteria provided in GENESIS, resulted in a noisy W_s response, and design 6 and design 15 appear to have particularly large errors.

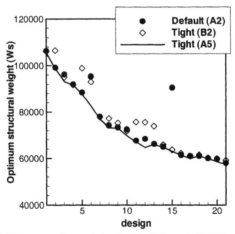

Figure 2: W_s response along a design line for different GENESIS parameters.

Since we suspected a convergence problem, we tightened convergence criteria. Out of the six different sets of the GENESIS control parameters used (Kim *et al.*, 2000), only three are discussed here. GENESIS has two nested iteration loops. In the outer loop an approximation to the original optimization problem is generated, which is solved in the inner loop via modified method of feasible direction (MMFD). The sequence of approximate optimizations is continued until there is no further change of the design variables or no further change of the objective function. When the outer loop convergence criteria were tightened from the default, W_s response was still noisy (Case B2). We found that an inner loop control parameter, ITRMOP, helps most to improve accuracy of the structural optimization. ITRMOP controls the number of times the inner loop convergence criteria are to be satisfied consecutively before the inner loop is terminated. By default, ITRMOP = 2, and when ITRMOP was increased to five, W_s response was much less noisy (Case A5) as seen in Figure 2. One important observation is that most of the erroneous GENESIS runs have heavier W_s, because a prematurely terminated minimization will end up with a higher objective function value, provided that no constraint is violated. The MMFD algorithm used in GENESIS first tries to find a feasible design, and none of the GENESIS runs produced an infeasible design for the HSCT problem.

To study the error in W_s from structural optimization, we used a mixed experimental design of 126 HSCT configurations, intended to permit fitting quadratic or cubic polynomial models for the five variable HSCT design problem. Optimization error, e, is defined as

$$e = W_s - W_s', \tag{1}$$

where W_s is the calculated optimum and W_s' is the true optimum, which is unknown for many practical engineering optimization problems. To estimate W_s', we need to perform high-fidelity optimization runs that can be expensive. Our strategy to estimate W_s' was to repeat GENESIS runs with six different sets of convergence criteria (including two high-fidelity cases with ITRMOP = 5), and take the best. This way, the optimization error, e, was calculated for each of the 126 HSCT configurations. Then, the mean and standard deviation of e were estimated for each case of different GENESIS parameters,

$$\hat{\mu}_{data} = \frac{\sum_{i=1}^{n} e_i}{n} = \bar{e}, \quad \hat{\sigma}_{data} = \sqrt{\frac{\sum_{i=1}^{n} (e_i - \bar{e})^2}{n-1}}, \tag{2}$$

where n (= 126) is the sample size. Table 2 shows that the average error was 4.86% for Case A2. The average error was increased to 5.63% for Case B2 although the outer loop convergence criteria were tightened, while Case A5 had a negligible average error of only 0.2%. In terms of computational cost, Case A5, a high-fidelity case, was more than twice as expensive as the low-fidelity cases, Case A2 or B2. The results show that the optimization error is attributed to convergence difficulties, and choice of appropriate convergence parameters is not trivial.

Table 2: Effects of GENESIS convergence criteria on W_s error.

	Case A2	Case A5	Case B2
Mean of optimization error, $\hat{\mu}_{data}$	3931 *lb.*	162 *lb.*	4553 *lb.*
(Percentage error w.r.t. mean W_s)	(4.86%)	(0.20%)	(5.63%)
Standard dev. of optimization error, $\hat{\sigma}_{data}$	7071 *lb.*	966.6 *lb.*	5991 *lb.*
Average CPU time on a SGI Origin	78.1 sec.	156.7 sec.	61.4 sec.

3.2 Probabilistic Modeling of Optimization Error

In the framework of robust design studies, uncertainty variables are often modeled using statistical distributions. With multiple structural optimization runs available, we can obtain a data driven model of the optimization error by fitting a probability distribution. We use the maximum likelihood estimation (MLE) for distribution fit (Law and Kelton, 1982, pp. 189-192). In MLE, we find a vector of distribution parameters, $\boldsymbol{\beta}$, to maximize the likelihood function, $l(\boldsymbol{\beta})$, which is a product of the probability density function, f, over the sample data x_i ($i = 1, ..., n$),

$$l(\boldsymbol{\beta}) = \prod_{i=1}^{n} f(x_i; \boldsymbol{\beta}). \tag{3}$$

The quality of fit is checked via χ^2 goodness-of-fit test (Law and Kelton, 1982, pp. 194-198), which is essentially a comparison of histograms between data and fit. The test results will be given in terms of the p-value. A p-value near one implies a good fit and a small chance that the data is inconsistent with the distribution. Conversely, a small p-value implies a poor fit and a high chance that the data is inconsistent with the distribution.

Considering the one-sidedness of the optimization error, we selected the Weibull distribution, which is defined by a shape parameter, α, and a scale parameter β. The probability density function (PDF) of the Weibull distribution is

$$f(x) = \begin{cases} \alpha \beta^{-\alpha} x^{\alpha-1} \exp\left(-\left(\frac{x}{\beta}\right)^{\alpha}\right) & if \; x \geq 0 . \\ 0 & if \; x < 0 \end{cases} \tag{4}$$

Once we get fitted parameters, α and β, estimates of mean and standard deviation of e can be calculated from

$$\hat{\mu}_{fit} = \frac{\beta}{\alpha} \Gamma\left(\frac{1}{\alpha}\right), \quad \hat{\sigma}_{fit} = \frac{\beta^2}{\alpha}\left\{2\Gamma(\frac{2}{\alpha}) - \frac{1}{\alpha}\left[\Gamma(\frac{1}{\alpha})\right]^2\right\}, \tag{5}$$

where Γ is the gamma function.

3.3 Direct Fit of Optimization Error

A straightforward approach to finding a probabilistic model of the optimization error is to fit a model distribution to e, calculated from Eq. 1. This approach is denoted a *direct fit*. The Weibull model was fitted to Cases A2 and B2, the low fidelity optimizations, and the results are summarized in Table 3. For both cases, p-values of the χ^2 test indicated acceptable fits with a 5% confidence level. The average errors estimated from the fit, $\hat{\mu}_{fit}$, were in good agreements with $\hat{\mu}_{data}$: -2.1% and 0.4% of discrepancies for Case A2 and Case B2, respectively. The estimates of standard deviation from the fits, $\hat{\sigma}_{fit}$, were also in good agreement with $\hat{\mu}_{data}$. Figures 3 and 4, comparing cumulative frequencies of e between the data (bars) and the direct fit (solid line), indicate that the Weibull model is well suited for the optimization errors for both Case A2 and Case B2. As a result, we obtained data driven probabilistic models for the optimization error.

Table 3: Estimation of optimization error via direct fit.

	Case A2	Case B2
$\hat{\mu}_{fit}$, *lb.*	3850	4553
(discrepancy w.r.t $\hat{\mu}_{data}$)	(-2.1 %)	(0.4 %)
$\hat{\sigma}_{fit}$, *lb.*	6894	6023
(discrepancy w.r.t $\hat{\sigma}_{data}$)	(-2.5 %)	(0.5 %)
α (estimate of shape parameter)	0.5912	0.7682
β (estimate of scale parameter)	2520	3915
p-value of χ^2 test	0.0925	0.8327

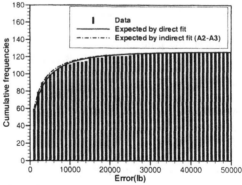

Figure 3: Comparison of cumulative frequencies of optimization errors of Case A2.

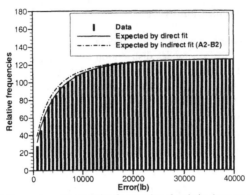

Figure 4: Comparison of cumulative frequencies of optimization errors of Case B2.

3.4 Indirect Fit of Optimization Error

The direct fit approach can be expensive because W_s' is estimated from higher-fidelity optimization runs. Moreover, high-fidelity optimizations are not always available. Instead, we may have optimization runs with two different convergence settings, where one is not particularly better than the other. Then, the probability distribution of the difference of optimal values from the two different convergence settings, instead of the optimization error itself, can be estimated (Kim, *et al.*, 2000).

For two optimization results, W_s' with convergence setting #1 and W_s^2 with convergence setting #2, model the optimization errors as random variables s and t,

$$s = W_s' - W_s'$$ (6)
$$t = W_s^2 - W_s'.$$

The difference of s and t is defined as the *optimization difference x,*

$$x = s - t = (W_s' - W_s') - (W_s^2 - W_s') = W_s' - W_s^2.$$ (7)

If s and t are independent, the probability density function (PDF) of x can be obtained by a convolution of the PDF functions $g(s; \beta_1)$ and $h(t; \beta_2)$,

$$f(x; \beta_1, \beta_2) = \int_{-\infty}^{\infty} g(s; \beta_1) h(s - x; \beta_2) ds.$$ (8)

Note that the optimization difference x is easily calculated from W_s' and W_s^2 that are readily available. We denote this approach an *indirect fit.*

The indirect fit was performed on the pair of Cases A2 and B2, two low fidelity cases. The χ^2 test on the optimization difference indicated a reasonable fit with a *p*-value of 0.197. As results of the indirect fit, we estimate the mean and standard deviation of the optimization error of each of the two cases involved. Table 4 shows that the estimates of mean error by the indirect fit, $\hat{\mu}_{fit}$, have reasonable agreements with $\hat{\mu}_{data}$: -13.7% and -12.6% discrepancies for Case A2 and Case B2, respectively. The estimates of standard deviation, $\hat{\sigma}_{fit}$, are also in a reasonable match with $\hat{\sigma}_{data}$. In Figures 3 and 4, the cumulative frequencies predicted by the indirect fit were compared to data and the results of the direct fits. Although the indirect fit is a little less accurate than the direct fit, the indirect approach is computationally more efficient because it does not require expensive higher-fidelity optimization runs. The results demonstrate the usefulness of the probabilistic model of the optimization error; the Weibull model allows us to estimate well the mean and standard deviation of the error from two sets of low-fidelity optimizations.

Table 4: Estimation of optimization error via indirect fit

	Case A2	Case B2
$\hat{\mu}_{fit}$, *lb.*	3394	3981
(discrepancy w.r t $\hat{\mu}_{data}$)	(-13.7 %)	(-12.6 %)
$\hat{\sigma}_{fit}$, *lb.*	7393	5729
(discrepancy w.r.t $\hat{\sigma}_{data}$)	(4.5 %)	(-4.4 %)
α (estimate of shape parameter)	0.509	0.710
β (estimate of scale parameter)	1756	3187

4. ESTIMATING RS BIAS ERROR

Response surface approximations fit numerical or physical experimental data with an analytical model that is usually a low-order polynomial. The error ε in the approximation is divided into bias (modeling) errors due to the differences between the model and the true response and into noise error. The noise errors are assumed to be uncorrelated and normally distributed random variables with zero mean and standard deviation σ, which is the same at all points. For the N data points matrix form of the collected responses can be represented,

$$\mathbf{y} = \mathbf{X}_1\boldsymbol{\beta}_1 + \boldsymbol{\varepsilon}, \tag{9}$$

where $\mathbf{X}_1$ is the matrix[1] whose terms in the row associated with the design point $\mathbf{x}$ are formed by monomials used in the RS model, and $\boldsymbol{\beta}_1$ is the coefficient vector associated with the monomials.

4.1 Mean Squared Error Criterion

As a measure of the error, and hence the uncertainty in the approximation, we use the mean squared error, *MSE*

$$MSE = Var[\hat{y}(\mathbf{x})] + (Bias[\hat{y}(\mathbf{x})])^2, \tag{10}$$

---

[1] The subscript '1' used for β_1 and X_1 refers to the set of monomials included in the model. Later, a subscript '2' will be used for the set of monomials needed to complete the model in order to obtain the true response

where $\hat{y}(\mathbf{x})$ is the prediction by the fitted model at design point $\mathbf{x}$. The first term in Eq. 10 represents the variance error due to noise and the second term bias error due to inadequate modeling.

This error expression is usually integrated over the design space, and the integral is minimized by choosing experimental designs that control the effect of one or both types of error. Here instead we characterize the error in the predictions of an RS approximation already constructed and determine the design regions where RS prediction may suffer due to either or both types of error. Therefore we use Eq. 10 to investigate the variation of *MSE* from point to point.

Now we consider the case where the true response at $\mathbf{x}$ is given as

$$y(\mathbf{x}) = \xi_1 \beta_1 + \xi_2 \beta_2, \tag{11}$$

where ξ_1 is the vector of fitting monomials calculated at point $\mathbf{x}$, and ξ_2 are terms missing from the assumed model. Since we usually do not know the true response, we often assume it to be a higher order polynomial. We can then write the mean of the true response as

$$E(\mathbf{y}) = \mathbf{X}_1 \beta_1 + \mathbf{X}_2 \beta_2, \tag{12}$$

where $\mathbf{X}_2$ is similar to $\mathbf{X}_1$, but due to the terms present in the true response that are not included in the fitting model. Following the derivation in Papila and Haftka (2001), Eq. 12 can be rewritten as

$$MSE(\mathbf{x}) \cong \sigma^2 \xi_1^! (\mathbf{X}_1^! \mathbf{X}_1)^{-1} \xi_1 + \beta_2^! \mathbf{G} \beta_2. \tag{13}$$

where $\mathbf{G}$ is a matrix formed by design matrices $\mathbf{X}_1$, $\mathbf{X}_2$, and fitting monomial vector ξ_1.

We look for points where the bias error is large. We now assume that we know what terms are missing from our model, but we do not have enough data to calculate the corresponding coefficients, β_2. If we can estimate the size of β_2 we can formulate a constrained maximization problem for a bound on the magnitude of the mean squared error that may be experienced at any given design point for the worst possible β_2 of that magnitude. The maximization problem yields the following eigenvalue problem (Papila and Haftka, 2001),

$$\mathbf{G} \beta_2 + \lambda \beta_2 = 0, \tag{14}$$

for which the maximum eigenvalue characterizes the maximum possible bias error associated with the assumed form of the true model. The corresponding eigenvector defines the missing polynomial coefficients for a given design point that result in the largest bias error when fitted only with the assumed model. The polynomial may be different at each point although the magnitude of the missing coefficient vector is constrained. So the eigenvalue calculated does not reflect the true polynomial corresponding to the data (as the data is insufficient to calculate β_2), but instead different polynomials of the same order (with different β_2) with the largest error for all possible polynomials considered.

4.2 Example: Cubic Polynomial in Two-variables

To demonstrate the use of the eigenvalue estimate of bias error we studied first a cubic polynomial in two-variables as the true model, and a quadratic polynomial as the fitted model with noise of σ in the data. The terms missing in the RS approximation and their coefficients are given as Eq. 15,

$$\xi_2^! = \begin{bmatrix} x_1^3 & x_1^2 x_2 & x_1 x_2^2 & x_2^3 \end{bmatrix}$$

$$\beta_2 = \begin{bmatrix} \beta_{21} & \beta_{22} & \beta_{23} & \beta_{24} \end{bmatrix}^T$$

(15)

Note that the eigenvalue problem in Eq. 14 does not depend on the response data. It depends only on the experimental design and the assumed true model. We first used a three-level factorial design to form the design matrices X_1 and X_2. Then we solved the eigenvalue problem, Eq. 14 with $\beta_2^! \beta_2 = 1$, at 21×21 mesh points over the design region. The maximum eigenvalues and corresponding eigenvectors were determined. Figure 5 shows a contour plot of the eigenvalues. It is seen that high error are expected at center of the boundaries of the design region: (-1,0), (1,0), (0,-1) and (0,1), and low values at (0,0), (-0 8,-0.8), (-0.8,0.8), (0.8,-0.8) and (0.8,0.8).

We now use a specific example where the quadratic portion is $50 + 3x_1 + 5x_2 + x_1^2 + 2x_1 x_2 + x_2^2$ and study sets of β_2 that are the eigenvectors associated with the maximum eigenvalues obtained at nine data points of the experimental design. The data were contaminated with random, normally distributed noise of zero average and standard deviation $\sigma = 0.3$ in order to simulate more general problems with both noise and modelling error We use the absolute residual as the error measure There are four distinct eigenvectors to investigate. Since the eigenvalue contours are based on different third order polynomials instead of a single one, we cannot expect exact agreement for any single polynomial. Instead, a given polynomial is expected to have a positive coefficient of correlation between the absolute residuals and the maximum eigenvalues Table 5 summarizes the coefficient of correlations, r, for the four polynomials The maximum of the four polynomials results in $r = 0.830$.

Table 5: Coefficient of correlations between the absolute residual and the maximum eigenvalues from Eq 14 for cubic polynomial example with noise

No	β_{21}	β_{22}	β_{23}	β_{24}	r
1	0	0.707	0.707	0	0.595
2	0	-0.707	0.707	0	0.380
3	0	0	-1	0	0.485
4	0	-1	0	0	0.307
Maximum of four	-	-	-	-	0.830

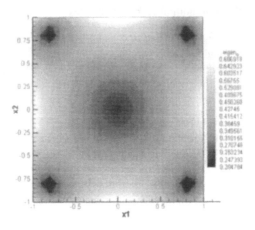

Figure 5: Maximum eigenvalue contour from Eq 14 for two-variable cubic polynomial, but fitted with quadratic model

Figure 6 shows the absolute residual (the maximum of the residuals from the four polynomials) contours when noise is present in the data. With the maximum of the residuals from the four polynomials, the coefficient of correlation is high as also seen by the qualitative agreement of Figure 5 and Figure 6.

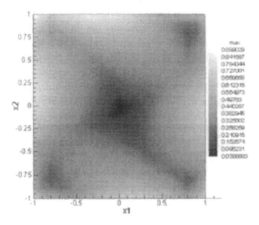

Figure 6: Absolute residual contours – with noise for cubic polynomial example (polynomial defined in Table 5)

4.3 Example: HSCT Two-variable Optimal Wing Structural Weight Problem

The two-variable HSCT problem is used as another example of RS modeling error. We generated optimal wing structural weight (W_s) data following Papila and Haftka (2000) for the nine data points from a three-level factorial design and the 24 supplementary points (Papila and Haftka, 2001), to obtain a good approximation of the true W_s model, which is not exactly known, unlike the polynomial example.

A more accurate, but expensive optimization setting with ITRMOP = 5 is used in order to estimate the true function form since it generates less numerical noise. The default GENESIS setting is used for demonstration of the present approach on RS approximations based on noisy data together with bias error.

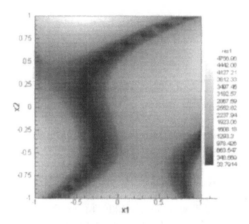

Figure 7: Absolute residual contours – HSCT quadratic wing weight equation

As in the polynomial example, we assume that the true function is a cubic polynomial, but we fit a quadratic approximation to the 9 data points. Since the experimental design and the model are the same as the polynomial example of the previous section, the maximum eigenvalue contours (Figure 5) are identical. We do not have the true values at the mesh points, but since the cubic approximation for W_s obtained by the accurate setting results on 33 points is an accurate approximation, we assume that this approximation is the true function. We then calculated the absolute residuals at all mesh points by subtracting the prediction by the quadratic RS constructed based on the nine points from the prediction of the assumed true function. The coefficient of correlation between the maximum eigenvalues and the absolute residuals is $r = 0.282$. The contours are shown in Figure 5 and Figure 7. There are shifts of the locations of high bias error compared to locations determined via eigenvalues, but they occur on the boundaries of the design region as predicted. The shifts are larger than for the polynomial example, but positive correlation is still achieved. We note that unlike the previous example, the cubic true weight function is just an assumption. Finally, the coefficient of correlation between the prediction variance (Figure 8) and the absolute residual is $r = 0.164$.

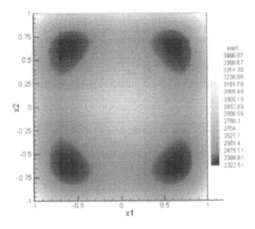

Figure 8: Estimated Standard Deviation ($= \sqrt{Var[\hat{y}(x)]}$) for two-variable wing problem

5. CONCLUDING REMARKS

A Weibull model was successfully fit to the errors due to incomplete convergence in optimal wing structural weight of the HSCT. An indirect approach using differences between two optimization results was proposed. The indirect fit enabled us to estimate the average errors of low-fidelity optimizations without performing expensive high-fidelity optimizations. The results demonstrated the usefulness of the probability model of the simulation error. As a result, we obtained a data-driven probability model of the simulation errors.

We presented an approach for identifying regions where large errors in response surface models are expected due the inadequacy of the fitting model. The approach leads to an eigenvalue problem with the largest eigenvalue indicating the worst error that may be experienced at a point. The examples showed that the eigenvalues may be helpful for identifying regions of high bias error. In particular, we found positive correlation between the maximum eigenvalues and the absolute residuals.

REFERENCES

Balabanov, V., Giunta, A. A., Golovidov, O., Grossman, B., Mason, W. H., Watson, L. T., and Haftka, R. T., 1999, Reasonable design space approach to response surface approximation, *Journal of Aircraft*, **36**, pp. 308-315

Kim, H., Haftka, R. T., Mason, W. H., Watson, L. T, and Grossman, B., 2000, A study of the statistical description of errors from structural optimization, AIAA 2000-4840, 8th AIAA/USAF/NASA/ISSMO Symposium on Multidisciplinary Analysis and Optimization, Long Beach, CA.

Knill, D. L., Giunta, A. A., Baker, C. A., Grossman, B., Mason, W. H., Haftka, R. T., and Watson, L. T., 1999, Response surface models combining linear and Euler aerodynamics for supersonic transport design, *Journal of Aircraft*, **36**, pp. 75-86.

Law, A. M. and Kelton, W. D., 1982, *Simulation Modeling and Analysis*, (New York: McGraw Hill).

McCullers, L. A., 1997, *Flight Optimization Systems*, Release 5.92, NASA Langley Research Center.

Myers, R. H. and Montgomery, D. C., 1995, *Response Surface Methodology: Process and Product Optimization Using Designed Experiments*, (New York: John Wiley & Sons, Inc).

Papila, M. and Haftka, R. T., 2000, Response surface approximations: noise, error Repair and modeling errors, *AIAA Journal*, **38**(12), pp. 2336-2343.

Papila, M. and Haftka, R. T., 2001, Uncertainty and response surface approximations, 42nd AIAA/ASME/ASCE/AHS/ASC Structures, Structural Dynamics, and Materials Conference, AIAA-2001-1680.

VR&D, Inc., 1998, *GENESIS User Manual*, Version 5.0, Colorado Springs, CO.

Analysis and Control of MEMS in the Presence of Uncertainty

George C. Johnson[1], Andrew Packard[1] and
Panayiotis Papadopoulos[1]

ABSTRACT

Uncertainties in material properties and structural dimensions are unavoidable at
any length scale, but they are particularly important as devices become smaller.
This work seeks to identify and quantify the sources and effects of the inherent
uncertainty in micro-electro-mechanical systems (MEMS). Methods for assessing the effects of this uncertainty on overall system response will be developed.
Quantitative bounds on the effect of uncertainty for any given design will be accessible and, once this is achieved, design variables can be chosen that minimize
the effects of the uncertainties. An example of robust control applied to MEMS
response is provided to demonstrate the potential benefits of this approach.

INTRODUCTION

Uncertainty is inherent in all systems. Geometric tolerances and material variability limit the extent to which the response of any particular device can be predicted from design specifications. As the geometric features of a structure or
system become smaller, the role of uncertainty becomes more significant. This
research will focus on identifying the sources of uncertainty in micro-electro-
mechanical systems (MEMS), quantifying the effects of uncertainty on system
behavior, and developing numerical schemes for predicting and controlling system response in the presence of uncertainty.

MEMS are systems involving both electrical and mechanical elements
whose smallest features are on the order of micrometers. A wide range of MEMS
devices have been developed recently, including sensors, micromotors, switches,
and biodevices (Kovacs, 1999; Senturia, 2000; Weigerink and Elwenspoek,
2001). Processes for creating such devices are similarly broad. The most common current approach for integrated MEMS devices (those with both electrical
and mechanical elements created at the same time from the same processes) involves polycrystalline silicon as the structural element. However, many other
structural materials are being introduced. These include single crystal silicon, as

---

[1] Department of Mechanical Engineering, University of California at Berkeley

in bulk-etched or silicon on insulator (SOI) devices; metals, most commonly found in LIGA devices; and polymers (Madou, 1997; Maluf, 1999)

The mechanical properties of these materials are not well characterized at the length scales involved in MEMS. This is partly due to the difficulty in making measurements at the micron scale, but is also partly due to the inherent variability of the materials and structures that are used for such characterization.

Geometric tolerances are a very significant issue in MEMS. For most processes, a geometric uncertainty of ±0.1 μm is as good as can be expected. While at the macro-level, this is a small portion of the structural dimension, the same is not true for MEMS. Consider a typical MEMS beam with 2 μm x 2 μm cross section and 100 μm length. The expected 0.1 μm uncertainty represents a relative uncertainty in width and thickness of 5% and in length of 0.1%. The relative uncertainty in cross sectional area is then on the order of 7% (assuming the width and thickness errors are uncorrelated), while the relative uncertainty in beam stiffness, which scales as the cube of the beam width, is on the order of 16%. At the macro-scale, this would be an unacceptable level of uncertainty. However, at the micro-scale, it is a fact of life. Unfortunately, the presence and effect of this uncertainty has not received the attention it deserves.

Complicating the issue of examining uncertainty is the fact that the material properties involved may themselves be uncertain. Suppose that the beam in the example above were made of polycrystalline silicon with average grain diameter of 0.5 μm. This implies that only about 15 to 20 grains would define any given cross section of the beam. These grains are likely to have random orientation governed by some distribution (texture), meaning that each cross section is defined by a relatively small number of randomly oriented grains. The cross sectional properties that derive from the crystalline properties are then expected to be random variables. Again, this unavoidable fact becomes more significant as the dimensions of the structure become smaller.

The nature of the geometric and material uncertainty are varied in terms of source and their associated length scale. One aspect of the variability occurs at a local scale, having correlation lengths for the random variables on the order of microns. Examples of this type of variability are surface roughness, side-wall etch profile, and grain size. A second length scale involves longer-range process variations across the region being processed (a wafer, in the case of silicon processing). These variations have correlation lengths on the order of mm to cm. Variations in thickness and residual stress across a silicon wafer exemplify this type of behavior. A third area of concern involves variations between nominally identical runs of a fabrication process, or even between different wafers in the same process run.

Each type of variation is important and deserves attention. Our research addresses each aspect of this problem, with focus on designing and controlling structures in the face of the inherent uncertainties resulting from each source. Every process (silicon-based, LIGA, etc.) will have its own unique set of issues. Our experimental work focuses on silicon-based MEMS, but we anticipate that the results will be sufficiently broad to be applicable to other processes, provided the uncertainties are sufficiently well identified and characterized.

MEMS applications in systems requiring high reliability demand close attention to the sources, effects and mitigation of uncertainty. Therefore, our research addresses robust schemes for the design, analysis and control of complex MEMS systems in the presence of uncertainty in both the as-built system and in the operating environment.

OVERVIEW OF POLYSILICON MEMS

The processes used to fabricate MEMS with polycrystalline silicon as the structural material are those used to create integrated circuits. A series of thin films are deposited, patterned and etched in a sequence that leaves the structural films attached to the substrate at only a few points, but free over much of the region. See Madou (1997) and Senturia (2000) for more details.

The fabrication sequence described here is a simple two-mask process involving a single sacrificial layer and a single structural layer. More complicated processing sequences provide for multiple structural layers and allow the fabrication of far more complex structures. First, a sacrificial layer, often a glass, is deposited on the substrate. Regions of this film are removed by patterning and etching to provide areas (etch holes) where the structural layer can attach to the substrate. Figure 1(a) shows a sacrificial layer with a single etch hole providing access to the substrate. The structural film is then deposited over the remainder of the sacrificial layer and the exposed substrate. (See Figure 1(b).) The structure itself is defined by patterning and etching the structural film. Finally, the structure is freed by another etch process in which the sacrificial layer is removed, but the structural layer is largely unaffected. (See Figure 1(c).) An example of a complicated system fabricated using this basic process flow is shown in Figure 2.

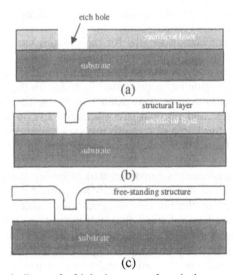

Figure 1. Schematic diagram for fabrication process for a single structural MEMS layer

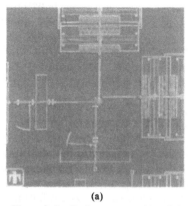
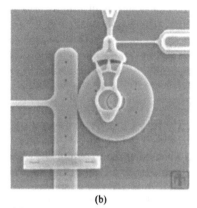

(a)   (b)

Figure 2. Sandia dynamometer (a) and close-up image of the central gear. (Images may be found at
http://mems.sandia.gov/micromachine/images2.html.)

Geometric uncertainties result from every step of this process. Each layer is deposited in a manner designed to provide uniform and well controlled thickness over the entire substrate. In reality, the thickness is not absolutely uniform over the substrate, and is rarely exactly the same from process run to process run. Thickness variations, which may reflect themselves as surface roughness, can affect the thickness and surface topology of all of the layers above them.

The steps used to create holes in the sacrificial layer and define the structural elements involve lithography and reactive ion etching of those regions that are not protected by the photoresist. The ideal etch profile is one in which straight lines in the design are straight on the as-built structure, and the side-walls are smooth and vertical. Errors in focusing the light through the mask onto the unexposed photoresist can result in features that are either wider or narrower than designed. Variations in etching of the structural material can result in side-walls that are neither vertical nor smooth. Figure 3 shows examples of such features found in two typical polysilicon processes (Jones, 1999).

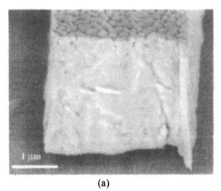

(a)   (b)

Figure 3. Scanning electron micrographs of two polysilicon beams. Material (a) has a finer surface texture and smaller grains than material (b)

The material in Figure 3(a) was deposited at 580C, while the material in Figure 3(b) was deposited at 650C. The work of Krulevitch (1994) has shown that the grain structure of these two materials is quite different, with the low-temperature deposition leading to small, equiaxed crystals and the high-temperature deposition leading to larger, columnar crystals. This grain structure is reflected in the surface topology.

EXAMPLES OF SYSTEM-LEVEL UNCERTAINTY IN MEMS

We provide three examples of observed variability in simple MEMS devices – two representing system failure and one associated with a device for measuring material properties. Failure may be associated with either an element of the system actually breaking (fracture) or with the moving parts of the system failing to remain free to move (stiction).

As part of his doctoral research at Berkeley, Jones (1999) measured the fracture strength of a range of different "flavors" of polycrystalline silicon. The device used for these measurements allowed a large number of beams to be tested in a relatively short period of time, thereby providing sufficient data to determine not only on the mean failure strength, but also the distribution that governs these measurements.

The structure used consists of a rigid central shuttle to which six cantilever beams are attached. The shuttle is moved by an off-chip probe, causing the base of the beams to move with it. The free end of each beam is restrained from motion by a small contact point that is initially several microns away from the beam. Figure 4 shows an optical micrograph of a structure during testing. In this test, the two shortest beams have failed, while the longer beams remain attached to

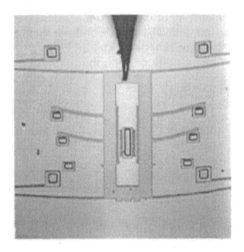

the

Figure 4. Optical image of fracture test structure during testing. The shortest two beams have already broken off of the central shuttle.

shuttle. This micrograph clearly shows the large deflections that can be supported by MEMS beams. The displacement of the shuttle is monitored optically, along with the integrity of each beam. The shuttle deflection is recorded as each beam breaks and this tip deflection is used with nonlinear beam analysis (the elastica problem) to determine the strain state associated with each failure.

Figure 5 shows the cumulative distribution of failure for two different runs of the MUMPs (Multi-User MEMS Processes) foundry service. These plots show the fraction of beams that had failed at a strain at or below the indicated value and clearly demonstrate the variation in fracture strength associated with a given process. Shown on each plot is also a fit of the cumulative Weibull distribution function to the experimental data. In both cases, the Weibull curve reasonably models the observations, though only in an empirical sense. That is, there is no analytic justification for this particular distribution function, nor for the magnitude of the parameters.

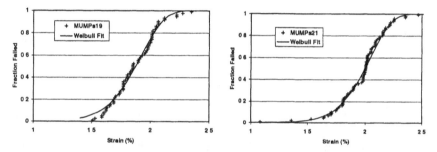

Figure 5. Strain to failure distribution for two MUMPs process runs.

The second example involves a different type of system failure common in MEMS – stiction. The large surface area to volume ratios of many MEMS devices can result in the unintended adhesion (stiction) of the compliant elements of the system to the substrate. Whether due to capillary, van der Waals or electrostatic forces, the magnitude of the attraction between the film and the substrate is often too great for the natural restoring forces of the system to overcome. Adhesion between surfaces also plays an important role in friction.

One way of quantifying stiction is by building an array of beams of varying compliance and measuring the deflected shape of each beam. The ends of longer, more compliant beams are more likely to be stuck to the substrate than are the shorter, stiffer beams. Thus, it is common to report stiction vs. beam length for materials subject to various processes. Figure 6 shows the data of Srinivasan, *et al.* (1998) for the effect of two different hydrophobic surface treatments on stiction of cantilever beams. In these experiments, beams of varying length were electrostatically pulled down to make contact with the substrate. Upon release of the actuation force, the beams were examined to see which remained adhered to the substrate. Though not shown in this figure, all beams that had the usual hydrophilic oxide layer and were longer than 150 µm remained attached to the substrate. The effect of the surface treatments was to allow beams up to 1 mm long to release from the substrate. However, some beams as short as 400 µm are stuck to the substrate, with the likelihood of a beam being stuck increasing with increasing length. Again, an empirical Weibull distribution is provided for each set of data.

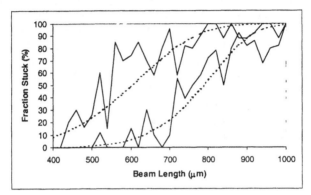

Figure 6. Fraction of cantilever beams remaining attached to the substrate after release of an external force. The two sets of data are for two separate hydrophobic coatings (See Srinivasan, *et al* , (1998).)

A final example of the uncertainty inherent in MEMS response involves the variability exhibited by a micromechanical strain gage (Lin, *et al.*, 1997) This passive device, shown in Figure 7, has a relatively long test beam which will extend or contract according to the sign of the strain in the film. Attached to the end of the test beam is a short "slope beam" to which an indicator beam is attached. As the test beam extends due to the strain, the slope beam deforms as shown and the indicator beam rotates. The deflection of the tip of the indicator beam is, for small strains, proportional to the magnitude of the strain.

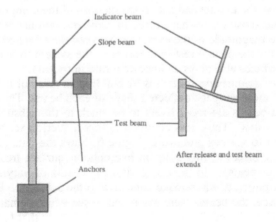

Figure 7. Micromechanical strain gage of Lin, *et al.* (1997)

A series of strain gages were examined from a single processing run, each indicating a strain reading with a resolution of 0.2%. As with the previous examples, the various devices showed considerable variation. The cumulative probability diagram for all of the strain gages is shown in Figure 8. None of the strain gages showed an apparent strain of less than 1%, and all showed an apparent strain of less than 2.4%. A Weibull distribution provides a qualitative description of the observed variation in strain.

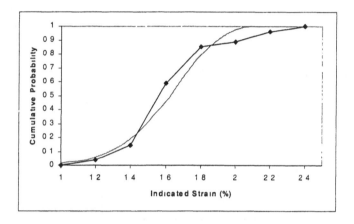

Figure 8. Distribution of apparent strain using the strain gage of Figure 7.

The three examples presented here clearly indicate the level of uncertainty that MEMS can exhibit at the system level. Our work seeks to understand this system-level behavior in terms of the uncertainties resulting from the fabrication (surface topography, grain structure, etch characteristics) and to develop methods of accommodating the response variability through robust control schemes.

ROBUST CONTROLLER DESIGN AND ITS APPLICABILITY TO MEMS

As suggested above, little has been done to quantify the magnitude of the various uncertainties, or to understand their effects on system-level response of MEMS. On the other hand, the effects of uncertainty have been address in some detail is in the design of robust control algorithms (Smith and Packard, 1996). Our work seeks to examine the applicability of this work in robust design to MEMS.

In this context, let us consider a system whose performance we wish to control. Let f represent a cost function whose value is to be minimized under appropriate choice of design or control variables Γ. Complicating the choice of Γ is the presence of uncertainties, denoted Δ. The approach taken here is to minimize the worst-case cost, as in

$$\min_{\Gamma} \max_{\Delta} f(\Gamma, \Delta) \cdot \qquad (1)$$

While easily stated, even simple cost functions typically lead to complicated analytical expressions requiring considerable computational effort. In this work, we focus on a specific form of the cost function and ranges of possible control and uncertainty that are rich enough to describe complex and relevant problems, but which have a mathematical structure permitting efficient computational methods (Zhou, *et al.*, 1996; McCloskey, *et al.*, 2000).

The structure used is that of linear fractional transformation (LFT), represented in block-diagram form as shown in Figure 9. The output vector e of this system is the error resulting from a set of excitations d and the design and uncertainty variables Γ and Δ. Mathematically, the LFT of M by Γ and Δ is written as

$$e = \left[M_{11} + M_{12} \begin{bmatrix} \Gamma & 0 \\ 0 & \Delta \end{bmatrix} \left(I - M_{22} \begin{bmatrix} \Gamma & 0 \\ 0 & \Delta \end{bmatrix} \right)^{-1} M_{21} \right] d \, , \tag{2}$$

where M_{ij} are components of a known matrix.

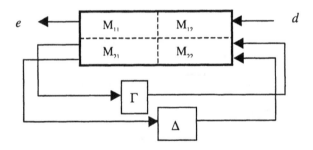

Figure 9. Block diagram of the Linear Fractional Transformation of M by Γ and Δ.

The cost function associated with this problem can then be identified as the gain from d to e. The robust design problem is then the mini-max problem

$$\min_{\Gamma} \max_{\Delta} \left\| M_{11} + M_{12} \begin{bmatrix} \Gamma & 0 \\ 0 & \Delta \end{bmatrix} \left(I - M_{22} \begin{bmatrix} \Gamma & 0 \\ 0 & \Delta \end{bmatrix} \right)^{-1} M_{21} \right\| . \tag{3}$$

Two simpler problems are contained within this structure: nominal design if there is no uncertainty ($\Delta = 0$) and worst-case analysis if there are no design or control variables ($\Gamma = 0$).

The advantages to constructing the robust design problem in this way is that there exist efficient algorithms that permit evaluation of this mini-max problem with reasonable computing power. In certain cases, this min-max design problem can be solved exactly. In other cases that are more structured, algorithms have been developed to provide rigorous bounds to the solution, while not necessarily leading to an exact solution.

An Example of Robust Control of MEMS

As an example of the applicability of this approach to MEMS, we consider the control of the simple structure shown in Figure 10, consisting of a series of comb-drive/sensor elements connected by beams. All of the beams are nominally identical, as are all of the masses. However, the uncertainty in the width and thickness of the material used to construct the beams leads to uncertainty in the effective rigidity.

The objective in this analysis is to displace each of the six mass elements the same amount. Simple analysis of this MEMS structure indicates that some of the forces must act in the direction of the displacement, while others must act in the opposite direction. In order to allow the application of static forces in either direction, both sides of the comb drive must be available for excitation. This means that a single comb drive cannot be used for both actuation and sensing of the motion. Thus, we choose to apply voltages to half of the comb drives (masses 1, 3 and 5), while reserving the others (masses 2, 4 and 6) for measurement of the displacement.

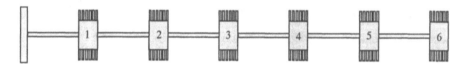

Figure 10. A six degree-of-freedom MEMS structure that is to be excited in such a way as to displace all six masses the same amount.

Since only some of the combs are actuated, it will be impossible to achieve the desired uniform displacement. The goal is to select the actuation mode that will allow the displacement to be most uniform (in a least squares sense). Four cases are considered - nominal and robust controllers that involve open-loop or feedback control. The open-loop (OL) controllers specify the magnitudes and directions of the forces based on analysis of the system, either without considering the presence of uncertainty (leading to the nominal OL controller) or including the uncertainty (leading to the robust OL controller). In designing the feedback (or closed-loop, CL) controllers, we admit that a small amount of noise will be present in the measurements. Again, both nominal and robust CL controllers were considered.

The analysis was performed for beams whose nominal dimensions are: 100μm length, 3μm width and 2μm thickness. The material is taken to be polycrystalline silicon with mass density of 2330 kg/m^3 and elastic (Young's) modulus of 167 GPa. The comb-drive elements are assumed to be effectively rigid, so their size does not factor into this static analysis. The target displacement for each of the masses is 1 μm.

Only the effective beam stiffness was taken to be uncertain, with the limit of the variation being ±15% of nominal. This corresponds to an uncertainty of ±5% in beam width, or for the beams considered here, an uncertainty of ±150nm.

This is well within the range that is reasonable for MEMS fabricated using $2\mu m$ design rules.

Let w_i, $i = 1, 2, \ldots 6$, denote the displacement of the ith mass, and let $F_j, j = 1, 3, 5$, denote the force acting on the mass indicated by the subscript. The displacement vector is determined from the force vector as

$$w = G_\Delta F , \tag{4}$$

where the presence of the uncertainty is explicitly shown in the 6-by-3 compliance matrix through the symbol Δ. (Note that Δ contains the uncertainties in each of the six beams. The nominal case is indicated by $\Delta = 0$.)

For the OL control problem, the three forces are the design (or control) variables used to minimize the cost function,

$$f = \left\| (w - w_{target}) \right\| = \left\| (G_\Delta F - w_0) \right\| \tag{5}$$

where w_{target} is the vector of desired displacements. The force vector associated with the nominal OL controller results from the minimization of Equation (5) in the absence of uncertainty,

$$\min_F \left\| (G_0 F - w_{target}) \right\| \Rightarrow F_{OL_{nominal}} . \tag{6}$$

The robust OL controller is obtained by minimizing the worst-case cost associated with all possible choices of uncertainty in beam stiffness,

$$\min_F \max_\Delta \left\| (G_\Delta F - w_{target}) \right\| \Rightarrow F_{OL_{robust}} . \tag{7}$$

The force vectors associated with the nominal and robust open-loop control of this structure are quite different. Specifically, these two vectors are

$$F_{OL_{nominal}} = \left\{ \begin{array}{c} 3.9 \\ -0.68 \\ 0.9 \end{array} \right\} \mu N, \quad F_{OL_{robust}} = \left\{ \begin{array}{c} 2.0 \\ -0.20 \\ 0.01 \end{array} \right\} \mu N . \tag{8}$$

Under nominal conditions, the robust controller is much worse than the nominal controller, having a cost function nearly twice as great. However, under conditions other than nominal, the robust scheme performs better than the nominal scheme, which can lead to displacement errors in excess of 100%. As will be shown later, neither of these open-loop strategies provide satisfactory control of the system when uncertainty exists in the beams' stiffness.

To improve upon the response of the system, we introduce a simple feedback control scheme in which the force vector F_j, $j = 1, 3, 5$, consists of a nominal force F_{0_j} plus a term linear in the measured displacements $\hat{w}_i$, $i = 2, 4, 6$, through a 3-by-3 gain matrix K as

$$F = F_0 + K(\hat{w} + \eta), \qquad (9)$$

where η represents the noise in the measurements. For this analysis, the noise level is taken to be 5% of nominal or ±50nm. The object in the closed-loop analysis is to minimize the cost function through appropriate choice of F_0 and K.

In order to express the cost function explicitly in terms of the design variables, as in Equations (6) and (7), the 3-by-3 compliance submatrix $\hat{G}_\Delta$ relating the measured displacements to the forces is introduced, so that

$$\hat{w} = \hat{G}_\Delta F. \qquad (10)$$

The cost function for the most general problem addressed is then

$$f = \left\| G_\Delta (I - K\hat{G}_\Delta)^{-1}(F_0 + K\eta) - w_{target} \right\|. \qquad (11)$$

The nominal CL controller minimizes the maximum of Equation (11) over all choices of measurement noise η, assuming no uncertainty. The robust CL controller minimizes the maximum of Equation (11) over all η and Δ within the specified ranges,

$$\min_{F_0,K} \max_{\eta,\Delta} \left\| G_\Delta (I - K\hat{G}_\Delta)^{-1}(F_0 + K\eta) - w_{target} \right\| \Rightarrow F_{CL_{robust}}, K_{CL_{robust}}. \qquad (12)$$

Note the similarity between this expression and that given in Equation (3) for the general robust control problem.

Both closed-loop schemes produce deflections at nominal stiffness that are virtually indistinguishable from those of the nominal open-loop case. The robust controller does suffer very slightly at nominal conditions, but this is more than made up for in its ability to deal with the presence of uncertainty.

Figure 11 shows the deflections for all four cases under nominal conditions. It is clear from this figure that the robust OL controller has considerably worse response at nominal stiffness than any of the other schemes.

Figure 12 shows the results of a Monte Carlo simulation involving calculation of the system displacements for different choices of beam uncertainty and measurement noise. All four control schemes are shown for each mass.

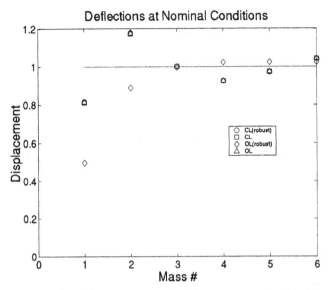

Figure 11. Deflections of the 6 masses under each controller with no uncertainty in beam stiffness.

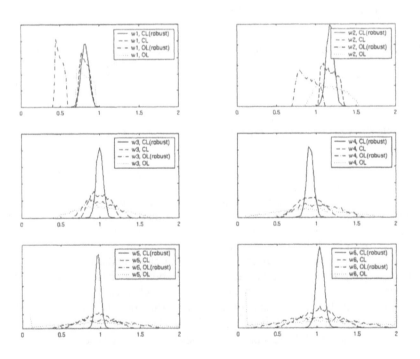

Figure 12. Variation in the displacements of the six masses for the range of beam stiffness and displacement measurements within the prescribed ranges of uncertainty and noise.

The three schemes that provide the same performance at nominal conditions show approximately the same spread in response for mass 1. The robust OL actuation scheme exhibits the same variation about it's nominal value as the others, but as noted above, the nominal response for this case is considerably worse than for the other cases.

Masses that are further from the fixed end are more difficult to control under all actuation schemes except for the robust CL approach. Focusing on mass 6, these simulations show that the open loop schemes only occasionally result in deflections within 10% of the desired amount. The nominal CL approach is slightly better, but the range of expected deflection of this mass is ±50% of the target. Only the robust CL scheme maintains the same variance about it's nominal response for all of the masses. A summary of the four cases, given in terms of the nominal and worst-case cost functions, is given in Table 1. This clearly shows the significant improvement in performance under all conditions associated with the robust closed-loop control scheme.

Table 1. Nominal and worst-case cost functions f for the four control schemes.

Control Scheme	Nominal f	Worst-Case f
Nominal Open Loop	0.27	2.5
Robust Open Loop	0.52	1.5
Nominal Closed Loop	0.27	0.93
Robust Closed Loop	0.27	0.29

CONCLUSIONS

Uncertainty in micro-electro-mechanical systems is a significant factor, at both the microscopic and the system levels. However, the tools of robust control design may prove to be vital in developing effective strategies for coping with this inherent uncertainty. The examples provided in this paper indicate the extent of the uncertainty in MEMS materials and systems, and suggest that appropriate use of robust design schemes are applicable.

Future work will address a range of topics. Specific work will focus on the characterization of local uncertainties associated with surface topography, developing an understanding of the relationships between local material and geometric uncertainty and the resulting system-level uncertainty, examination of the applicability of robust design schemes for micro-electro-mechanical systems, and development of new and more efficient computational schemes for evaluating system response and controller design.

REFERENCES

Jones, P.T., 1999, *The Fracture Strength of Brittle Films used for MEMS Devices*, Dissertation. University of California at Berkeley.

Kovacs, G. T. A., 1998, *Micromachined Transducers Sourcebook*, (New York: McGraw-Hill).

Krulevitch, P. A., 1994, *Micromechanical Investigations of Silicon and Ni-Ti-Cu Thin Films*, Dissertation. University of California at Berkeley.

Lin, L. W.; Pisano, A. P.; Howe, R. T., 1997, A micro strain gauge with mechanical amplifier, *Journal of Microelectromechanical Systems*, **6**, pp. 313-321.

Madou, M. J., 1997, *Fundamentals of Microfabrication*, (Boca Raton, FL: CRC Press).

Maluf, N., 1999, *An Introduction to Microelectromechanical Systems Engineering*, (Boston: Artech House).

McCloskey, R; Packard, A; Sipila, J., 2000, Branch and bound computation of the minimum norm of a linear fractional transformation over a structured set, *IEEE Transactions on Automatic Control*, **45**, pp. 369-375.

Senturia, S. D., 2000, *Microsystem Design*, (Boston: Kluwer Academic).

Smith, R and Packard, A., 1996, Optimal Control of Perturbed Linear Static Systems, *IEEE Trans. Autom. Control*, **41**, pp. 579 – 584.

Srinivasan, U.; Houston, M.R.; Howe, R.T.; Maboudian, R., 1998, Alkyltrichlorosilane-based self-assembled monolayer films for stiction reduction in silicon micromachines, *Journal of Microelectromechanical Systems*, **7**, pp. 252 -260

Weigerink, R. J. and Elwenspoek, M., *Mechanical Microsensors*, (Berlin: Springer Verlag).

Zhou, K., Doyle, J. and Glover, K., 1996, *Robust and Optimal Control*, (New York: Prentice-Hall).

Parallel Algorithms for Large-Scale Simulation-Based Optimization

O. Ghattas[1] and L. T. Biegler[1]

1 ABSTRACT

We consider variants of Successive Quadratic Programming (SQP) applied to Simultaneous Analysis and Design (SAND) implementations for PDE-based engineering modeling systems. In this study we consider in particular large scale PDE solvers for fluid flow, reactions and interactions with solid surfaces. We sketch the SQP algorithm and options that need to be compatible with the information obtained from the modeling system. Here we consider four levels of implementation and highlight each level with challenging PDE-based optimization problems. These implementation levels also need to be supported with object oriented software platforms that take advantage of advanced computing environments and lead to flexible modifications in solution and optimization algorithms. The benefits of this approach will be demonstrated through numerical results on challenging optimization problems with PDE models.

2 INTRODUCTION

Engineering modeling systems form the cornerstone of analysis and design in a broad range of disciplines. In computational mechanics and the analysis of transport systems, a wide variety of PDE modeling systems and products are available. In addition to solving discretized partial differential equations, most of these also have capabilities for mesh generation, options for construction of finite element bases, and a wide variety of linear iterative solvers and preconditioners.

These engineering systems represent hundreds of man-years of software development. However, virtually all of them were developed for analysis and not for design optimization. As a result, there remain some interesting challenges in leveraging this software investment in order to adapt these systems for optimization. In principle, such systems require smooth functions and first (and

[1] Carnegie Mellon University, Pittsburgh, PA 15213, email:
Omar_Ghattas@navier.cml.cs.cmu.edu, biegler@cmu.edu

possibly second) derivatives to be accessed from the modeling equations by the nonlinear programming (NLP) algorithm. Moreover, in the solution of nonlinear systems, many modeling systems apply a Newton-based approach to converge the finite element equations. In some implementations, an exact Jacobian could be constructed for the Newton step. More often, this matrix is simplified so that some nonlinear terms are neglected. Since modeling systems aim to converge the nonlinear equations efficiently, with suitable preconditioners for the iterative solution of linear systems that comprise Newton steps, an inexact Jacobian is often satisfactory for engineering analysis. However, the absence of accurate Jacobian information presents a major hurdle for the implementation of efficient gradient-based NLP algorithms. Finally, even if accurate Jacobian information is computed, the Jacobian matrix is rarely constructed explicitly. Hence, it may be difficult to make its elements available to the NLP algorithm for further calculation.

This paper explores the interface of widely used PDE modeling systems with Successive Quadratic Programming (SQP) NLP solvers. Proper implementation requires a strong interaction between the modeling system and the NLP algorithm. To motivate this discussion, we consider four levels of implementation to NLP algorithms. At the most basic level, we consider a black-box or Nested Analysis and Design (NAND) implementation. This interface requires very little interaction of the NLP solver with the PDE model. Gradients for the NLP solver are obtained by finite difference, the NLP has few variables for design optimization. Also the modeling system, which is converged for every NLP iteration, requires little modification for the interface. On the other hand, the black-box implementation can suffer from repeated and time-consuming solution of the PDE model and intermediate convergence failures of the model.

At the other extreme we consider an open implementation which allows complete access to the discretization of the PDE model and its first and second derivatives. This approach leads to full flexibility for optimization along with a SAND (Simultaneous Analysis and Design) optimization strategy. With this implementation one can apply fast NLP algorithms, but to a much larger problem that includes all of the discretized model and decision variables. In the development of math programming algorithms, full information from an open implementation is always assumed to be available. In practice, however, a fully open implementation requires many engineering modeling systems to be rewritten, with little reuse of existing components.

In addition, one can consider two intermediate approaches with implementations that are tailored to the engineering modeling system. These two approaches are distinguished by whether accurate Jacobian elements of the PDE model can be made available to the NLP algorithm. In the *direct tailored approach* reduced gradients and search directions for SQP are calculated directly using Newton steps from the PDE modeling system and the "sensitivity" of Newton steps to the decision variables. On the other hand, the *adjoint tailored approach* requires reduced gradients and search directions for SQP, calculated using *both* the model Jacobian and its transpose. One advantage of this approach is the availability of multiplier estimates.

The paper outlines advances made to the SQP algorithms as well as implementation strategies for PDE-based optimization. In the next section we summarize the SQP algorithm and introduce the optimization problem. In section 3 we survey the Lagrange-Newton-Krylov-Schur (LNKS) method, a fully open approach based on a full space SQP algorithm. This approach takes advantage of preconditioners in the KKT matrix as well as parallelism in large-scale linear systems. An application to a challenging wing control problem with finite element models of the fluid flow demonstrates this approach. To complement this strategy, a tailored approach is presented in Section 4, which deals with a reduced space SQP (rSQP) algorithm. This approach requires less information from a detailed modeling system and consequently, can be interfaced more easily to existing PDE-based modeling tools. While the tailored approach is not as efficient as the LNKS method, it is nevertheless a SAND method that is suitable for optimization problems with few (say, < 100) decision variables. Finally, Section 5 concludes the paper and presents directions for future work.

2. SAND OPTIMIZATION

We consider optimization problems that take the form:

$$\text{Min } f(x) \text{ s.t. } c(x)=0, x\in [x^L, x^U] \tag{1}$$

where $x\in \mathbf{R}^n$, $f: \mathbf{R}^n \rightarrow \mathbf{R}$ and $c: \mathbf{R}^n \rightarrow \mathbf{R}^m$ are generally assumed to be smooth functions. Here we also assume n, m >> n-m. While the equations $c(x) = 0$ may be straightforward to construct for PDE models, in many modeling systems these equations are not constructed explicitly and their functional form is generally not available through the NLP interface. Similarly gradients of $f(x)$ and $c(x)$ may not be available to the interface, although we may assume that exact derivatives are often generated and used internally in the modeling system. Also, for this study we assume that second derivatives are not available from the modeling system.

For PDE optimization, the NLP method of choice is usually some variant of Successive Quadratic Programming (SQP), but this method is applied differently for each implementation. At an iterate x_k, SQP generates a search direction d_k by solving:

$$\text{Min } g(x_k)^T d + 1/2\, d^T W(x_k)\, d \tag{2}$$
$$\text{st } c(x_k) + A(x_k)^T d = 0,$$
$$x_k + d \in [x^L, x^U]$$

where $g(x)$ denotes the gradient of $f(x)$, $W(x, \lambda)$ denotes the Hessian of the Lagrangian function $L(x, \lambda) = f(x) + \lambda^T c(x)$ and $A(x)$ denotes the n x m matrix of constraint gradients:

$$A(x) = [\nabla c_1(x), ..., \nabla c_m(x)]. \tag{3}$$

A new iterate is then computed as: $x_{k+1} = x_k + \alpha_k d_k$, where α_k is a step length parameter chosen in a line search procedure which tries to ensure global convergence. For example, this can be done by enforcing sufficient reduction in a merit function, e.g., an exact penalty function

$$\Phi(x) = f(x) + \gamma \|c(x)\| \tag{4}$$

where $\gamma > 0$ is a penalty parameter.

For fully open implementations, a large NLP is considered directly by the SQP algorithm. All equations and variables are accessible and gradient calculations are straightforward. Nevertheless, care must be taken to solve these problems with a large-scale implementation of SQP. For this purpose, both sparse full-space methods and reduced Hessian variations of SQP have been proposed and demonstrated for PDE applications.

For black box (NAND) implementations, SQP is applied only to a subset of the problem variables, as the model equations and variables, a subset of x, are eliminated implicitly. As a result, the size of the optimization problem remains small (typically less than 100 variables) and no large-scale extensions are required for SQP. On the other hand, because the model is solved repeatedly and finite difference derivatives are required, black box optimization is more costly than with an open implementation. We therefore consider an ideal SAND implementation first, that takes full advantage of the first and second derivative information from the model as well as preconditioners derived from PDE structures for solving linear subsystems of the KKT matrix.

3. LAGRANGE-NEWTON-KRYLOV-SCHUR OPTIMIZATION

The LNKS method was first introduced by Biros and Ghattas (2001) and was inspired by domain-decomposed Schur complement PDE solvers. In such techniques, reduction onto the interface space requires exact subdomain solves, so one often prefers to iterate within the full space while using a preconditioner based on approximate subdomain solution (Keyes and Gropp, 1987). In our case, the decomposition is into states and decisions, as opposed to subdomain and interface spaces. A basic component of this method is the Krylov-Schur solver -the inner iterative solver and the preconditioner that accelerate the computations of a Newton step for the KKT optimality conditions. We introduce and analyze these preconditioner(s) in Biros and Ghattas (2001), where we also examine the parallelizability and scalability of the LNKS algorithm. Here we present a summary of this method and demonstrate its effectiveness on a large-scale optimization problem.

We consider the NLP (1) without bound constraints and exploit the structure of the problem by partitioning x into state variables $x_s \in \mathbf{R}^m$, and decision variables $x_d \in \mathbf{R}^{n-m}$. The Lagrangian function is used to convert the constrained optimization to a system of nonlinear equations. Here the first order optimality conditions or Karush-Kuhn-Tucker (KKT) conditions are given by:

$$\nabla L(x^*, \lambda^*) = g(x^*) + A(x^*)\lambda = 0 \tag{5}$$
$$c(x^*) = 0$$

The KKT optimality conditions (5) define a system of nonlinear equations and the Jacobian of this system is termed the KKT matrix. Assuming sufficient smoothness, and an initial guess sufficiently close to a solution, Newton steps obtained by the above system converge quadratically to the solution (x^*, λ^*) (Fletcher, 1987). Nevertheless, this full space Newton approach also presents difficulties: a descent direction is not guaranteed, second derivatives are required, and the KKT system itself is difficult to solve. The size of the KKT matrix is more than twice that of the forward PDE problem, and it is expected to be very ill-conditioned, not only from the forward problem but also from the different scales between first and second derivatives submatrices. Moreover, the KKT matrix is indefinite; mixing negative and positive eigenvalues is known to slow down Krylov solvers. Therefore, a good preconditioner is essential to make the method efficient.

In LNKS we use a proper Newton method to solve for the KKT optimality conditions. To compute the Newton step we solve the KKT system using an appropriate Krylov method. At the core of the algorithm lies the preconditioner for the Krylov method: an inexact version of the QN-RSQP algorithm. To derive the preconditioner we rewrite the Newton step for (5) in a block-partitioned form:

$$\begin{bmatrix} W_{ss} & W_{sd} & A_s \\ W_{ds} & W_{dd} & A_d \\ A_s^T & A_d^T & 0 \end{bmatrix} \begin{bmatrix} p_s \\ p_d \\ \lambda \end{bmatrix} = - \begin{bmatrix} g_s \\ g_d \\ c \end{bmatrix} \tag{6}$$

There are a number of ways to solve (6). For instance, reduced space SQP (rSQP) is equivalent to a block-row elimination: given p_d, solve the last block of equations for p_s, then solve the first to find λ, and finally solve the middle one for p_d, the search direction for the decision variables. Therefore rSQP can be written as a particular block-LU factorization of the KKT matrix:

$$\begin{bmatrix} W_{ss} A_s^{-T} & 0 & I \\ W_{ds} A_s^{-T} & I & A_d A_s^{-1} \\ I & 0 & 0 \end{bmatrix} \begin{bmatrix} A_s^T & A_d^T & 0 \\ 0 & W_z & 0 \\ 0 & W_{sd} - W_{ss} A_s^{-T} A_d^T & A_s \end{bmatrix} \tag{7}$$

Note that these factors are permutable to block triangular (LU) form and that W_z is the reduced Hessian with respect to x_d, given by:

$$W_z = W_{ss} + S^T W_{ss} S - S^T W_{sd} - W_{ds} S \tag{8}$$

where $S = A_s^{-T} A_d^T$. Based on the Schur-type factorization we use the following preconditioner for the KKT system:

$$\begin{bmatrix} 0 & 0 & I \\ 0 & I & A_d A_s^{-1} \\ I & 0 & 0 \end{bmatrix} \begin{bmatrix} A_s^T & A_d^T & 0 \\ 0 & W_z & 0 \\ 0 & 0 & A_s \end{bmatrix} \tag{9}$$

The key components of the preconditioner are A_s^{-1} and W_z, the preconditioners for the forward problem and the reduced space (or decision space) equations, respectively. In particular, a natural choice for W_z is a BFGS like method, which is commonly used in QN-RSQP methods. For an analysis of this preconditioner and more details on the derivation of the algorithm, see Biros and Ghattas (2001). Biros and Ghattas (1999) also provide theoretical and numerical evidence that these preconditioners work very well.

3.1 A Case Study for LNKS

Consider the active flow control of the wing represented in Figure 1. The optimization problem is given by:

Objective function:

$$f(u, p, b) = \frac{1}{2} \int_{\Omega} (\nabla u + \nabla u^T) : (\nabla u + \nabla u^T) \, d\Omega + \frac{\rho}{2} \int_{\Omega} b \cdot b \, d\Omega$$

Strong form of the Navier-Stokes equations:

$$-\nu \nabla \cdot (\nabla u + \nabla u^T) + (\nabla u)u + \nabla p + b = 0 \text{ in } \Omega$$
$$\nabla \cdot u = 0 \text{ in } \Omega$$
$$u = 0 \text{ on } \Gamma$$

where u is the velocity field, p is the pressure and b is the vector of body forces. The decision variables are the velocities at trailing edge of the wing and the objective function is the vorticity in the flow field. Note the differences in Figure 1 between the case with no control and active control (optimized).

No Controls   *Active Controls*
Figure 1: Elimination of Wing Tip Vortices

states controls	preconditioning	Newton iter	average KKT iter	time (hours)
117,048	QN-RSQP	161	....	32.1
2,925	LNKS-EX	5	18	22.8
(32 procs)	LNKS-PR	6	1,367	5.7
	LNKS-PR-TR	11	163	1.4
389,440	QN-RSQP	189	....	46.3
6,549	LNKS-EX	6	19	27.4
(64 procs)	LNKS-PR	6	2,153	15.7
	LNKS-PR-TR	13	238	3.8
615,981	QN-RSQP	204	....	53.1
8,901	LNKS-EX	7	20	33.8
(128 procs)	LNKS-PR	6	3,583	16.8
	LNKS-PR-TR	12	379	4.1

Table 1: Method Comparison and Isogranular Efficiency

Table 1 summarizes the results and shows how the LNKS method with exact (EX), preconditioner (PR) and inexact (TR) solves compares with a standard rSQP method. Here very large problems are solved with over 8900 decision variables and 615,000 state variables and equations., and we see that the best LNKS method is consistently over an order of magnitude faster than the SAND rSQP method. Moreover, in the isogranular results (where the problem size per processor remains constant), the scale-up of results is excellent. Further evidence of the near linear scalablity properties (η with implementation, algorithmic and total scale-ups) for LNKS can be seen in Table 2.

CRAY T3E-900

procs	agr Gflops	its	time	speedup	η_i	η_{k_t}	η
16	0.81	38	18,713	1.00	1.00	1.00	1.00
32	1.55	39	10,170	1.84	0.95	0.97	0.92
64	3.14	40	5,985	3.13	0.97	0.95	0.92
128	4.86	40	3,294	5.68	0.75	0.95	0.71

SGI ORIGIN 2000

procs	agr Gflops	its	time	speedup	η_i	η_{k_t}	η
16	1.09	37	13,512	1.00	1.00	1.00	1.00
32	2.13	40	6,188	1.84	0.96	0.93	0.89
64	6.08	38	3,141	3.13	0.96	0.97	0.93
128	7.90	39	1,402	5.68	0.87	0.95	0.83

Table 2: Scale-up of LNKS on parallel processors

4. TAILORED RSQP METHODS

The LNKS method exploits the properties of Newton's method and incorporates preconditioners that are essential for the solution of linear systems derived from PDE models. On the other hand, few existing modeling systems provide second derivatives; often exact first derivatives are not available either. For these tasks we consider a tailored approach based on rSQP methods. With *tailored* methods we combine the best features of both open and black box approaches. Here we assume that the model variables and equations are solved with Newton's method within the PDE solver and at each SQP iteration we require only one iteration of Newton's method for these equations. Hence the model equations are not solved repeatedly as in the NAND approach. Again we consider a partition of variables as in Section 3 as well as the linear system in (6).

One can show through symmetric transformations of (6) and simplifications of the multiplier calculations (Alkaya et al., 2000) that the rSQP method is equivalent to:

$$
\begin{bmatrix} 0 & 0 & A_s \\ 0 & W_z & A_d \\ A_s^T & A_d^T & 0 \end{bmatrix} \begin{bmatrix} p_s \\ p_d \\ \lambda \end{bmatrix} = - \begin{bmatrix} g_s \\ g_d + w \\ c \end{bmatrix} \tag{10}
$$

where $w = -(A_d A_s^{-1} W_{ss} - W_{ds}) A_s^{-T} c$ and W_z is given by (8). We term the system in (10) the *adjoint* tailored approach. Solving (10) as part of a Newton method will lead to quadratic convergence in x and the elements p_s, p_d and λ can be solved in sequence from (10). A further simplification in rSQP methods is to approximate $B_k \sim W_z$ using a quasi-Newton method (e.g., BFGS) and to set $w = 0$. In this way, second derivatives are not required from the modeling system, but first derivatives need to be supplied and accessed directly. On the other hand, a further simplification of (10) leads to:

$$
\begin{bmatrix}
0 & 0 & I \\
0 & B_k & A_d A_s^{-1} \\
I & A_s^{-T} A_d^{T} & 0
\end{bmatrix}
\begin{bmatrix}
p_s \\
p_d \\
A_s \lambda
\end{bmatrix}
= -
\begin{bmatrix}
g_s \\
g_d \\
A_s^{-T} c
\end{bmatrix}
\tag{11}
$$

Equation (11) represents the *direct* tailored approach. Note that the matrices A_s and A_d need not be accessed directly. Instead (11) can be solved using only the Newton step from the forward problem, $-A_s^{-T} c$, and the "sensitivity" matrix of this step, $S = A_s^{-T} A_d^{T}$. Both of these terms can be computed and supplied from within the PDE solver.

The direct approach can be linked to any PDE-based modeling system that uses Newton solvers for the forward problem. The matrix S can be obtained at the cost of n-m back-solves, which can also be done in parallel. Therefore, the direct approach may be preferred for problems with few decision variables ((n-m) small). On the other hand, the adjoint method does not require the S matrix and becomes advantageous when (n-m) is large. Finally, from (11) we see that the multipliers λ are not calculated explicitly with the direct approach. Instead, a multiplier-free algorithm has been developed (Biegler et al., 1995) that retains desirable global and local convergence properties. Moreover, the rSQP algorithm developed for the direct tailored approach can be adapted to deal with inequality constraints in (1). Here the following quadratic program is solved:

$$
\begin{aligned}
\text{Min} \quad & (g_d - S^T g_s)^T p_d + 1/2\, p_d^T B_k\, p_d \tag{12} \\
\text{st} \quad & x_{s,k} - A_s^{-T} c_k + S\, p_d \in [x_s^{L}, x_s^{U}] \\
& x_{d,k} + p_d \in [x_d^{L}, x_d^{U}]
\end{aligned}
$$

in the reduced space. Additional characteristics of this method can be found in Biegler et al. (1997). Advanced features of this algorithm as well as a detailed numerical comparison can be found in Biegler and Waechter (2001). Finally, because the tailored approach can be interfaced as a SAND method to existing modeling systems, we have developed a flexible object oriented implementation for this purpose, as described next.

4.1 rSQP++: An Object Oriented Implementation

An new object-oriented (OO) framework for building Successive Quadratic Programming (SQP) algorithms, called rSQP++, has been implemented in C++ (Bartlett and Biegler, 2001). The rSQP++ framework is designed to incorporate many different SQP algorithms and to allow external configuration of specialized linear algebra objects such as matrices and linear solvers. In particular, data-structure independence has been recognized as an important feature missing in current optimization software. In addition, it is possible for the client to modify the SQP algorithms to meet other specialized needs without having to touch any of the source code within the rSQP++ framework.

There are several challenges in trying to build a framework for SQP (as well as for many other methods) that allows for maximal sharing of code, and at the same time is understandable and extendible. Specifically, three types of variability are encountered in the design of rSQP++: (a) Algorithmic variability, (b) implementation variability and (c) NLP specific specializations.

(a) First, we need to come up with a way of modeling and implementing iterative algorithms that will allow for steps to be reused between related algorithms and for existing algorithms to be extended. This type of higher level algorithmic modeling and implementation is needed to make the steps in our rSQP algorithms more independent so that they are easier to maintain and to reuse. A framework called *GeneralIterationPack* has been developed for these types of iterative algorithms and serves as the backbone for rSQP++.

(b) The second type of variability is in allowing for different implementations of various parts of the rSQP algorithm. There are many examples where different implementation options are possible and the best choice will depend on the properties of the NLP being solved. One example is whether to represent the sensitivity matrix, $S = A_s^{-T} A_d^T$ explicitly or implicitly. Another example is the implementation of the Quasi-Newton reduced Hessian $B_k \sim W_z$. The choice of whether to store B_k directly or its factorization (and in what form) depends on the choice of QP solver used to solve (11) or (12). Yet another example is allowing different implementations for the QP solver.

(c) A third source of variability is in how to allow users to exploit the special properties of an application area. Abstract interfaces to matrices have been developed that serve as the foundation for facilitating the type of implementation and NLP specific linear algebra variability described above. In addition, these abstract interfaces help manage some of the algorithmic variability such as the choice of different reduced space decompositions.

Figure 2 shows a coarse grained UML object diagram for a rSQP++ algorithm configured and ready to solve an NLP. At the core is a set of algorithmic objects. The rSQPAlgorithm object acts as the hub for the algorithm; its main job is to fire off a set of steps in sequential order and perform major loops. One or more rSQPStep objects perform the actual computations in the algorithm. The rSQPStep objects operate on iteration quantity objects IterQuantity that are stored in the rSQPState object. In addition to simple linear execution of an algorithm, more sophisticated control strategies can be performed. This design allows step classes to be shared in many different related algorithms and also provides for modifications of the algorithm by adding, removing and replacing rSQPStep and IterQuantity objects. In other words, the behavior of the algorithms is not fixed and can be modified at runtime. In this way, users can modify the rSQP algorithms without touching any of the base source code in rSQP++.

Also shown in Figure 2 are DecompositionSystem and NLP objects. These objects provide the keys to specializing the linear algebra for a particular NLP. The DecompositionSystem object abstract choice of reduced space decomposition away from the rSQP algorithm. The DecompositionSystemVarReduct node subclass is for variable reduction decompositions. A BasisSystem object is used to abstract the variable reduction matrices including the basis matrix A_s.

The DecompositionSystemVar-, ReductOrthogonal and DecompositionSystemVarReductCoordinate subclasses implement the orthogonal and coordinate decompositions. The NLP interface is used to abstract the application. The base NLP interface provides basic information such as variable bounds and the initial guess and computes f(x) and c(x). The NLPFirstOrderInfo specialization is for NLPs that can compute g(x) and A(x). The matrix A(x) is represented as an abstract matrix object of type MatrixWithOp and can therefore be implemented by any appropriate means. Through this matrix interface, the optimization algorithm can perform only simple operations like matrix vector multiplication v = op(A) u. It is only in conjunction with a compatible BasisSystem object that the algorithm can perform all the needed computations. The NLPSecondOrderInfo interface is for NLPs that can compute the Hessian of the Lagrangian, W, which is also abstracted as a MatrixWithOp object. In this way the core rSQP++ code is independent of the specialized data-structures and solvers for an NLP. By configuring the algorithm with NLP and BasisSystem objects and MatrixWithOp objects for A(x) and possibly W, specialized data structures and linear algebra for an NLP can be accommodated.

The NLPFirstOrderInfo interface assumes that matrix vector multiplications with A(x) and its transpose can be performed. The BasisSystem interface assumes that linear systems involving the basis matrix A_s and its transpose can be solved for arbitrary right hand sides. For many applications, these requirements can not be met. For these applications, the TailoredApproach interface is defined and it is used by the algorithm to extract the bare minimum information (i.e. g(x), $A_s^{-T} c$ and $S = A_s^{-T} A_d^T$). With this information, both coordinate and orthogonal reduced space decompositions can be used. A more detailed introduction to rSQP++ as well as additional features can be found in Bartlett and Biegler (2001) and in Bartlett (2001). In particular, this object oriented program has been linked to the MPSalsa PDE modeling system at Sandia National

Laboratory. Next we describe a case study that demonstration of this approach on a challenging PDE-based design problem.

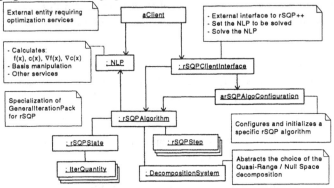

UML Object Diagram: Instantiation of rSQP++ components ready to solve an NLP

Figure 2: Coarse-grained Object Diagram

4.2 A Case Study for the Tailored Approach

The rSQP optimization algorithm has been linked to the MPSalsa parallel reacting flows code with the goal of developing SAND methods for use with large-scale PDE simulations. MPSalsa is an unstructured grid finite element code that uses a fully coupled Newton method to solve the PDEs governing fluid flow, heat transfer, and non-dilute mass transfer. In this case study, we present results for optimization of a Chemical Vapor Deposition reactor for growing thin films of Gallium Nitride (GaN). GaN is used in blue light emitting diodes and other photonic devices. The quality of the film is highly dependent on the uniformity of the growth rate at different positions in the reactor and our objective is therefore to minimize nonuniformities in the film.

The reactor is shown in Figure 3. Here we use an axisymmetric (2D) model. A mixture of trimethylgallium, ammonia, and hydrogen gases ($Ga(CH_3)_3$ or TMG, NH_3, and H_2) enter the top of the reactor, flow over the disk, which is heated, and then flow down the annular region out the bottom of the mesh. At the heated disk, the $Ga(CH_3)_3$ and NH_3 react to deposit a GaN film and release three molecules of methane (CH_4). The main parameter used in this paper is the inlet velocity of the gases, V. The objective function measures the uniformity of the growth rate of GaN over the disk. For this we choose a smooth approximation (KS function) to the L_∞ norm of the differences in growth rates. The equations for fluid flow consist of a momentum balance (the Navier-Stokes equations), the total mass balance (continuity equation) and an energy equation. The diffusive flux term (Multicomponent Dixon-Lewis Formulation) includes transport due to concentration gradients and thermal diffusion (Soret effect). At the disk surface, a complicated set of chemical reactions are approximated well by a transport limited model. In this case, the growth rate of GaN on the surface (as well as the consumption of $Ga(CH_3)_3$ and NH_3, and the production of CH_4) is proportional to the concentration of trimethylgallium $(Ga(CH_3)_3)$ at the surface. Numerous physical properties in the above equations are dependent on the local temperature and composition. In the MPSalsa code, we use the Chemkin library and database format to obtain these physical properties. These terms add considerable nonlinearity to the problem. Additional details of this model can be found in Salinger et al. (2001).

The above system of coupled PDEs is solved with the MPSalsa code, which uses a Galerkin/least-squares finite element method to discretize these equations over the spatial domain. This code is designed for general unstructured meshes in 2D and 3D, and runs on massively parallel computers. The discretized system for this problem contains 31995 unknowns. A fully coupled Newton's method is used to robustly calculate steady-state solutions. While analytic Jacobian entries are supplied for derivatives with respect to the solution variables and the density, derivatives of the other physical properties are only calculated with the numerical Jacobian option. The resulting linear system at each iteration is solved using the Aztec package of parallel, preconditioned iterative solvers. Using the rSQP algorithm described above, we obtain the uniform profile shown in Figure 3 (profile b). Using multiple Intel PIII (500 MHz) processors, the tailored SAND approach requires only 1.56 CPU hours. On the other hand, when one uses the NAND optimization approach in the Dakota solver, a CPU time of 22 hours is required (van Bloemen Waanders et al., 2001).

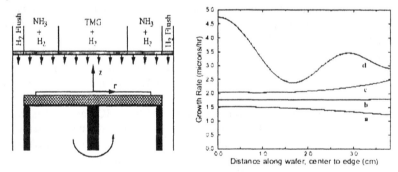

Figure 3: Schematic of CVD Reactor and Growth Profiles

5 CONCLUSIONS

This paper surveys recent advances in the development of simultaneous (SAND) optimization strategies. These methods are much more efficient than NAND or black box strategies but require careful consideration of implementation to existing PDE modeling tools. Here we consider various SQP implementations to exploit features of existing PDE modeling systems and discuss four such levels for SAND algorithms.

In particular, we summarize a fully open, full-space SQP strategy (LNKS) and demonstrate how this SQP performs on a large fluid flow problems with active control around a wing. LNKS allows full flexibility and exploitation of preconditioned Krylov solvers. It is especially well suited to problems with many decision variables. A prototype version of LNKS has also be developed and tested extensively in the Veltisto package (Biros and Ghattas, 2001).

On the other hand, if less information is available from the PDE model, a Reduced Space SQP (rSQP) approach can be applied for SAND optimization. Here we have developed and refined the rSQP algorithm that is best suited to problems with few decision variables and can be interfaced easily on existing Newton-based modeling systems. As a result, it can then inherit initialisation schemes, supporting routines, model structure and parallelism from these systems. This approach has been implemented in an object oriented program (rSQP++) and has been benchmarked on challenging nonlinear systems with reacting flows.

Future work will deal with a number of extensions to PDE-based SAND optimization. First, we are working on the incorporation of inequality constraints using both active set and barrier methods. In particular, the latter strategies have the advantage of directly exploiting the structure of the KKT matrix and avoiding the combinatorial problem of choosing the active set of constraints. Also, the barrier method has been prototyped in the IPOPT code (Waechter and Biegler, 2001) for both full space and reduced space versions of SQP.

A challenging issue that we are also investigating is the adaptivity of FEM mesh for Shape Optimization. Recent work of Malcevic (2001) has led to efficient strategies for dynamic remeshing. We plan to use these in the design of SQP-based shape optimization strategies. Moreover, we will extend these SQP approaches to deal with Optimization under Uncertainty with Multiperiod formulations (so-called *Robust Optimization*) (see Rooney and Biegler, 2001). Finally, with our advances in optimization strategies we intend to target challenging applications and sophisticated modeling tools at Sandia, including the SIERRA package.

6. ACKNOWLEDGEMENTS

Funding from the National Science Foundation under Grant 9732301 and from the Computer Science Research Institute (CSRI) at Sandia National Laboratory is gratefully acknowledged. We also pleased to acknowledge our Sandia colleagues W. Hart, A. Salinger, B. van Bloemen Waanders and our students, R. Bartlett, G. Biros, G. Itle and I. Malcevic involved in this project.

7. REFERENCES

Alkaya, D., S. Vasantharajan and L. T. Biegler, "Generalization of a Tailored Approach for Process Optimization" *I & EC Research* , 39, 6, pp. 1731-1742 (2000)

Bartlett, R. A., PhD Thesis, Chemical Engineering Department, Carnegie Mellon University, Pittsburgh, PA (2001)

Roscoe A. Bartlett and Lorenz T. Biegler, "rSQP++ : An Object-Oriented Framework for Successive Quadratic Programming" First CSRI Workshop on PDE-based Optimization, Santa Fe, NM (2001)

Biegler, L. T., Claudia Schmid, and David Ternet, "A Multiplier-Free, Reduced Hessian Method For Process Optimization," *Large-Scale Optimization with Applications,* p. 101, IMA Volumes in Mathematics and Applications, Springer Verlag (1997)

L. T. Biegler and Andreas Waechter, "SQP SAND Strategies that Link to Existing Modeling Systems" First CSRI Workshop on PDE-based Optimization, Santa Fe, NM (2001)

George Biros and Omar Ghattas, "Parallel Lagrange-Newton-Krylov-Schur methods for PDE-constrained optimization." (2001) Submitted for publication.

George Biros and Omar Ghattas, "Parallel {N}ewton-{K}rylov algorithms for {PDE}-constrained optimization." in *Proceedings of SC99, The SCxy Conference series*, Portland, Oregon, November 1999. ACM/IEEE.

Fletcher, R., *Practical Methods of Optimization*, Wiley, New York (1987)

Malcevic, I., PhD Thesis, Civil and Environmental Engineering Department, Carnegie Mellon University, Pittsburgh, PA (2001)

Rooney, W. C., and L. T. Biegler, "Nonlinear Confidence Regions for Design Under Uncertainty," accepted for publication, *AIChE J.* (2001)

A.G. Salinger, R.P. Pawlowski, J.N. Shadid, B.van Bloemen Waanders, R. Bartlett, G.C. Itle, and L. Biegler, "rSQP Optimization of Large-Scale Reacting Flow Applications with MPSalsa," First CSRI Workshop on PDE-based Optimization, Santa Fe, NM (2001)

B. van Bloemen Waanders, A. Salinger, R. Pawlowski , L. T. Biegler, R. A. Bartlett, "Simultaneous Analysis and Design Optimization of Massively Parallel Simulation Codes using an Object Oriented Framework," *Tenth SIAM Conference on Parallel Processing for Scientific Computing,* March 2001

Waechter, A. and L. T. Biegler, "Line Search Filter Methods for Nonlinear Programming," Technical Report, CAPD, Carnegie Mellon University (2001)

Modeling and Simulation for Design Under Uncertainty

Sankaran Mahadevan[1]

1 ABSTRACT

This paper presents a generalized framework and computational algorithms for the uncertainty-based design of large engineering systems. The methods compute and use sensitivity information on the drivers of system performance, reliability, cost etc., facilitating systematic design and life cycle management decisions based on the criterion of maximum expected utility. A Bayesian framework is developed for the inclusion of physical, informational, and model uncertainties. The framework includes model validation and confidence assessment procedures and integrates model-based and test-based methods for reliability estimation. The methods are demonstrated for application to a composite helicopter rotor component.

2 INTRODUCTION

The design of any engineering system requires the assurance of its reliability and quality. Due to the uncertainties in the system characteristics and demand, such assurance cannot be given with certainty. One common approach is to quantify the reliability of a design in probabilistic terms. Traditional design approaches have largely relied on testing to quantify the reliability, and safety factors based on experience and judgement to achieve the desired reliability. Such empirical design practices in many cases are not linked to the physics of the system behavior in terms of design, manufacturing, and environmental variables. The linkage to total cost measures such as life cycle cost is more qualitative and experience-based than quantitative. Concurrency and "total quality" issues in design are currently not supported through quantitative links between reliability, uncertainty, design variables, and production processes and controls. Furthermore, the use of testing to quantify the uncertainty and reliability is prohibitively expensive for large, complex systems. Therefore, the development of a modeling and simulation-based methodology to rationally incorporate the uncertainties in engineering design is becoming more and more attractive. The simulation-based methodology facilitates quantitative evaluation, more effective use of experiments, and rational engineering specifications on production processes and variables.

---

[1] Vanderbilt University

Uncertainty in engineering analysis and design arises from several sources (Oberkampf et al, 1999). Some of the "known" sources are: (1) *Physical uncertainty or inherent variability:* The demands on an engineering system as well as its properties always have some variability associated with them, due to environmental factors and variations in operating conditions, manufacturing processes, quality control etc. Such quantities are represented in engineering analysis as random variables, with statistical parameters such as mean values, standard deviations, distribution types etc. estimated from observed data. (2) *Informational Uncertainty:* This includes several types of uncertainty associated with the type of information available: statistical uncertainty due to small number of samples, imprecise information, etc. The accuracy of the statistical distribution parameters depends on the amount of data available. Thus the distribution parameters themselves are uncertain, and have to be treated as random variables. On the other hand, information may be imprecise or qualitative, and it is not easy to treat this type of uncertainty through random variables. (3) *Modeling Error:* This results from approximate mathematical models of the system behavior, and from numerical approximations during the computational process. For new and complex engineering systems, this type of uncertainty is not quantifiable *a priori*.

This paper discusses strategies to include all these types of uncertainty in a computational framework for decision-based design. The framework should include the quantification of design reliability or risk, cost, utility, etc. and their sensitivity to the uncertainties in the system. Alternative design options with various distribution parameters as design variables, and the associated life-cycle cost and utility consequences, should be included. A systematic procedure needs to be incorporated for confidence assessment of model prediction, acceptance or rejection of the system model, and for combining model prediction with development tests and in-service information during the life cycle of the system. The following sections present several methods to achieve these objectives.

3 UNCERTAINTY QUANTIFICATION

Various methods are available to compute the uncertainty in the system response due to uncertainties in the input quantities. Probabilistic techniques have been pursued extensively, by modeling the inherent variability through random variables. These methods, along with Bayesian techniques, can also consider statistical uncertainty that derives from lack of adequate statistical data. Non-probabilistic methods have been pursued to deal with imprecise information. In addition, the errors due to mathematical modeling and numerical approximation and discretization need to be quantified. Thus a comprehensive strategy is needed for computing the total uncertainty in modeling and simulation, and to determine the contribution of each type of uncertainty to the overall uncertainty. The discussion in this paper is restricted to probabilistic approaches.

As a first step in probabilistic modeling, Monte Carlo simulation has been used in many studies. However, the basic Monte Carlo method is too time-consuming to achieve acceptable accuracy for low probability events. Several efficient sampling schemes and variation reduction techniques such as Latin Hypercube Sampling and adaptive importance sampling (e.g., Mahadevan and Dey, 1997) have been developed.

An attractive alternative to Monte Carlo simulation is to use analytical approximations that combine probability theory and optimization methods (e.g., Rackwitz and Fiessler, 1978; Cruse et al, 1988). These are based on first-order or second-order approximations of the system performance constraint equations (limit states), and analytical estimation of the reliability through the identification of the most probable combination of the random variables that cause the system to be at the limit state.

A significant advantage of these analytical methods is the quantification of the sensitivity of the system reliability to the uncertain quantities. Design performance sensitivities to changes in mean, variance and truncation of the variables can be estimated, and the engineer can seek to maximize the expected utility by selecting and changing those variables to which the expected utility is more sensitive. For a system with multiple components, these sensitivities can be computed at both levels - individual component, and overall system (Mahadevan, 1997). However, these methods cannot handle discrete variables. Therefore, efficient sampling methods, and branch and bound technique need to be incorporated for discrete random variables.

Most studies have only used the aforementioned solution techniques to incorporate the first type of uncertainty, namely, physical or inherent variability. Mehta et al. (1992) have used Bayesian techniques to consider statistical and model uncertainties in developing confidence bounds on the output through a nested computation, using either Monte Carlo or analytical methods. These confidence intervals can be combined with probabilistic sensitivity information to increase the system's robustness (i.e., reduce the sensitivity of system performance to variations in operational conditions).

Ditlevsen (1982) represented model uncertainty as an additional random vector field to be added to problems with physical variability and statistical uncertainty, such that the entire uncertainty estimation is done through probabilistic analysis. The data on the additional random vector was obtained through extensive calibration work. A similar approach has been taken by several other researchers (Der Kiureghian, 1989; Sorensen and Thoft-Christensen, 1983) and design specification authorities, by including a model error correction factor. In the absence of large amounts of data, model uncertainty needs to be included through other qualitative or quantitative measures.

Several studies have focused on the estimation of discretization error in finite element analysis. Alvin (1999) used the response surface technique to include discretization error in non-deterministic analysis. The response surface is estimated from computer experiments, which vary both uncertain physical parameters and the fidelity of the computational mesh. This response surface is then used to propagate the variabilities in the continuous input parameters, while the mesh size is taken to zero, its asymptotic limit.

4 DECISION-MAKING UNDER UNCERTAINTY

The goal of engineering analysis is to provide information for decision making. Uncertainties are present in all facets of engineering decision, and some amount of risk is unavoidable. A systematic, rational framework is provided by the decision tree approach, combined with the concepts of utility theory (Ang and Tang, 1984). The optimum alternative according to the maximum expected utility value (EUV) criterion is

$$E(U_{opt}) = \max_i \left\{ \sum_j P_{ij} U_{ij} \right\} \tag{1}$$

where U_{ij} is the utility of the *j*th consequence associated with alternative *i*, P_{ij} is the probability of the *j*th consequence associated with alternative *i*, and U_{opt} is the utility of the optimum among *n* design alternatives. Note that the utility U_{ij} can also be expressed as a function of monetary value, if necessary.

   The decision tree approach also lends itself easily to prior and posterior analyses, thus facilitating a Bayesian approach for decision making with additional information. As the engineering system is being developed, additional information in the form of research, new literature, expert opinion, laboratory and field test data etc. may become available. The added cost of obtaining new information can also be included in the decision analysis. This may be used to make decisions regarding (1) whether additional information should be obtained (in this regard, the "value of information" can be included in the analysis), and (2) optimal sample size for statistical characterization of the random variables in the problem.

   Design under uncertainty has been implemented in the context of reliability-based optimum design in many studies (e.g., Rao, 1984; Frangopol, 1985; Mahadevan, 1992; Kowal, 1997; Pu et al, 1997). This has typically employed the minimization of expected weight or expected cost as the objective function, and component and system-level reliability requirement as constraints. The design variables are almost always the mean values of some of the random variables. However, in the context of modeling and simulation-based life cycle engineering, life-cycle cost (or lifetime utility) becomes an important consideration. This should include issues such as manufacturing cost, deployment cost, usage cost, repairs etc.

   When the design variables are random, decision-based design can be based on: (a) mean values, (b) standard deviations (e.g. manufacturing tolerances), (c) truncation of distributions, etc. Different alternatives have different reliability, cost and utility consequences. Since the sensitivity of failure probability to changes in the mean, variance, truncation etc. can be easily quantified, it is quite straightforward to include relevant distribution parameters as design variables. This sensitivity is computed, using the first order reliability approximation $P_{ij} = \Phi(-\beta_{ij})$, as (Madsen, Krenk and Lind, 1986)

$$\frac{\partial P_{ij}}{\partial \theta_k} = \varphi(-\beta_{ij}) \frac{\partial \beta_{ij}}{\partial \theta_k} \tag{2}$$

where β_{ij} is referred to as the reliability index, computed as the minimum distance from the origin to the limit state in the space of equivalent uncorrelated standard normal variables; θ_k is the kth distribution parameter, ϕ and Φ are the PDF and CDF of a standard normal variable. While this is easy to compute, the determination of the cost (or utility) sensitivities to changes in the uncertain design variables is not straightforward; it is dependent on the specific processes and materials used by the manufacturer. However, it is not necessary to determine the total manufacturing costs of the design, but merely the differential costs associated with specific changes in the design. The change in the product reliability can be estimated as,

$$\Delta P_{ij}\Big|_{\theta_k=\theta_{k0}} = \sum_{k=1}^{n}\left(\frac{\partial P_{ij}}{\partial \theta_k}\right)_{\theta_k=\theta_{k0}} \Delta\theta_k \tag{3}$$

where $\Delta\theta_k$ is the change in parameter k, and n is the number of changes in primitive variable distribution characteristics. The change in the product utility can be similarly estimated by replacing P_{ij} with U_{ij} in the above equation. With these relationships, a constrained optimization problem formulation and appropriate search algorithms can be used to obtain the optimum design.

5 GENERALIZED BAYESIAN FRAMEWORK FOR UNCERTAINTY MANAGEMENT

The use of model-based simulation for uncertainty quantification introduces several types of uncertainty. Consider the computation of failure probability in reliability analysis:

$$P(g(\mathbf{X}) < 0) = \int_{g(\mathbf{X})<0} f_{\mathbf{X}}(\mathbf{x})\ d\mathbf{x} \tag{4}$$

Every term on the right hand side of this equation is uncertain. There is inherent variability represented by the random variables $\mathbf{X}$, statistical uncertainty represented by the uncertainty in f (the joint probability density function of $\mathbf{X}$), and model uncertainty in the computation of $g(\mathbf{X})$, the limit state function that relates the system response to the performance capacity.

A Bayesian framework as shown in Figure 1 is proposed in this paper to incorporate the different kinds of uncertainty encountered in every stage of the system life cycle. The main advantage of the Bayesian approach is that it can combine the prior information and the objective information from numerical data. The prior information could be either objective or subjective.

Several applications of this Bayesian framework have been explored at Vanderbilt University for reliability updating and system health management by integrating model-based prediction with test or inspection information.

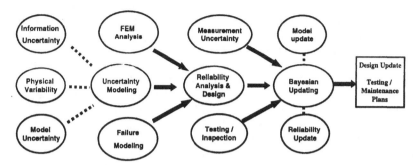

Figure 1. Bayesian framework for uncertainty management

Mahadevan, Zhang and Smith (2001) have extended this methodology to system reliability reassessment, through the use of Bayesian networks. Multiple failure sequences and correlations between component failures are considered and incorporated in the Bayesian network construction. Both forward and backward propagation formulae are developed. In the case of complicated structures, the Bayesian network is combined with the branch-and-bound method to consider only the significant failure sequences. Since most real systems involve multiple components and multiple failure sequences, the Bayesian network facilitates a comprehensive framework for system-level reliability management.

In the following sections, the application of this Bayesian framework is demonstrated for two important problems in system design and development, namely, reliability test planning and model validation.

6 INTEGRATION OF COMPUTATION AND TESTING

Let the vector of random variables $\mathbf{X}$ describe inherent variability, and the vector of random variables θ describe statistical and model uncertainties, which includes parameters that define the probability model of $\mathbf{X}$ and the limit state function $g(\mathbf{X})$. The uncertainty in θ is addressed through the definition of a prior distribution $f(\theta)$ based on limited data, past experience or expert opinion. For a time-dependent reliability problem, the failure probability at a specified time t, i.e., the CDF of life at t, is obtained as follows:

$$F_T(t,\theta) = \int_{g(X,\theta,t)\leq 0} f_{X,\theta}(X,\theta)dX \tag{5}$$

With uncertainty in θ, the failure probability at t is not deterministic, but is a random variable with its own distribution over [0,1]. The distribution of failure probability is evaluated through a nested reliability analysis. Well known reliability techniques, e.g., Monte Carlo simulation, first-order reliability method (FORM), second-order reliability method (SORM) etc. can be used to compute this integral. Note that this is the prior distribution for Bayesian analysis.

The fatigue life of a mechanical component is usually modeled using a lognormal or a Weibull distribution. The use of lognormal distribution makes it possible to develop closed-form analytical solutions. In the case of the Weibull distribution, there is no conjugate distribution for the parameters; therefore, all the posterior calculations have to be done numerically.

Before the tests, the parameters and confidence bounds of the failure probability distribution can be estimated using nested reliability analysis and regression. After the life test data is available, these parameters are updated. In traditional reliability testing, tests need to be continued till a desired confidence level is achieved. If the desired confidence interval width is taken as the criterion for terminating the testing, the integration of reliability computation and testing can be used to reduce costs through adaptive test planning. After each test, the Bayes estimators for the distribution parameters are recalculated, and the number of tests remaining to be performed to achieve the desired confidence interval width is updated Refer Mahadevan and Zhang (2001) for the mathematical details of this procedure.

This method is applied to the probabilistic life prediction of a composite helicopter rotor hub component, shown in Figures 2 and 3. The failure of the component due to fatigue delamination is considered. Dey et al (1999) consider the combined effect of tension and bending loads, and use a strain energy release rate method to model delamination failure. A virtual crack-closure technique (VCCT) method (Raju, 1987) is used to calculate the total strain energy release rate (G) at the delamination tip. Delamination onset is assumed to occur when the calculated G exceeds G_{crit} derived from material coupon delamination tests (Murri et al, 1997). A non-linear ANSYS finite element model has been used to determine the local forces and displacements needed to calculate G, and these G values are used with the Response Surface Method to approximate the response of the specimen.

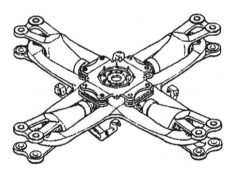

Figure 2. Rotor hub assembly

The corresponding limit state function is written as

$$g = G_{crit} - 175.344 \times \gamma \times (0.569 - 0.0861E_{11} + 0.0231P - 0.117\theta$$
$$- 0.000546P^2 + 0.003760^2 + 0.0046P\theta)$$

(6)

where γ represents the uncertainty and approximation in the calculation of G, including the discretization error in finite element analysis and approximation in the response surface equation. This equation is used in the first order reliability analysis (FORM) with the probabilistic characterizations of the evolved random variables to calculate the probability of failure at different values of t.

The critical strain energy release rate G_{crit} is a function of load cycles N and is also a random variable. Its mean value as a function of N and its standard deviation are obtained from material property tests and regression analysis. The statistical uncertainty in the determination of the mean and standard deviation of G_{crit}, which is caused by the limited statistical data, is characterized and used in the nested reliability analysis of Equation (6). The CDF of log life with mean value and 95% confidence bounds at different life values are shown in Figure 4.

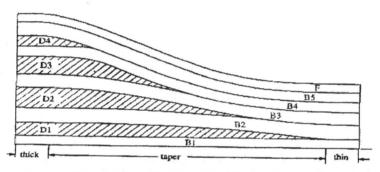

Figure 3. Half of the symmetric section of tapered composite test specimen

Table 1. Uncertainty characterization

Input parameter	Mean value	Standard deviation	Distribution type
Applied angle (θ, degree)	$\sim N(10, 1.0)$	$\sim N(1.2, 0.12)$	normal
Applied load (P, kN)	$\sim N(124.6, 12.46)$	$\sim N(13.35, 1.335)$	normal
Modulus (E_{11}, $10^9 N/m^2$)	48.34	0.63	normal
Model uncertainty parameter (γ)	-	-	uniform [0.9,1.1]

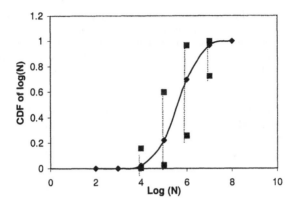

Figure 4. CDF of log life with statistical uncertainty

Next we have fatigue test data on 12 tapered laminate specimens, which is provided by Bell helicopter. Figure 5 shows the variation of 95% confidence interval width of mean life with gradually available test data. The results obtained by using the integration method are compared with the results by using "tests-only" approach. The confidence interval width obtained from the integration with prior computation is always smaller than the estimation only from tests, or we can say, less tests need to be done to obtain the same confidence interval width using the integration method than using the "tests-only" approach. For example, to achieve a confidence interval width of 2.0 for mean life, 12 tests are needed in the "tests-only" approach. However, with the integration of reliability computation, only 9 tests are needed. Thus, efficient and adaptive test planning can be implemented, where the testing is stopped when the confidence interval width reaches the desired value.

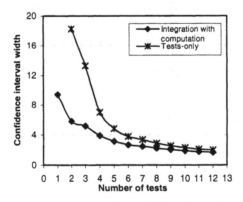

Figure 5. Variation of confidence interval width with number of tests

7 MODEL VALIDATION

In the case of model uncertainty, when there exist several possible models to describe a phenomenon, the Bayesian approach can be used to include all the candidate models by assigning model weight (the probability of each model being correct) and integrating the effects of all the models. When there is observation/data available, the model weights may be updated. This approach has been used to account for probability distribution type uncertainty and linear regression model uncertainty problems in statistics (Edwards 1984, Guedes Soares 1988, Draper, 1995, Volinsky 1997), and for mechanical model uncertainty in reliability prediction (Zhang and Mahadevan, 2000).

In many real engineering applications, usually there is only a single mathematical/mechanical model available. The soundness of the model is usually not 100% certain, but the model is assumed to be correct since there is no other choice. This uncertainty can be reduced through a deeper investigation of the mechanics, or, if there is some empirical information available on the system behavior, the model may be assessed and a decision may be made regarding acceptance or rejection of the model. This section investigates the latter issue. There are two basic assumptions in this kind of model assessment: 1) The validity of a model is judged only through its output, assuming that the investigation of its basic assumption and mechanics is unavailable; 2) The data used in the model prediction and the observation are assumed to be correct, and the computational errors are neglected; therefore, the inconsistency between the observed data and the model prediction is caused solely by the adoption of the wrong model.

This problem may be solved using classical hypothesis testing at a specified significance level. However, it has been argued that this method is difficult to interpret and sometimes misleading (Edward et al 1963, Leamer 1978, Berger and Sellke 1987). An alternative method is to use Bayesian hypothesis testing.

Bayesian hypothesis testing is investigated here for the validation of mechanical reliability computation models. Both time-independent and time-dependent problems are considered. The methodology is derived for problems both with and without statistical uncertainty. For problems without statistical uncertainty, the reliability output is expressed through a failure probability or life distribution. For such problems, the proposed method develops the computation of a Bayes factor, which can be used for acceptance/rejection decision-making. For problems with statistical uncertainty, the reliability output is expressed through a distribution function of the failure probability or life distribution. For such problems, the Bayes factor itself becomes a random variable. Therefore, the proposed method uses the probability of the Bayes factor exceeding a specified value as the decision criterion for model acceptance/rejection.

Consider two competing models M_i and M_j for the same system. Their prior probabilities to be the correct model are denoted by $P(M_i)$ and $P(M_j)$. Using Bayes' theorem, when an event/data Y is observed, the relative posterior probabilities of two hypotheses are obtained as:

$$\frac{P(M_i|observation)}{P(M_j|observation)} = \left[\frac{P(observation|M_i)}{P(observation|M_j)}\right]\left[\frac{P(M_i)}{P(M_j)}\right] \tag{7}$$

The term in the first set of square brackets on the right hand side is called "Bayes factor" (Jeffreys 1961). This Bayes factor can be viewed as the "weighted" likelihood ratio of M_i to M_j. The data are said to favor M_i relative to M_j if the Bayes factor exceeds one, that is, if the observed data is more likely under hypothesis M_i than it is under hypothesis M_j. Other criteria based on logarithmic transformation of the Bayes factor have also been suggested (Jeffreys, 1961).

For the case when only one model is available, the alternative model is the case that the model M is not correct. Therefore, the Bayes factor is $P(observation|M \text{ is correct})/P(observation|M \text{ is not correct})$. For time-independent problems without statistical uncertainty, the Bayes factor may be computed as

$$B = \frac{P(data|M \text{ is correct})}{P(data|M \text{ is not correct})} = \frac{x_0^k(1-x_0)^{n-k}}{\int_0^1 x^k(1-x)^{n-k} f(x|M \text{ is not correct})dx}$$
$$= x_0^k(1-x_0)^{n-k}(n+1)C_k^n \tag{8}$$
$$= (n+1)P(k|x_o,n)$$

where n is the number of tests, k is the number of failures, and x_0 is the failure probability estimate from the model.

For time-independent problems with statistical uncertainty, the Bayes factor is obtained as

$$B = \frac{P(data|M \text{ is correct})}{P(data|M \text{ is not correct})} = \frac{\int x(\theta)^k[1-x(\theta)]^{n-k} f(\theta)d\theta}{\int_0^1 x^k(1-x)^{n-k} f(x|M \text{ is not correct})dx} \tag{9}$$

The Bayes factor in Equation (7) is a random variable, and therefore the probability of $B>1$ (favoring model M) may be computed using the limit state-based reliability approach.

For time-dependent problems without statistical uncertainty, the life range may be divided into several intervals, and Equation (8) repeated for each interval. Therefore, several Bayes factors are obtained. For time-dependent problems with statistical uncertainty, Equation (9) is extended to the computation of $P(B > 1)$ for several intervals. It is easy to make the decision to accept or reject the model, if all intervals provide similar inference. However, when some Bayes factors favor model M while others indicate the opposite, the interpretation of these results and decision-making becomes qualitative or subjective. Such decision would likely include factors such as the cost of performing further test/investigation on the model, the consequences of adopting the wrong model, etc.

The Bayes factor approach is applied to assess the helicopter rotor component life prediction model of the previous section. The fatigue test data is divided into 6 intervals. The corresponding Bayes factors are calculated for both cases of with and without statistical uncertainties, and the results are shown in Table 2. It is seen that all the six Bayes factors in Column 4 are larger than 1, leading to acceptance of the life prediction model. However, if statistical uncertainty is considered, it is seen that $P(B_2>1)$ is not very high, which partly reflects the large statistical uncertainty involved in the model prediction. Thus there is less confidence in accepting the model.

Table 2. Bayes factors in the 6 life intervals

$\log(N)$ interval	k_i	Predicted x_{0i}	B_i (Equation 5, no statistical uncertainty)	$P(B_i>1)$ (with statistical uncertainty)
[4.0, 4.5]	1	0.08	4.987	87.37%
[4.5, 5.0]	3	0.14	2.019	69.85%.
[5.0, 5.5]	2	0.21	3.580	79.3%
[5.5, 6.0]	3	0.25	13.421	97.2%
[6.0, 6.5]	2	0.155	3.83	81.4%
[6.5, 7.0]	1	0.105	4.835	85.35%

Future work in this direction needs to treat model acceptance test planing as a decision theory problem, requiring the consideration of additional test costs, cost/risk consequences of accepting the current model, etc.

8 CONCLUSION

Integrated computational and test-based methods have been presented in this paper to incorporate all three types of uncertainty for design and certification analyses of complex engineering systems. The proposed framework includes the quantification of design reliability or risk and cost, and their sensitivity to the uncertainties in the system.

Bayesian methods have been developed to quantify the modeling and information uncertainty, including the uncertainty in mechanical and statistical model selection and the uncertainty in distribution parameters. The Bayesian methodology is combined with hypothesis testing to develop a model validation procedure. An adaptive method is developed to determine the number of tests needed to achieve a desired confidence level in the reliability estimates, by combining prior computational prediction and test data.

These methods have been demonstrated for the probabilistic life prediction of a composite helicopter rotor component under fatigue delamination, and combined with test data for uncertainty and reliability quantification, confidence level simulation, and model validation.

9 ACKNOWLEDGEMENT

The research reported in this paper was supported by funds from Sandia National Laboratories, Albuquerque, NM (Contract No. BG-7732), under the

Sandia/NSF Life Cycle Engineering Program (Project Monitors: Dr. David Martinez, Dr. Steve Wojkiewicz). The support is gratefully acknowledged. Sandia is a multi-program laboratory operated by Sandia Corporation, a Lockheed Martin Company, for the U.S. Department of Energy under Contract DE-AC04-94AL85000.

The example problem was first supported by funds from the U.S. Army Research Office STTR Contract Nos. 37448-EG-ST1 and DAA G55-98-C-0037 (Project Manager: Dr. Gary Anderson). The project was in collaboration with Brentwood Technologies (Dr. Animesh Dey, Dr. Robert, Tryon) and Bell/Textron Helicopter (Dr. Carl Rousseau). The support and contributions of these colleagues are gratefully acknowledged.

10 REFERENCES

Alvin, K. F., 1999, "A Method for Treating Discretization Error in Nondeterministic Analysis" *Proceedings of AIAA/ASME/ASCE/AHS/ASC Structures, Structural dynamics, and Materials Conference and Exhibit*, pp. 3061-3083.

Ang, A. H.-S. and Tang , W. H., 1984, *Probability Concepts in Engineering Planning and Design*. Volume II, (New York: John Wiley & Sons).

Berger, J. O., and Sellke, T., 1987, "Testing a point null hypothesis: the irreconcilability of *P* values and Evidence", *Journal of American Statistical Association*, **82**(397), pp. 112-122.

Box, G. E. P. and Tiao, C., 1973, *Bayesian Inference in Statistical Analysis*. Addison-Wesley, Reading, Massachusetts.

Cruse, T. A., Wu, Y. T., Dias, J. B., and Rajagopal, K. R., 1988, "Probabilistic Structural Analysis Methods and Applications", *Computers & Structures*, **30**, pp. 163-170.

Der Kiureghian, A., 1989, "Measures of structural safety under imperfect states of knowledge", *Journal of Structural Engineering*, ASCE, **115**(5).

Dey, A., Mahadevan, S., Tryon, R., Wang,Y. and Rousseau, C., 1999, "Fatigue reliability prediction of tapered composites", *Paper AIAA-99-1586*, 40th AIAA/ASME/ASCE/AHS/ASC Structures, Structural Dynamics and Materials (SDM) Conference, (St. Louis, MO), pp. 2935-2941.

Ditlevsen, O., 1982, "Model Uncertainty in Structural Reliability", *Structural Safety*, Vol 1, pp.73-86.

Draper, D., 1995, "Assessment and propagation of model uncertainty", *Journal of the Royal Statistical Society Series B*, **57**(1), pp. 45-97.

Edwards, G., 1984, "A Bayesian Procedure for Drawing Inference from Random Data" *Reliability Engineering*, **9**, pp.1-17.

Everett, R. A., Bartlett, F. D. and Elber, W., 1990, *Probabilistic Fatigue Methodology for Six Nines Reliability*, NASA Tech Memo 102757.

Frangopol, D. M., 1985, "Multicreteria Relaibility-Based Structural Optimization", *Structural Safety* 2, pp.154-159,.

Guedes Soares, C., 1997, "Quantification of model uncertainty in structural reliability", In: (Guedes Soares, C, editor.) *Probabilistic Methods for Structural Design*: 17-38. (Netherlands: Kluwer Academic Publishers).

Jeffreys, H., 1961, *Theory of Probability* (3rd. ed.), (London: Oxford University Press).

Kowal, M. T., 1997, "Mechanical System Reliability and Cost Integration", *Reliability-Based Mechanical Design*, ed. T. A. Cruse, (New York: Marcel Dekker).

Leamer, E. E., 1978, *Specification Searches*, (New York: John Wiley & Sons).

Madsen, H. O., Krenk, S. and Lind, N. C., 1986, *Methods of Structural Safety*, (New Jersey: Prentice-Hall Inc).

Mahadevan, S., 1992, "Probabilistic Optimum Design of Framed structures", *Computers and Structures*, **42**(3), pp.365-274.

Mahadevan, S., 1997, "System Reliability Analysis," *Reliability-Based Mechanical Design*, ed. T. A. Cruse, (New York: Marcel Dekker).

Mahadevan, S. and Dey, A., 1997, "Adaptive Monte Carlo Simulation for Time-Variant Reliability Analysis of Brittle Structures", *AIAA Journal*, **35**(2), pp.321-326.

Mahadevan, S., and Zhang, R., 2001, "Fatigue test planning using reliability and confidence simulation", *International Journal of Materials and Product Technology*, Vol. 16, Nos. 4-6, pp. 317-332.

Martz H. F. and Waller, P. A., 1982, *Bayesian Reliability Analysis*, (John Wiley & Sons).

Mehta, S., Cruse, T. A., and Mahadevan, S., 1992, "System Certification", *Reliability Technology* – 1992. *The Aerospace Division*, Ed. T. A. Cruse, ASME, **28**.

Murri,G. B., O'Brien, T. K. and Rousseau, C. O. , 1997, "Fatigue Life Prediction of Tapered Composite Laminates", *American Helicopter Society*, 53rd Annual Forum, Apr-May, Virginia Beach.

Oberkampf, W., Deland, S. M., Rutherford, B. M., Diegert, K.V. and Alvin, K. F. , 1999, "A New Methodology for the Estimation of Total Uncertainty in Computational Simulation" *Proceedings of 40th AIAA/ASME/ASCE/AHS/ASC Structures, Structural dynamics, and Materials Conference and Exhibit*, pp. 3-3060.

Pu, Y., Das, P. K., and Faulkner, D. , 1997, "A Strategy for Reliability-Based Optimization", *Engineering Structures*, **19**, pp. 276-282.

Rackwitz, R. and Fiessler, B., 1978, "Structural reliability under combined random load sequence", *Computer and Structures*, **9**(5), pp. 484-494.

Raiffa, H. and Schlaifer, R., 1961, *Applied Statistical Decision Theory*, (Cambridge MA: Harvard University Press).

Raju, I. S., 1987, "Calculation of strain energy release rates with higher order and singular finite elements", *Engineering Fracture Mechanics,* **28**, pp. 251-274.

Rao, S. S., 1984, "Multiobjective Optimization in Structural Design with Uncertain Parameters and Stochastic Processes", *AIAA Journal*, **22**, pp. 1670-1678.

Sørensen, J. D. and Thoft-Christensen, P., 1983, "Model uncertainty for bilinear hysteretic systems", *System Modeling and Optimization*, ed. Thoft-Christensen, (Springer-Verlag), pp. 585-596.

Thoft-Christensen, P., and Murotsu, Y., 1986, *Application of Structural System Reliability Theory*. (Berlin: Springer Verlag).

Tvedt, L. , 1990, "Distribution of Quadratic Forms in Normal Space-Application to Structural Reliability", *Journal of Engineering Mechanics*, ASCE, **116**(6), pp.1183-1197.

Volinsky, C. T., et al., 1997, "Bayesian model averaging in proportional hazard models: assessing the risk of a stroke", *Journal of Royal Statistical Society*, **46**(4), pp. 433-448.

Zhang, R., and Mahadevan, S., 2000, "Model uncertainty and Bayesian updating in reliability-based inspection", *Structural Safety*, **22**(2),pp. 145-160.

Zhang, R., and Mahadevan, S., 2001, "Integration of computation and testing for reliability estimation", *Reliability Engineering and System Safety* (in print).

11 SUBJECT INDEX

Uncertainty analysis, probability, statistics, modeling, simulation, engineering design, reliability, Bayes statistics, optimization.

Very Efficient Simulation for Engineering Design Problems with Uncertainty

Chun-Hung Chen[1]

1 INTRODUCTION

The process of design or decision making is usually modeled as a design or optimization problem. However, because of complexity, stochastic relations, and so on, not all real-world problems can be represented adequately in the existing model forms. Attempt to use analytical models for such systems usually require so many simplifying assumptions that the solutions are likely to be inferior or inadequate for implementation. Often, in such instances, the only alternative form of modeling and analysis available to decision maker is simulation.

Stochastic simulation technology, such as discrete-event simulation and Monte Carlo simulation, has matured over the past decade and is now commonly used to evaluate large-scale real systems with complex stochastic behavior. Simulation allows one to more accurately specify a system through the use of logically complex, and often non-algebraic, variables and constraints. This capability compliments the inherent limitation of traditional optimization. However, the added flexibility often creates models that are computationally intractable. To obtain a good statistical estimate for a design decision, a large number of simulation runs is usually required for each design alternative. This is due to the stochastic features and the slow convergence of a performance measure estimator relative to the number of runs (Fabian 1971, Kushner and Clark 1978). Furthermore, many alternative designs must be simulated in order to find a good design. The total computation cost for simulation-based approaches is too expensive (Law and Kelton 1991). New Techniques must be developed to address the computational issues.

There exists a large literature on innovative methods for improving the efficiency of simulation experiments. Bratley et al. (1987) and Fishman (1996) provide a comprehensive presentation of recent developments of simulation methodologies. Some methods exploit the fact that the required number of simulation runs decreases when the simulation variance is reduced. Experimental as well as theoretical result for some special cases have shown that some sort of dependence among experiments for different designs can increase the chance of selecting the true best design. Glasserman and Vakili (1994) prove, for the case of additive performance measures of finite-state Markov chains, that association indeed increases the rate of convergence. Yang and Nelson (1991) and

[1] Department of Systems Engineering & Operations Research, George Mason University, Fairfax, VA 22030

Glasserman and Yao (1992) show that the schemes of common random numbers and control variates are helpful in obtaining better confidence intervals for various selection procedures when the performance measure is obtained by averaging i.i.d. random variables with normal distributions. Glynn (1994) and Heidelberger (1993) deal with rare event problems by developing an importance sampling scheme. Control Variates (Nelson, 1990) attempts to take advantage of correlation between certain random variables to obtain variance reduction. The method of applying variance reduction techniques usually depends on the particular model of interest.

On the other hand, several researchers have shown that allocating simulation runs in an uneven manner can significantly reduce the total number of runs required to identify the best candidate design. Previous work includes Dudewicz and Dalal (1975) and Rinott (1978) who develop two-stage procedures for allocating runs to designs and several papers (Nelson 1995, Bechhofer et al. 1995) that extend the technique to general ranking and selection problems. The major disadvantages of existing approaches is that they utilize only the information of variance to control simulation experiments. While simulation efficiency is improved, the time savings is not significant and so the applicability is still limited.

Our approach extends the current literature in three important ways:
- Whereas much of the optimal design literature focuses on deterministic optimization models, our research introduces a nonlinear stochastic optimization framework that incorporates uncertainty, variability, and risk preferences.
- While most existing variance reduction techniques for simulation are designed for a particular class of special-structure problems, the proposed methodology is applicable to a much broader class of simulation problems.
- Our novel technique utilizes the information of both variance and the profile of the design space to *intelligently* determine the most efficient way of performing simulation. This results in drastic reduction in computation time.

This chapter is organized as follows: In the next section, we formulate the engineering design problem. Section 3 introduce the novel idea of optimal computing budget allocation (OCBA) and presents an asymptotic allocation rule for OCBA. The performance of the technique is illustrated with a series of numerical examples in Section 4. Section 5 concludes the chapter.

2 ENGINEERING DEISGN WITH UNCERTAINTY

In this section, we offer a general formulation of engineering design problems. Broadly stated, the problem is to identify a design that optimizes the utility function of the design engineer, where the utility function and design specifications may be complex, stochastic, or not precise. We formulate the problem using the following notation.

θ_i the system design parameter vector for design i, $i = 1, 2,..., k$,

ξ a random vector that represents uncertain factors in the systems.

$f(\theta_i, \xi)$ the system performance measure, a function of the decisions variables θ_i, and the uncertainty ξ. Since θ_i is random, f is also a random variable or random vector.

$U(\cdot)$ the utility function which has multiple attributes. Typical examples include a mean-variance function (Markowitz 1959) or the von Neumann-Morgenstern utility function (Fishburn 1970, Hazelrigg 1996). If only extreme case analysis is of interest, say $P\{U(f(x, y_s))>C\}$, where C is a critical value, then the first term of the objective function becomes $E[I_{U(f(x,y))>C}]$ where I is the indicator function: $I_B=1$ when B is true, 0 otherwise.

The design problem is then stated as,

$$\text{\textit{Minimize}} \quad -E[U(f(\theta_i, \xi)] \tag{1}$$

For notation simplicity, we define

$$L(\theta_i, \xi) \equiv -U(f(\theta_i, \xi) \tag{2}$$

Thus the design problem can be stated as

$$\min_{\theta_i \in \Theta} J(\theta_i) \equiv E[L(\theta_i, \xi)] \tag{3}$$

where Θ, the search space, is an arbitrary, huge, structureless but finite set; J, and the performance criterion which is the expectation of L, the sample performance. Note that for the complex systems considered in this paper, $L(\theta, \xi)$ is available only in the form of a complex calculation via simulation. The system constraints are implicitly involved in the simulation process, and so are not shown in (3). The standard approach is to estimate $E[L(\theta_i, \xi)]$ by the sample mean performance measure

$$\bar{J}_i \equiv \frac{1}{N_i} \sum_{j=1}^{N_i} L(\theta_i, \xi_{ij}), \tag{4}$$

where ξ_{ij} represents the j-th sample of ξ and N_i represents the number of simulation samples for design i. Denote by

σ_i^2 the variance for design i, i.e., $\sigma_i^2 = \text{Var}(L(\theta_i, \xi))$. In practice, σ_i^2 is unknown beforehand and so is approximated by sample variance.

b the design having the smallest sample mean performance measure, i.e., $\bar{J}_b \leq \min_i \bar{J}_i$,

$\delta_{b,i} \equiv \bar{J}_b - \bar{J}_i$.

As N_i increases, $\bar{J}_i$ becomes a better approximation to $J(\theta_i)$ in the sense that its corresponding confidence interval becomes narrower. The ultimate accuracy of this estimate cannot improve faster than $1/\sqrt{N}$. Note that each sample of $L(\theta_i, \xi_{ij})$ requires one simulation run. A large number of required samples of $L(\theta_i, \xi_{ij})$ for all designs may become very time consuming.

While the design with the smallest sample mean (design b) is usually picked, design b is not necessarily the one with the smallest unknown mean performance. *Correct selection* is therefore defined as the event that design b *is actually the best design* (i.e., with the smallest population performance). In the remainder of this paper, let "CS" denote Correct *Selection*. In engineering design, $P\{CS\}$ increases as more simulation has been performed. We gradually increase the number of simulation runs until $P\{CS\}$ is sufficiently high. There exists some literature about the estimation of $P\{CS\}$ (e.g., Bernardo and Smith, 1984, Chen, 1996, Inoue and Chick, 1998).

3 OPTIMAL COMPUTING BUDGET ALLOCATION

Our efficient simulation technique is the one that optimally allocates a computing budget to the designs under evaluation. Intuitively, to ensure a high probability of correctly selecting an optimal design, a larger portion of the computing budget should be allocated to those designs that are critical in the process of identifying good designs. In other words, a larger number of simulations must be conducted with those critical designs in order to reduce estimator variance. On the other hand, limited computational effort should be expended on non-critical designs that have little effect on identifying the good designs even if they have large variances. Overall simulation efficiency is improved as less computational effort is spent on simulating non-critical designs and more is spent on critical designs. The ideas are explained using the following simple example. Suppose we are performing simulations for 5 design alternatives in order to determine a design with minimum mean performance measure. First of all, we conduct some preliminary simulation for all 5 designs. Figure 1 gives an example of their 99% confidence intervals obtained from the preliminary simulation. Note that the uncertainty of estimation is due to the system's stochastic features.

As seen in Figure 1, while there is uncertainty in the estimation of the performance for each design, it is obvious that designs 2 and 3 are much better than other designs, if we intend to find a design with minimum performance measure. And so only designs 2 and 3 need to be further simulated to reduce estimation uncertainty in order to correctly identify the best design. By stopping simulations for designs 1, 4, and 5 earlier, we can save a lot of computation cost.

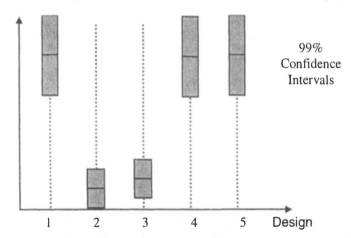

Figure 1 99% confidence intervals for five design alternatives after some preliminary simulation.

However, most cases are not as trivial as that in Figure 1. It is more common to see cases like another example shown in Figure 2, where some designs seem better, but not outstandingly better, than other designs. It is not straightforward to determine which designs can be stopped from the simulation experiment in general, and when they should be stopped. Our research will provide a systematic approach to this issue. Namely, we will determine when and which designs must be stopped for further simulation, or which designs should be further simulated, in a way that a highest simulation efficiency can be achieved.

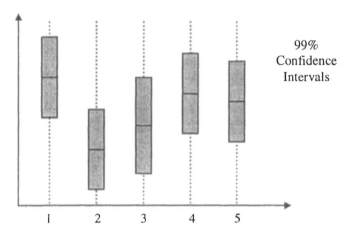

Figure 2. 99% confidence intervals after some preliminary simulation.

To be specific, we would like to allocate simulation trials to designs in a way that the simulation efficiency is maximized given that a desired simulation quality is obtained. We present a new optimal computing budget allocation (OCBA) technique to accomplish this goal. Stating this more precisely, we wish to choose N_1, N_2, $\cdots$, N_k such that $P\{CS\}$ is maximized, subject to a limited computing budget T,

$$\max_{N_1, \cdots, N_k} P\{CS\}$$
$$s.t. \ N_1 + N_2 + \cdots + N_k = T,$$
$$N_i \in N, i = 1, \ldots, k. \tag{5}$$

Here N is the set of non-negative integers and $N_1 + N_2 + \cdots + N_k$ denotes the total computational cost assuming the simulation times for different designs are roughly the same. Chen et al. (2000) offer an asymptotic allocation for the simulation budget allocation given in (5).

Theorem 1. Given a total number of simulation samples T to be allocated to k competing designs whose performance is depicted by random variables with means $J(\theta_1), J(\theta_2), ..., J(\theta_k)$, and finite variances $\sigma_1^2, \sigma_2^2, ..., \sigma_k^2$ respectively, as $T \to \infty$, $P\{CS\}$ can be asymptotically maximized when

$$(a) \quad \frac{N_i}{N_j} = \left(\frac{\sigma_i / \delta_{b,i}}{\sigma_j / \delta_{b,j}} \right)^2 , \, i, j \in \{1, 2, ..., k\}, \text{ and } i \neq j \neq b,$$

$$(b) \quad N_b = \sigma_b \sqrt{\sum_{i=1, i \neq b}^{k} \frac{N_i^2}{\sigma_i^2}} \, .$$

where N_i is the number of samples allocated to design i, $\delta_{b,i} = \bar{J}_b - \bar{J}_i$, and $\bar{J}_b \leq \min_i \bar{J}_i$.   #

Remark 1. In the case of $k = 2$ and $b = 1$, Theorem 1 yields

$$N_1 = \sigma_1 \sqrt{\frac{N_2^2}{\sigma_2^2}} \, ,$$

Therefore,

$$\frac{N_1}{N_2} = \frac{\sigma_1}{\sigma_2} \, .$$

This evaluated result is identical to the well-known optimal allocation solution for $k = 2$.   #

With Theorem 1, we now present a cost-effective sequential approach based on OCBA to select the best design from k alternatives with a given computing budget. Initially, n_0 simulation replications for each of k design are conducted to get some information about the performance of each design during the first stage. As simulation proceeds, the sample means and sample variances of each design are computed from the data already collected up to that stage. According to this collected simulation output, an incremental computing budget, Δ, is allocated based on Theorem 1 at each stage. Ideally, each new replication should bring us closer to the optimal solution. This procedure is continued until the total budget T is exhausted. The algorithm is summarized as follows.

A Sequential Algorithm for Optimal Computing Budget Allocation (OCBA)

Step 0. Perform n_0 simulation replications for all designs; $l \leftarrow 0$; $N_1^l = N_2^l = \cdots = N_k^l = n_0$.

Step 1. If $\sum_{i=1}^{k} N_i^l \geq T$, stop.

Step 2. Increase the computing budget (i.e., number of additional simulations) by Δ and compute the new budget allocation, $N_1^{l+1}, N_2^{l+1}, \cdots, N_k^{l+1}$, using Theorem 1.

Step 3. Perform additional $\max(0, N_i^{l+1} - N_i^l)$ simulations for design i, for $i = 1, ..., k$.
$l \leftarrow l + 1$. Go to Step 1.

In the above algorithm, l is the iteration number. As simulation evolves, design b, which is the design with the smallest sample mean, may change from iteration to iteration, although it will converge to the optimal design as the l goes to infinity. When b changes, Theorem 1 is directly applied in step 2. However, the older design b may not be simulated at all in this iteration in step 3 due to extra allocation to this design in earlier iterations.

In addition, we need to select the initial number of simulations, n_0, and the one-time increment, Δ. Chen et al. (1999) offers detailed discussions on the selection. It is well understood that n_0 cannot be too small as the estimates of the mean and the variance may be very poor, resulting in premature termination of the comparison. A suitable choice for n_0 is between 5 and 20 (Law and Kelton 1991, Bechhofer et al. 1995). Also, a large Δ can result in waste of computation time to obtain an unnecessarily high confidence level. On the other hand, if Δ is small, we need to the computation procedure in step 2 many times. A suggested choice for Δ is a number bigger than 5 but smaller than 10% of the simulated designs.

4 NUMERICAL TESTING AND COMPARISON WITH OTHER ALLOCATION PROCEDURES

In this section, we test our OCBA algorithm and compare it with several different allocation procedures by performing two numerical experiments: one has 10 designs in the design space; the other has as many as 100 designs.

4.1 Different Allocation Procedures

In addition to the OCBA algorithm, we test several procedures and compare their performances. Among them, equal allocation represents the use of direct simulation, and Rinott is highly popular in simulation literature. We briefly summarize the compared allocation procedures as follows.

Equal Allocation

This is the simplest way to conduct simulation experiments and has been widely applied. The simulation budget is equally allocated to all designs. Namely, we simulate all design alternatives equally, that is, $N_i = T/k$ for each i. The performance of equal allocation will serve as a benchmark for comparison.

Rinott Procedure

The two-stage procedure of Rinott (1978) has been widely applied in the simulation literature (Law and Kelton 1991). Unlike the OCBA approach, the two-stage procedures are developed based on the classical statistical model. See Bechhofer et al. (1995) for a systematic discussion of two-stage procedures. In the first stage, all designs are simulated for n_0 samples. Based on the sample variance estimate (S_i^2) obtained from the first stage, the number of additional simulation samples for each design in the second stage is determined by:

$$N_i = \max(0, \lceil (S_i^2 h^2 / d^2 \rceil - n_0), \text{ for } i = 1, 2, \ldots, k,$$

where $\lceil \bullet \rceil$ is the integer "round-up" function, d is the indifference zone, h is a constant which solves Rinott's integral (h can also be found from the tables in Wilcox 1984). In short, the computing budget is allocated proportionally to the estimated sample variances. The major drawback is that only the information on variances is used when determining the simulation allocation, while the OCBA algorithm utilizes the information on both means and variances. As a result, the performance of Rinott's procedure is not as good as others. We do, however, include it in our testing due to its popularity in the simulation literature.

4.2 Numerical Experiments

In all the numerical illustrations, we estimate $P\{CS\}$ by counting the number of times we successfully find the true best design (design 1 in this example) out of 10,000 independent applications of each selection procedure. $P\{CS\}$ is then obtained by dividing this number by 10,000, representing the correct selection frequency.

Experiment 1. 10 Designs

There are ten design alternatives. Suppose $L(\theta_i, \xi)) \sim N(i, 6^2)$, $i = 1, 2, .., 10$. We want to find a design with the minimum mean. It is obvious that design 1 is the actual best design. In the numerical experiment, we compare the convergence of $P\{CS\}$ for different allocation procedures. We have $n_0 = 10$ and $\Delta = 20$.

Different computing budgets are tested. Figure 1 shows the test results using OCBA and the other two different procedures discussed in section 4.1. We see that all procedures obtain a higher $P\{CS\}$ as the available computing budget increases. However, OCBA achieves a same $P\{CS\}$ with a lower amount of computing budget than other procedures. In particular, we indicate the computation costs in order to attain $P\{CS\} = 99\%$ for different procedures in Figure 3. Our OCBA can further reduce the simulation time by 75% for $P\{CS\}=99\%$.

It is worth noting that Rinott's procedure does not perform much better than the simple equal allocation. This is because Rinott's procedure determines the number of simulation samples for all designs using only the information of sample variances. On the hand, Rinott's procedure is much slower than OCBA. This is because when determining budget allocation, OCBA exploits the information of both sample means and variances, while Rinott's procedure does not utilize the information of sample means. The sample means can provide the valuable information of relative differences across the design space.

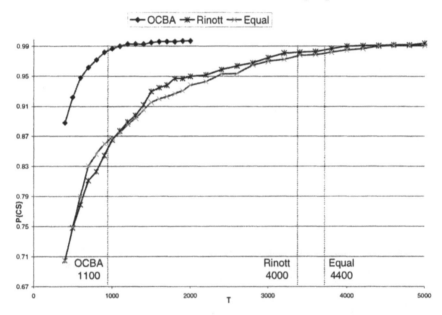

Figure 3 .*P*{CS} vs. *T* using three different allocation procedures for experiment 1. There are 10 designs for simulation. The computation costs in order to attain *P*{CS} = 99% are indicated.

Experiment 2. 100 Designs

This is a variant of experiment 1, but with a much bigger design space. To see the performance of the OCBA algorithm within a bigger design space, we increase the number of designs to 100. $L(\theta_i, \xi)) \sim N(i/10, 1^2)$, $i = 1, 2, .., 99, 100$. Note that we have the range of the means for these 100 designs the same as those in earlier 10-design experiments, which is from 1 to 10.

Figure 4 depicts the simulation results. We see that OCBA is more than one order of magnitude more efficient than the compared procedures. The higher efficiency is obtained because a larger design space gives the OCBA algorithm more flexibility in allocating the computing budget.

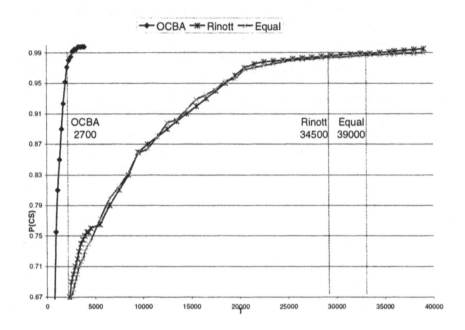

Figure 4. $P\{CS\}$ vs. T using three different allocation procedures for experiment 2. There are 100 designs for simulation. The computation costs in order to attain $P\{CS\} = 99\%$ are indicated.

5 COLCLUSIONS

We present a highly efficient procedure to identify the best design out of k (simulated) competing designs. The purpose of this technique is enhance the efficiency of simulation experiments in engineering design. The objective is to maximize the simulation efficiency, expressed as the probability of correct selection within a given computing budget. Our procedure allocates replications in a way that optimally improves an asymptotic approximation to the probability of correct selection. We also compare tow other allocation procedures, including a popular two-stage procedure in simulation literature. Numerical testing shows that our approach is much more efficient than all compared methods. Comparisons with the crude ordinal optimization show that our approach can achieve a speedup factor of 3~4 for a 10-design example. The speedup factor is even higher with the problems having a larger number of designs.

6 ACKNOWLEDGEMENTS

This work has been partially supported by NSF under Grants DMI-0002900, DMI-0049062, and by George Mason University Research Foundation. The author would like to thank Drs. Ken Chong, George Hazelrigg, and Vicente Romezo their valuable comments and suggestions.

7 REFERENCES

Bechhofer, R. E., Santner, T. J. and Goldsman, D. M., 1995, *Design and Analysis of Experiments for Statistical Selection, Screening, and Multiple Comparisons*, John Wiley & Sons, Inc.

Bratley, P., Fox, B. L. and Schrage, L. E. 1987, *A Guide to Simulation*. 2nd ed. Springer-Verlag.

Chen, C. H. 1996, A Lower Bound for the Correct Subset-Selection Probability and Its Application to Discrete Event System Simulations, *IEEE Transactions on Automatic Control*, **41**, pp. 1227-1231.

Chen, C. H., Wu, S. D. and Dai, L. 1999, Ordinal Comparison of Heuristic Algorithms Using Stochastic Optimization, *IEEE Transactions on Robotics and Automation*, **15**, pp. 44-56.

Chen, C. H., Lin, J. Yücesan, E. and Chick, S. E. 2000, Simulation Budget Allocation for Further Enhancing the Efficiency of Ordinal Optimization, *Journal of Discrete Event Dynamic Systems: Theory and Applications*, **10**, pp. 251-270.

Dudewicz, E. J. and Dalal, S. R. 1975, Allocation of Observations in Ranking and Selection with Unequal Variances. *Sankhya*, **B37** pp. 28-78.

Fabian, V., 1971, *Stochastic Approximation, Optimization Methods in Statistics*, Edited by J. S. Rustagi, Academic Press, New York.

Fishburn, P. C., 1970, *Utility Theory for Decision Making*, Wiley.

Fishman, G. S. 1996, *Monte Carlo: Concepts, Algorithms, and Applications*, Springer-Verlag.

Glasserman, P. and Vakili, P. 1994, Correlation of Uniformized Markov Chains Simulated in Parallel, *Prob. Engi. Information Science*, **8**, pp. 309-326.

Glasserman, P. and Yao, D. D. 1992, Some Guidelines and Guarantees for Common Random Numbers, Management Science, **38**, pp. 884-908.

Glynn, P. W. 1994, Efficiency Improvement Technique, *Annals of Operations Research*.

Hazelrigg, G. A., 1996, *Systems Engineering: An Approach to Information-Based Design*, Prentice Hall.

Heidelberger, P. 1993, Fast Simulation of Rare Events in Queueing and Reliability Models, *Performance Evaluation of Computer and Communication Systems*, ed. L. Donatiello and R. Nelson, Springer Verlag, pp. 165-202.

Kushner, H. J., and Clark, D. S. 1978, *Stochastic Approximation for Constrained and Unconstrained Systems*, Springer-Verlag.

Law, A. M. and Kelton, W. D. 1991, Simulation Modeling & Analysis, McGraw-Hill, Inc.

Markowitz, H., 1959, *Portfolio Selection: Efficient Diversification of Investments*, John Wiley, New York.

Nelson, B. L. 1990, Control-Variates Remedies, *Operations Research*.

Nelson, B. L., 1995, Stochastic modeling: analysis & simulation. McGraw-Hill, Inc.

Rinott, Y., 1978, On Two-stage Selection Procedures and Related Probability Inequalities, *Communications in Statistics*, **A7**,pp. 799-811.

Yang, W. and Nelson, B. L. 1991, "Using Common Random Numbers and Control Variates in Multiple Comparison Procedures", Operations Research, **39**, pp. 583-591.

Decision-Theoretic Methodology for Performance-Based Design

T. Igusa[1], S. G. Buonopane[1], D. Q. Naiman[1] and B. R. Ellingwood*

1 INTRODUCTION

Performance-based design is the next-generation approach to design for civil engineering structures. Three major research issues associated with performance-based design are: (a) civil structures are built without the use of a prototype or full-scale testing and are one-of-a-kind systems; (b) the uncertainties associated with new technology, such as new materials or structural configurations, and external loads are high; and (c) the cost of failure can be many times larger than the cost of the original structure, and the societal impact is severe. Other engineering disciplines are also confronted with the need to reduce full-scale tests and design efficiently within a context of high uncertainties and high failure costs.

In this report of a recently awarded research project, the planned research tasks that are directed towards the above research issues are described. The description of the inter-related research tasks is grouped into decision-making, simulation, and test-bed analysis sections.

2 RESEARCH TASKS WITHIN A DECISION-MAKING FRAMEWORK

While there are many possible approaches to the problem briefly described in the introduction, it is believed that a Bayesian decision framework provides the most appropriate platform from which a unified set of techniques can be developed. The motivation for using this framework is that prior knowledge can be used in a systematic manner in lieu of full-scale measurements and that decision-making can proceed in a manner that accounts for high uncertainties and costs of failure. The complexity of a structural system, however, and the resulting high-dimensional design space makes a direct application of Bayesian decision theory to structural design a significant computational challenge. Some details on approaching the computational and information research issues in developing a Bayesian decision framework are given in the following.

2.1 Propagation of Uncertainty between Scales Using Bayesian Inference

There has been increasing interest in using high-resolution models to simulate structural mechanics systems. Advances in computing technology have made possible detailed simulation of material behaviour down to the sub-micron

[1] Johns Hopkins University; *now at Georgia Tech

level, provided that the material properties are not random. Nevertheless, straightforward simulation of material behaviour that incorporates randomness in the material properties will not be possible because of the large number of random variables that increases at finer levels of detail.

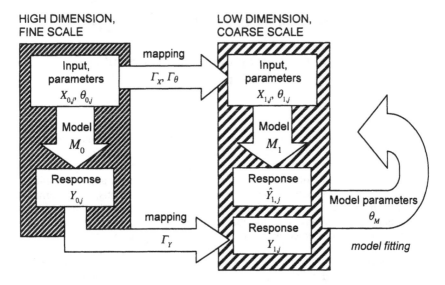

Figure 1 Standard Model-Fitting Approach to Multiple Scales

To state this more precisely, consider a model M_0 at a small scale. The model is represented by an operator that relates a set of input variables X_0 such as stresses and displacements prescribed on a boundary of a physical domain with a set of response variables Y_0 such as internal stresses and displacements. The model is in terms of a set of parameters, θ_0. Herein, it is assumed that θ_0 is random and that X_0 is deterministic. Mathematically, the model can be represented as

$$Y_0 = M_0(X_0, \theta_0) \tag{1}$$

If standard simulation techniques are used, then a set of n simulated parameters $\Theta_0 = \{\theta_{0,1}, \theta_{0,2}, ..., \theta_{0,n}\}$ would be generated, and using Equation (1), the corresponding simulated responses $\mathbf{Y}_0 = \{Y_{0,1}, Y_{0,2}, ..., Y_{0,n}\}$ would be generated by solving

$$Y_{0,j} = M_0(X_0, \theta_{0,j}) \quad \text{for } j = 1, ..., n \tag{2}$$

Inference on the probabilistic characteristics of the response would be made based on the simulated data set $\mathbf{Y}_0$.

The model M_0 would typically represent a small subcomponent of the entire system of interest, such as one of hundreds of connections in a building. To perform a realistic analysis of the system, it is necessary to condense the information contained in the detailed model M_0 and the associated high dimensional vectors θ_0, X_0 and Y_0 to develop a reduced model M_1 with relatively low dimensional parameter, input and response vectors θ_1, X_1 and $\hat{Y}_1$.

$$\hat{Y}_1 = M_1(X_1, \theta_1; \theta_M) \tag{3}$$

The hat is used because, in general, $\hat{Y}_1$ will include approximation errors associated with the fact that model M_1 is coarse relative to M_0. In the argument for M_1, there is an additional model-fitting parameter vector θ_M used to fit the model to the more detailed simulated model M_0, to experimental data (which will not be explored in this paper), or to both. For any fitting procedure involving model M_0, it is necessary to map the vectors θ_0, X_0 and Y_0 into the lower dimensional spaces corresponding to θ_1, X_1 and Y_1. These mapping relations can be represented as

$$\theta_1 = \Gamma_\theta(\theta_0), \quad X_1 = \Gamma_X(X_0), \quad Y_1 = \Gamma_Y(Y_0) \tag{4}$$

Typically, Γ_θ, Γ_X, and Γ_Y would involve local averaging, spatial fitting (such as a linear boundary displacement field used in model M_1 to a more general displacement field obtained in model M_0) or other dimension-reducing operation.

If model M_0 is not used, the parameter and input vectors θ_1 and X_1 can be given directly, or they can be given in terms of the vectors θ_0 and X_0 using Equation (4). If model M_0 is used, the input vector X_0 and the set of n simulated parameter and response vectors Θ_0 and Y_0 are mapped into the lower dimensional vectors and sets X_1, Θ_1 and Y_1 using Equation (4). The model-fitting parameter vector θ_M is adjusted so that a norm of the difference

$$\hat{Y}_1 - Y_1 = M_1(X_1, \Theta_1; \theta_M) - Y_1 \tag{5}$$

is minimized in some statistical (e.g., least squares) sense. Given models M_0 and M_1, there are many statistical methods for determining the model-fitting parameter vector θ_M. The standard model-fitting scheme is illustrated in Figure 1.

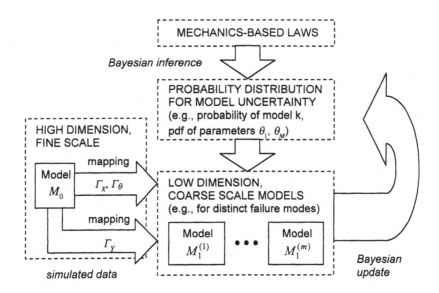

Figure 2 Bayesian Network for Uncertainty Propagation through Multiple Scales

If Bayesian statistical methods are used, then θ_M is described in terms of probability distributions. The basic problem of such methods is that the amount of computation required to generate Y_0 using model M_0 is prohibitively large, particularly when statistical information to describe the behaviour of the response in a neighbourhood away from the mean is required. In the work currently in progress by the writers, however, new approaches beyond what can be loosely described as generic statistical methods are being explored. The most important aspect of these new approaches is the integration of structural mechanics with statistical analysis. While the laws of mechanics are already included in model M_0, the laws are applied at a level of detail too fine to allow for direct interpretation of the results obtained from model M_1. What appears to be a more promising approach to the analysis of model M_1 is the use of a Bayesian inference engine, formulated through the use of Bayesian networks, to provide mechanics-based probabilistic information for the parameter vector θ_M.

The inference tools, integrated as shown in Figure 2, work as follows. The information of the physical component is given by input vector X_0 and simulated parameter vector θ_{0j}. The computational approach to obtain information to be used to formulate model M_1 is to generate the response vector Y_{0j} using the model M_0 in Equation (2) and map the results using Equation (4) to the data set $(X_1, \theta_1, \hat{Y}_{1,j})$. Due to the computationally intensive process required in Equation (2), the alternative is to use a Bayesian network to make an inference on the response corresponding to the input vector X_0 and simulated parameter vector θ_{0j}. The flow of information in this second approach may have more than one pattern, as indicated in Figure 2.

2.2 Influence of Model Uncertainty on the Design Decision Process

To facilitate the introduction of new technology into structures, designers need models that can predict structural behaviour and failure limits. This is particularly important in engineered structures that are developed with the lack of, or limited full-scale data and which are characterized by high costs of failure. A fundamental challenge that faces engineers developing simulation models of new technology is the development of comprehensive predictive capabilities for performance and failure. There is a predilection to continue to use old technology, because its performance characteristics have been time-tested in the field, and there is a reassuring degree of confidence in its models. From the designer point of view, the uncertainty in the predictive capabilities of models of new technology combined with high failure costs constitute a significant barrier to using new technology.

Model uncertainty is defined to be the uncertainty associated with whether one or more predictive models reflect the actual physical behaviour of the full-scale structure under field conditions. The uncertainty arises due to the assumptions and approximations needed to formulate any simulation model. The research work outlined in the previous subsection is intended to reduce model uncertainty through the power of simulation at detailed scales. Two additional approaches to reducing model uncertainty are through the use of laboratory tests and the use of expert opinion. Since these two approaches are well established within Bayesian decision theory, they will be used but not further developed in the planned research effort. While model uncertainty can be reduced using the above approaches, it cannot be completely eliminated. This is not a fundamental problem because, within a Bayesian decision framework, all uncertainties are incorporated in the decision-making process to produce a design with the highest expected utility.

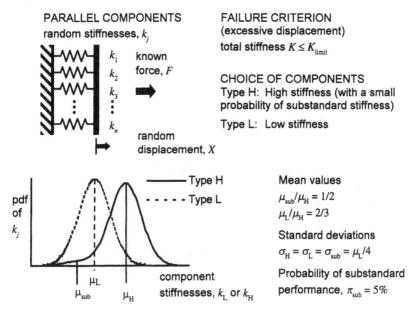

Figure 3 Example of Trade Off between Performance and Uncertainty

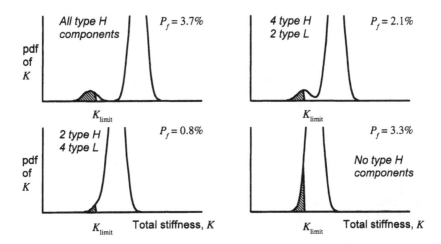

Figure 4 Probability of Failure, P_f, for Different Combination of Components

The quantification of uncertainty in one or more models can be expressed mathematically using a hierarchical Bayes analysis, using essentially a nested sequence of priors. The issue of how to quantify model uncertainty accurately remains a basic problem in information theory, and has led many researchers to develop alternative methods for describing uncertainty such as fuzzy sets, Dempster-Schafer measures, and non-monotonic logic. In the perspective of design, however, these alternative methods do not appear promising because they violate certain basic axioms of rational decision-theory (Hazelrigg, 1999).

In the current research, the relationship between model uncertainty and structural design is being explored in depth. Two issues are of interest: (a) designs that are not highly sensitive to model uncertainty of structural components and (b) trade-off between high-performance new technology with relatively large model uncertainty and medium- to low-performance old technology with small or no model uncertainty.

As an example of the trade-off issue, consider the simplest problem of a parallel system, as shown in Figure 3, in which the probability of failure is given as a simple inequality, Prob[$K \leq K_{\text{limit}}$], in which the performance variable $K = K_1 + \ldots + K_n$ is a sum of random variables representing the performance of n parallel components. Keeping the number of components n fixed, the designer is faced with a choice of using either a high- or low-performance component. In the low-performance component, there is negligible modelling uncertainty, while in the high-performance component, there is a small but non-negligible probability that the performance, measured by the stiffness in this example, is substandard. The randomness of the relative performance of the two choices of components is quantified in this example as follows. For the low-performance component, the stiffness k_L, is a Gaussian random variable with mean μ_L and standard deviation σ_L. For the high-performance component, the stiffness k_H, is modelled as a mixture of Gaussian random variables (Titterington, Smith, and Makov, 1985) with probability density

$$p(k_H) = (1-\pi_{\text{sub}}) \, \phi(k_H; \mu_L, \sigma_L) + \pi_{\text{sub}} \phi(k_H; \mu_{\text{sub}}, \sigma_{\text{sub}})$$

Here, μ_H and σ_H are the mean and standard deviation of the stiffness which are originally modelled for the high-performance component, π_{sub} is the probability that the component performs in a substandard mode, and μ_{sub} and σ_{sub} are the mean and standard deviation of the stiffness in the substandard mode. The shape of the probability distributions for the high- and low-performance components are indicated in Figure 3 for the case where the standard deviations are all equal and the other parameters are as indicated.

For the case of six parallel components, the probability distribution of the total stiffness K for four of the possible design choices, corresponding to the number type H components equal to $n_H = 6, 4, 2$, and 0, are shown in Figure 4. It can be seen that the lowest probability of failure is where there is a mixture of two high-performance and four low-performance components. In the current research, the generalization of such results to more general configurations than the parallel system in Figure 3 is being explored (Igusa, Buonopane and Ellingwood, 2001).

2.3 Development of Design & Analysis Strategies for Practical Optimization

Optimal design of large, complex structures with uncertainty in the physical properties and simulation model is computationally intractable. For a simplified case of a structure with N components each with M design alternatives, the space of possible designs is M^N. If the uncertainties associated with each component were quantified by P random variables each with Q possible discrete (or discretized) values, then the expected cost for each possible design would require Q^P analyses of the structure. For an exhaustive search for the optimal design under uncertainty, $M^N Q^P$ analyses are needed. Indeed, there are numerous methods for vastly improving the search for the optimal design and the computation of expected cost. Nevertheless, even with the use of such highly efficient methods, the number of structural analysis would remain to be prohibitive even with the computational tools available to the research community.

There are two steps in the research plan to address these issues. The first is to effectively reduce the dimension of the design space through (a) proper assignments of design utilities and (b) development of system-level structural design strategies rather than component-by-component design optimization. An example of a proper assignment of design utilities is given first. Highly irregular construction patterns, which occupy a major portion of the original design space, lead to high possibilities of construction error, which may result in unsafe structures. By assigning low utilities to such irregular construction patterns, and by investigating upper bounds of a design choice before embarking on a structural analysis, a large portion of the design space can be quickly dismissed before expending substantial computational effort. This is indicated by Figure 5.

Design choices for beam-to-column connections: A, B, C

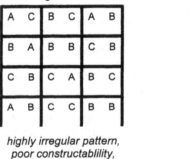 

highly irregular pattern,
poor constructablility,
substantially reduced utility

regular pattern,
good constructablility,
unreduced utility

Figure 5 Design Utilities for Connection Design in a 4-Story Steel Frame

System-level structural design strategies are defined as design rules that provide guidance to configuration at the global structural level rather than numerical values to dimensional and other physical properties at the component level. Some important structural design strategies are: static indeterminancy, provisions for bracing, and the use of stiffened cores within a building plan. The research focus will be on developing structural design strategies within the context of uncertainties in component properties.

The third, more challenging step in the research plan, is to reduce the space of random variables for each design choice. The most common approach to reducing the space of random variables is to invoke, wherever possible, the central limit theorem to justify the use of a single Gaussian random variable to quantify the average or cumulative effect of a large number of possibly non-Gaussian random variables. In mechanics problems, this approach is inherent in the averaged or effective medium theory, and is found to be valid for many spatially variable properties such as the Young's modulus. Another common approach is to use maximum- or minimum-value probability distributions such as the Weibull, Gumbel, and related distributions to quantify the maximum or minimum of a large number of random variables. In mechanics problems, this approach is suitable for strength or other limiting behaviour of parallel system configurations

There remain, however, many situations in which such powerful asymptotic approaches are inappropriate or highly inaccurate. Current investigation is on separating spatially local versus remote random variables, permitting randomness only within the local field and approximating the remote field with mean values. This reduces the space of random variables to a low dimensional region corresponding to the local field. This spatial decomposition of random variables has been successfully used in solid-state physics in the Claussius-Mossotti analysis of dielectric materials.

3. RESEARCH TASKS FOR IMPROVNG SIMULATION

In performance-based design, the issues of optimizing design parameters and computing reliability measures need to be addressed simultaneously. The problem is significantly more challenging from a computational point of view than the reliability problem itself, and the need for more powerful computational methods requires attention. The goal of this section is to provide an overview of directions for further investigation for improving simulation methodology with the purpose of making performance-based design more practical.

A mathematical framework for the problem of designing for improved structural reliability can be described as follows. The system under investigation is one for which generalized force F and displacements X are N-vectors related by a *global stiffness matrix K,* which is an $N \times N$ positive definite matrix. The stiffness matrix is itself a function of a parameter Θ about which there is uncertainty. The components of this parameter are quantities such as moments of inertia, moduli, areas, lengths, and while these components are unknown, their uncertainty is reflected by a specified probability distribution whose probability density is $\pi(\Theta)$. Generalized forces and displacements are linearly related in an elastic structural system by the equation $F=KX$. We will be focusing our attention for the time being on studying the reliability of the system under a fixed known force F, so that the displacement vector takes the form $X=K(\Theta)^{-1}F$.

There are many choices to be made in the level of generality in which we investigate designing for improved structural reliability. The type of structures we undertake to investigate could allow for a non-linear force-displacement relationship, but even the linear case is not well understood so that restricting attention to the linear case for now seems appropriate. Also, design modifications can lead to changes in the global stiffness matrix that are not of a parametric nature. In other words, the set of designs under consideration might not be naturally *nested,* so that the analysis would require a more complex family of global stiffness matrices. Again, this level of generality would be appropriate to allow for in the long run, but we choose the more restrictive model to begin with.

In order to discuss reliability we need to specify a *failure domain,* which is a set Ξ of displacements that we deem unacceptable for the structure. Assuming that we are interested in the probability of failure when the structure is subjected to a specific know force, the failure probability may be expressed in the form

$$p = P_\pi\left[\Theta \in X^{-1}(\Xi)\right] = \int I_{X^{-1}(\Xi)}\pi(\theta)d\theta \tag{6}$$

where $I_{X^{-1}(\Xi)}(\theta)$ denotes the indicator function of the set $X^{-1}(\Xi)$, the function of θ taking the value 1 for $\theta \in X^{-1}(\Xi)$ and 0 otherwise. The distribution of the random displacement variable X is rather unwieldy for analytic treatment. Even if, as is typically the case, the parameters Θ have simple multivariate distributional forms, e.g. multivariate normal, gamma, etc., it remains the case that the X is a complicated non-linear function of these parameter. For a typical system we hope to investigate, we anticipate N to be on the order of 1,000, so that finding a symbolic expression for the function $X = X(\theta)$ is entirely out of the question.

3.1 Strategies for Failure Probability Evaluation

We are left with the following strategies, and various combinations of them, for calculation of the failure probability: (1) obtaining analytic bounds, (2) naïve Monte Carlo simulation, (3) analytic approximation of the probability integral (e.g. Laplace's method of approximation), and (4) various techniques making naïve Monte Carlo simulation more effective, including speedups based on quick acceptance, importance sampling, and use of samples to speed up the inversion method. In this and the next sections, we give a brief overview of these methods.

It is often the case that the failure event can be decomposed as a union of a large number of more elementary events. This would be the case, for example, if failure of the system meant that one of the displacement components exceeded some threshold, so that the failure event for the system takes the form $\cup_{i=1}^n\{|X_i|>\tau_i\}$, where the τ_i are specified constants. One of the most common and simple approaches to bounding a probability of this form is to use what is common referred to as the Bonferroni upper bound (Bonferroni, 1935, 1936)

$$\sum_{i=1}^n P\left[|X_i| > \tau_i\right]. \tag{7}$$

This is useful in many probabilistic applications, because it is typically the case that events based on single components are simpler to handle. Unfortunately, in the types of large structural reliability problem, the random variables X_i are entangled in a complex non-linear fashion, so that computation of a single Bonferroni term is no simpler than computation of the whole failure probability. If by some means the Bonferroni bound was to become tractable then recent strategies for sharpening it (Naiman and Wynn, 1997, Naiman and Priebe, 2001) could prove useful.

Laplace's integral approximation (Bleistein and Handelsman, 1986) method gives approximations to integrals of the form

$$\mu = \int g(\theta)\pi(\theta)d\theta \tag{8}$$

when the integrand is a smooth function. This method has been used in structural engineering applications (Au and Beck, 1999).

Naive Monte Carlo simulation is carried out by generating samples $\Theta_1, \ldots, \Theta_n$ from the distribution π and estimating the failure probability using

$$\hat{p} = \frac{1}{n} \sum_{i=1}^{n} I_{X^{-1}(\Xi)}(\Theta_i) \tag{9}$$

This estimate is unbiased and has variance equal to $p(1-p)/n$ in which p is the (unknown) failure probability being estimated. The procedure is straightforward to implement provided that (a) it is a simple matter to generate samples from the distribution π, and (b) it is straightforward to determine whether or not a given Θ_i lies in $X^{-1}(\Xi)$. Condition (a) can be difficult to attain for a general multivariate distribution π, and it is typically the case that for complicated distributions one must rely upon approximate methods such as Monte Carlo Markov chain (MCMC) methods and other variants of the Metropolis-Hastings algorithm to generate random variates (Bremaud, 1999, Fishman, 1996, Gilks, et.al. 1996). On the other hand, it is often the case that when uncertainty is modelled for various components of an unknown parameter, many of the components are specified to be independent, so that generating samples from smaller dimensional distributions is all the is required. In this case exact and efficient methods for sampling the desired distributions are usually available (Devroye, 1986). For condition (b), it would usually be the case that once the displacement vector X_i is calculated from Θ_i (which will be computationally intensive as it involves inversion of the stiffness matrix), it is relatively inexpensive computationally to determine if X_i lies in Ξ.

Thus, it may be argued that even simple Monte Carlo simulation can be quite expensive to carry out in this context. The method is also inefficient, especially when a small probability p, which in structural engineering applications is of the order of 10^{-4} or less, is to be estimated (Fishman, 1996).

3.2. Improvements to Naïve Monte Carlo Simulation

A standard approach method to improving on naïve Monte Carlo evaluation is to find efficient algorithms for confirming failure or non-failure in special cases (Devroye, 1986). Such algorithms require analytic treatment of the failure set. Once these algorithms are found, they may be inserted into the simulation code to check for avoidance of the expensive (full matrix inversion) version of checking the failure condition. This approach is closely related to importance sampling. For example, easily calculated lower bounds on the eigenvalues of the stiffness matrix $K = K(\Theta)$ can be used to check for a bound on the norm of the displacement vector. Speedups can also be developed in an adaptive manner making use of newly discovered information about the failure region obtained from new samples.

Importance sampling is a general method for improving the efficiency of Monte Carlo simulation (Fishman, 1996). The key idea is that instead of approximating an expectation of the form in Equation (8) by sampling Θ according to π and taking $\hat{\mu} = g(\Theta)$, we can rewrite Equation (8) in the form

$$\mu = \int g^*(\theta) \pi^*(\theta) d\theta \tag{10}$$

where π^* is an alternative probability density function (only required to have the same support as π), and where $g^*(\theta) = g(\theta)\pi(\theta)/\pi^*(\theta)$, then estimate μ using $\tilde{\mu} = g^*(\Theta^*)$, where Θ^* is sampled from π^*. This estimator remains unbiased and the *art* of importance sampling is to find π^* in a given context that yields a substantial reduction in the variance of the estimator.

Importance sampling leads to another key method for improving Monte Carlo efficiency: the re-use of samples in determining the change in the reliability estimate that results from a change in the distribution π. One of the key issues we will need to grapple with in the design context is that of determining the effect on failure probability estimates that results from obtaining additional information about the model parameters, or from modifying some of these parameters so that their distribution changes.

Let us fix an initial probability distribution π for the unknown design parameters θ, and denote the failure probability by p_π in order to emphasize the dependence on the underlying distribution. Assume we have used naïve Monte Carlo sampling to determine this probability, so that we have collected samples $\Theta_1, ..., \Theta_n$ from π and we have calculated the corresponding displacements $X_i = X(\Theta_i)$, $i = 1,...,n$ to arrive at the estimator in Equation (9). Next, assume we wish to determine p_{π^*}, where π^* is some alternative probability distribution. We can then estimate this failure probability using

$$\hat{p}_{\pi^*} = \frac{1}{n}\sum_{i=1}^n I_\Xi(X(\Theta_i))\pi(\Theta_i)/\pi^*(\Theta_i) \tag{11}$$

The key benefit of using this estimate in the current setting is that almost all of the effort in the initial calculations is devoted to computing the displacements, and these do not need to be recomputed when we change the distribution. The calculations can be re-used over and over again for an unlimited number of alternative distributions.

Since the efficiency of the naïve Monte Carlo method depends critically on the solution of a large linear system of equations, it would be beneficial to devise an inversion algorithm that exploits all of the properties of these systems. In particular, the sparsity of the stiffness matrix leads to several candidate methods and code design decisions to create an efficient solver (Greenbaum, 1997). Many of the iterative algorithms currently in use are *conjugate gradient methods* (Hestenes and Stiefel, 1952), and are designed so as adapt to the structure of a *single* input matrix, but do not make use of the fact that this matrix is one of a family of sampled matrices. For our problem, it is envisioned that the algorithm would be called a large number of times, and one should expect a gain in efficiency could be achieved if the first several sample matrices are used in a *learning stage* to *train* an algorithm that is optimally tailored to handle the matrices appearing as input at a later stage.

4. FIRST STEP IN DEVELOPING A TEST BED PROBLEM

The behaviour of steel moment connections is one example in structural engineering where high-resolution modelling is necessary to capture many important aspects of behaviour. But the significant behavioural aspects must be mapped to a less detailed, building-level (reduced) model in order to analyze the behaviour of a typical building structure. In many cases, the most significant factor in determining the behaviour of the reduced model will be the mode of failure that occurs in the high-resolution model. Bayesian inference may provide a means to distinguish failure modes within the context of the reduced model.

A steel moment connection may fail in a number of modes—yielding or plastification of the flanges, local buckling in the compression flange of the beam, local buckling in the web of the panel zone, or fracture propagating from the weldments. For this example, consider the first two modes of failure. The reduced model will be based on the moment-rotation behaviour of the connection that is determined by local behaviour in the detailed model. Consider the simplified moment rotation-behaviours in Figure 6. The critical moments (M_y or M_b) may be significantly different, or even if they are nearly equal, the post-critical behaviour is typically vastly different. Therefore, determining which failure mode is most likely to occur may significantly contribute to our ability to describe the behaviour of the building-level system.

In the yielding and buckling cases, some distribution of the critical moments will exist due to randomness in material properties and physical dimensions. Such randomness would need to be explicitly included in the high-resolution model, but can be sufficiently represented in the reduced model by the distribution of the critical moment about its mean. Let $f(M_y; \mu_{M_y}, \sigma_{M_y})$ and $f(M_b; \mu_{M_b}, \sigma_{M_b})$ be unspecified probability density functions (pdfs) of yield moment and buckling moment. The variation due to properties which are not explicitly included in the reduced model (such as material properties or physical dimensions) are captured by the variances σ_{M_y} or σ_{M_b}. Other properties of the system that are critical to the failure mode response are included explicitly in the reduced model by the mean values μ_{M_y} and μ_{M_b} to be random variables as well. Here the mean values are determined by functional relationships based on the mechanics of the problem.

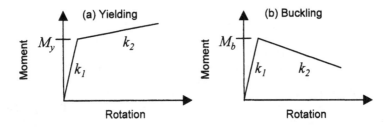

Figure 6. Two Failure Modes for Steel Moment Connection

Consider the mode of failure of yielding of the flanges. We identify two physical phenomena that significantly affect the observed (measured by experiment or high-resolution model) value of M_y — the presence of shear stresses and residual stresses in the flanges. Moment connections are often assumed to transfer shear through the bolted shear tab at the beam web; however, detailed finite element modelling has shown that a significant portion of the beam shear passes through the welds at the beam flanges since they may be significantly stiffer than the bolted shear tab. The presence of the large shear stresses in the flanges will reduce the apparent yield moment due to the biaxial stress state. Let M^* be the yield moment with no shear stress, V^* the shear required for yield with no moment, and V the current shear transferred through the beam flanges. The yield surface can be represented as a function $M_y = g_1(M^*, V^*, V)$. Residual stresses in the beam flanges may be present due to the manufacture of the beam but also due to the field welding. The field welding process will result in residual stresses that may be highly variable and may affect the observed yield moment of the connection. In this case we might propose a function of the form $M_y = M^* + g_2(R)$ where M^* is the yield moment neglecting the effect of residual stresses and g_2 is a function of R, some measure of residual stress magnitude, which takes on both positive and negative values. Combining these two effects into a single model for the mean of M_y results in

$$\mu_{M_y} = g_1(M^*, V^*, V) + g_2(R) \tag{12}$$

In general, V and R will be random variables so that μ_M will also be a random variable. The variance in the mean must also be defined by developing some realistic functional relationships relating it to the input variables M^*, V^*, V and R, or by assuming a constant value.

The flange-buckling mode of failure can be analyzed in a similar manner. Again we identify several physical phenomena that affect the observed buckling moment, M_b. Here we will consider imperfections in the flange, non-ideal rotational restraint provided by the beam web and weld to the column, and residual stresses in the flanges. If the beam flanges are initially distorted in some manner that approximates the local buckling mode, then the flanges may buckle at a significantly reduced moment. The function to describe this reduction would be a non-linear, monotonically decreasing function of two imperfection measures—magnitude, D, and wavelength, L—denoted as $M_b = h_1(D, L)$. Both D and L would be modelled as random variables. Analytical models of local buckling in beam flanges are typically based on plate solutions assuming boundary conditions that are rotationally free, simply supported or rigid. In the case of the beam flange, the beam web and weld to the column face will provide some partial rotational restraint that is difficult to represent with simple analytical plate solutions. Therefore, we might modify the buckling moment based on some rotational stiffness parameter, S, giving $M_b = M_b^* + h_2(S)$, where M_b^* is the analytical buckling moment based on idealized boundary conditions and the function h_2 modifies the buckling moment. In this case we are using a random variable to account for inability of the analytical model to represent the actual conditions, as well as randomness in the input properties of the model. Finally, residual stresses may reduce the buckling moment since local yielding can result in large local deformations in the beam flanges. Local yielding may also reduce the rotational restraint provided by the beam web or flange weld. In this case we include the residual stress measure as an argument in the functions h_1 and h_2, resulting in the following expression for the mean of the buckling moment

$$\mu_{M_b} = h_1(D, L, R) + h_2(S, R) \tag{13}$$

All of the input variables may be considered as random variables, and an expression for the variance of M_b must also be developed. Note that the inclusion of residual stress effects in both the yielding and buckling moments allows for interaction between these two failure modes.

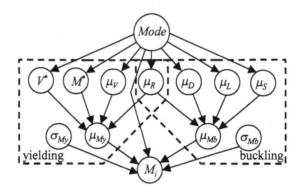

Figure 7 Bayesian Network for Steel Connection with Two Failure Modes

Figure 7 shows the proposed Bayesian network that accommodates the two failure modes considered and allows for the introduction of evidence from the high-resolution model. The follow variables are considered as random variables— V, R, D, L, S—and a prior distribution must be specified for each. Typically, the mean and variance of each variable would be satisfactory. For clarity of presentation, Figure 7 shows only the nodes for the mean values. The network as shown assumes that the variances of critical moments are constants. The network must accommodate evidence on the variable M_i, the moment at which the slope of the moment-rotation response changes from its initial linear portion. In this case we are making the simplifying assumption that the initial stiffness, k_i, is deterministic so that the randomness observed in the rotation, θ, will be a direct function of M_i. In general, k_i will be random and we will observe pairs of data (θ_i, M_i). The node *Mode* is a discrete random variable with two states corresponding to the two modes of failure. A prior distribution for *Mode* must be specified based on the probability of occurrence of each mode. After instantiation of evidence through the node M_i, a posterior distribution of the failure mode is obtained through a network calculation performed via an appropriate sampling algorithm, such as Markov Chain Monte Carlo (MCMC). One also obtains posterior distributions on the mean values of critical moment in each mode of failure, μ_{M_y} and μ_{M_b}.

In order to incorporate the results of the Bayesian network of the high-resolution model into the reduced model, the network would need to include explicitly those degrees-of-freedom on which the reduced model depended. In the steel moment connection example, those degrees-of-freedom would be forces and rotations applied to the connection. Given the values of these observed forces from a building-level analysis (or an interim analysis step) the expanded Bayesian network could then be used to make inferences, in a probabilistic sense, on the mode of failure.

5 CONCLUDING REMARKS

To address the research issues associated with performance-based design of large, complex structures, as outlined in the introduction, a set of research tasks related to a Bayesian decision framework and the associated simulation problem has been described. A test-bed problem, which provides practical focus to the research effort, is in its initial stages of development. While the current emphasis is on small- to medium-scale analysis, the eventual goal is to encompass small- to system-level scales.

This material is based upon work supported by the National Science Foundation at Johns Hopkins University under Grant Number DMI-0087032 and by the Sandia National Laboratory at Georgia Tech under contract A0356-15896. This research support is gratefully acknowledged.

6 REFERENCES

Au, S.K. and Beck, J.L., 1999, A new adaptive importance sampling scheme for reliability calculations, *Structural Safety*, **21**, pp. 135–158.

Bonferroni, C.E., 1935, Il calcolo delle assicurazioni su gruppi di teste, In *Studi in Onore del Professore Salvatore Ortu Carboni*, (Rome), pp. 13-60.

Bonferroni, C.E., 1936, Teoria statistica delle classi e calcolo delle probabilità, In *Pubblicazioni del R Istituto Superiore di Scienze Economiche e Commerciali di Firenze*, **8**, pp. 3–62.

Bleistein, N. and Handelsman, R., 1986, *Asymptotic Expansions of Integrals*, (New York: Dover).

Bremaud, P., 1999, *Markov Chains: Gibbs Fields, Monte Carlo Simulation and Queues* (New York: Springer).

Fishman, G., 1996, *Monte Carlo : Concepts, Algorithms, and Applications*, (New York: Springer-Verlag).

Devroye, L., 1986, *Non-Uniform Random Deviate Generation*, (New York: Springer Verlag).

Gilks, W.R., Richardson, S. and Spiegelhalter, D.J., 1996, *Markov Chain Monte Carlo in Practice* (New York: Chapman and Hall).

Greenbaum, A., 1997, *Iterative Methods for Solving Linear Systems, Frontiers in Applied Mathematics 17*, (Philadelphia: SIAM).

Hazelrigg, G.A. 1999, An axiomatic framework for engineering design. *Journal of Mechanical Design*, **121**, pp. 342–347.

Igusa, T., Buonopane, S.G. and Ellingwood, B.R., 2001, Reliability-based simulation for performance-based design. In *Proceedings of the 8th International Conference on Structural Safety and Reliability*, Newport Beach, edited by Corotis, R.B., (Rotterdam: A.A. Balkema), in press.

Hestenes, M.R. and Stiefel, E. 1952, Methods of Conjugate Gradients for solving linear systems, *Journal of Research of the National Bureau of Standards*, **49**, pp. 409–436.

Naiman, D.Q. and Priebe, C.E. 2001, Computing Scan Statistic p-Values using Importance Sampling, with Applications to Genetics and Image Processing, *Journal of Computational and Graphical Statistics*, **10**, to appear.

Naiman, D. and Wynn, H.P. 1997, Abstract Tubes, Improved Inclusion-Exclusion Identities and Inequalities, and Importance Sampling, *Annals of Statistics*, **25** pp. 1954–1983.

Titterington, D.M., Smith, A.F.M. and Makov, U.E., 1985, *Statistical Analysis of Finite Mixture Distributions*, (New York: John Wiley).

The Characterization and Management of Uncertainty in Model-Based Predictions

Roger G. Ghanem[1]

1 INTRODUCTION

This paper presents concepts and methods that are relevant to the task of quantifying and managing the confidence associated with model-based predictions. This includes the determination of requisite mitigating actions to achieve a target level of confidence. The lack of confidence is assumed to derive from uncertainty being present at various stages of the predictive process.

Uncertainty is ubiquitous in the natural, engineered, and social environments. Devising rationales for explaining it, strategies for its integration into scientific determinism and mitigating its consequences has been an active arena of rational endeavor where many scientific concepts have taken turn at fame and infamy. Far from being a static concept, uncertainty is the complement of knowledge, and as thus, continually adapts itself to knowledge, feeding on its evolution to redefine its claim over science.

Mechanics is a framework for applying deductive and mathematical reasoning to enhance our understanding of the physical world. Thus, far from being accidental, the interaction of mechanics and uncertainty is rather by design, as they both mold the physical world in a complementary fashion. The substance of this interaction is attested to by the simultaneous evolution of mechanics and rational models of uncertainty as embodied, for example, in the contributions of, among others, Gauss, Bernoulli, Euler, Legendre, Laplace, Rayleigh, Einstein, and Feynman. The driving forces behind each generation of these scientists reflect the contemporary state of technology.

The present state of science is increasingly shaped by the availability of significant computational resources able to resolve very complex models of physical phenomena, as well as the availability of sophisticated measurement tools that, combined with these computational resources, are capable of observing the same physical phenomena across an impressive range of length and time scales. With this science, comes the ability to manufacture complex, unique, and expensive systems, as well as, the ability to design systems whose successful performance relies on the designer accurately predicting their operational environment. Issues that merit consideration in this context are related to the ability of quantifying the confidence warranted from model-based predictions

---

[1]Department of Civil Engineering, The Johns Hopkins University, Baltimore, MD 21218, USA; Email: ghanem@jhu.edu

as well as the requisite resources to achieve a target confidence. Resources, in this context, refer to the combined resources of modeling, computing, and experimentation.

The problem can be conceptually framed in terms of events in the set generated by the parameters of the system under investigation, being mapped into events in the set generated by the state of the system. Variability in the set of basic events induces variability in the set of predictions, with the correspondence specified through a deterministic mapping, which is typically derived from first principles by relying on a mechanistic representation of the phenomena involved. Probabilistic statements about the parameters of the model are associated, through this mapping, with probabilistic statements about the predictions of the model.

Contributions to formulating and solving the problem can be made at a number of key steps. Firstly, events of interest to a given problem must be identified. These consist, for example, of failure events, of levels of exceedance, or of some death/birth events. A mathematical description of the set to which these predicted events belong must be adopted that is rich enough to discern between the various events of interest. Moreover, a corresponding, consistent, topology over the set to which the basic events belong must be identified that is commensurate with available instrumentation technology. Secondly, the measure on the various events associated with the system's parameter must be estimated. In a probabilistic framework such as the present one, this amounts to estimating the probability measure of the various events. The outcome of this task, and hence the confidence in the overall uncertainty quantification and propagation process, depends in no small measure on the choice of model and data used to calibrate the associated parameters. The ability to adaptively update the probabilistic characterization of the associated random variables is a critical feature of any useful framework. Thirdly, algorithms for propagating the uncertainty in the parameters into uncertainty in the predictions must be developed. This algorithm should clearly take into consideration the mechanistic model adopted for the problem, the accuracy with which numerical predictions of this model can be resolved, as well as the practical questions that the uncertainty analysis is trying to address. Obviously, this uncertainty propagation task can be performed, at least conceptually, by relying on the Monte Carlo simulation (MCS) paradigm. Adaptive sampling techniques where the adaptivity is customized and optimized by taking account of the underlying mechanics should be developed. These techniques should be endowed with error estimation capabilities, that are as much based on the mechanics of the problem as on statistics of the data.

Based on the above, a solution of the following form should be sought,

$$u = \hat{u} + \epsilon_h|_{p,data,model} + \epsilon_p|_{data,model} + \epsilon_d|_{model} + \epsilon_m$$

where $\hat{u}$ is some computed prediction of the solution, ϵ_h, ϵ_p, ϵ_d, and ϵ_m are error estimators that can be controlled by refining the numerical approximations, refining the probabilistic approach, refining the data, and refining the model, respectively. Obviously, these errors are not statistically independent of each others and the above additive scheme should be construed to involve the proper statistical conditioning, as indicated. Adaptive techniques for controlling ϵ_h are well established in deterministic computational mechanics. Recent formalisms are being developed that enable adaptation schemes for ϵ_p and ϵ_d. It is clear that the certification of the predicted solution $\hat{u}$ to be within a specified tolerance requires the rational control of all error terms. Selection between competing models can then be made by identifying that model which achieves its target error reduction within specified limits on computational and data collection resources.

It should be noted that the approach adopted in the present work consists of computing a mathematical characterization of the solution and the associated error estimates as random quantities. This characterization can subsequently be used in decision making by taking appropriate norms and integrals.

The next section presents a brief review of probabilistic concepts including the mathemtical characterization of uncertainty. This is followed by a demonstration of how these characterizations can be used in developing estimates of stochastic field variables.

2   REVIEW OF PROBABILISTIC CONCEPTS

The development presented in this paper hinges on the definition of random variables as *measurable functions* from the space of elementary events to the real line. As functions, approximation theory, as developed for deterministic functions, can be applied to random variables. The main question to be addressed, is the characterization of the solution to a physical problem where some parameters of the model have been modeled as stochastic processes. The answer to this question lies in the realization that in the deterministic finite element method, as well as most other numerical analysis techniques, a solution to a deterministic problem is known once its projection on a basis in an appropriate function space has been evaluated. It often happens, in deterministic analysis, that the coefficients in such a representation have an immediate physical meaning, which distracts from the mathematical significance of the solution. Carrying this argument over to the case involving stochastic processes, the solution to the problem will be identified with its projection on a set of appropriately chosen basis functions. A random variable, will thus be viewed as a function of a single variable, θ, that refers to the space of elementary events. Monte Carlo simulation can be viewed as a collocation along this θ dimension. Other approximations along this dimension are possible, and can be associated with different choices of basis functions in the corresponding space of random variables. This theoretical development is consistent with the identification of the space of second order random variables as a Hilbert space with the inner product on it defined as the operation of statistical correlation (Loeve, 1977). Second order random variables are those random variables with finite variance, they are mathematically similar to deterministic functions with finite energy.

2.1 Mathematical Characterization of Random Variables

2.1.1 Karhunen-Loeve Expansion

The Karhunen-Loeve expansion (Loeve, 1977) of a stochastic process $\alpha(\mathbf{x}, \theta)$, is based on the spectral expansion of its covariance function $R_{\alpha\alpha}(\mathbf{x}, \mathbf{y})$. Here, $\mathbf{x}$ and $\mathbf{y}$ are used to denote spatial coordinates, while the argument θ indicates the random nature of the corresponding quantity. The covariance function being symmetrical and positive definite, by definition, has all its eigenfunctions mutually orthogonal, and they form a complete set spanning the function space to which $\alpha(\mathbf{x}, \theta)$ belongs. It can be shown that if this deterministic set is used to represent the process $\alpha(\mathbf{x}, \theta)$, then the random coefficients used in the expansion are also orthogonal. The expansion then takes the following form,

$$\alpha(\mathbf{x}, \theta) = \overline{\alpha}(\mathbf{x}) + \sum_{i=1}^{\infty} \sqrt{\lambda_i} \xi_i(\theta) \phi_i(\mathbf{x}), \tag{1}$$

where $\overline{\alpha}(\mathbf{x})$ denotes the mean of the stochastic process, and $\{\xi_i(\theta)\}$ forms a set of orthogonal random variables. Furthermore, $\{\phi_i(\mathbf{x})\}$ are the eigenfunctions and $\{\lambda_i\}$ are the eigenvalues, of the covariance kernel, and can be evaluated as the solution to the following integral equation

$$\int_{\mathcal{D}} R_{\alpha\alpha}(\mathbf{x}, \mathbf{y}) \phi_i(\mathbf{y}) d\mathbf{y} = \lambda_i \phi_i(\mathbf{x}), \tag{2}$$

where $\mathcal{D}$ denotes the spatial domain over which the process $\alpha(\mathbf{x}, \theta)$ is defined. The most important aspect of this spectral representation is that the spatial random fluctuations have been decomposed into a set of deterministic functions in the spatial variables multiplying random coefficients that are independent of these variables. If the random process being expanded, $\alpha(\mathbf{x}, \theta)$, is gaussian, then the random variables $\{\xi_i\}$ form an orthonormal gaussian vector. The Karhunen-Loeve expansion is mean-square convergent irrespective of the probabilistic structure of the process being expanded, provided it has a finite variance (Loeve, 1977). Note that for this a number of useful forms, defined over regular geometric domains, an analytical solution of the integral eigenvalue problem has been obtained (Ghanem and Spanos, 1991). The monotony of the decay of the eigenvalues is guaranteed by the symmetry of the covariance function, and the rate of the decay is related to the correlation length of the process being expanded. Thus, the closer a process is to white noise, the more terms are required in its expansion, while at the other limit, a random variable can be represented by a single term. In physical systems, it can be expected that material properties vary smoothly at the scales of interest in most applications, and therefore only few terms in the Karhunen-Loeve expansion can capture most of the uncertainty in the process.

2.1.2   Polynomial Chaos Expansion

The covariance function of the solution process is not known apriori, and hence the Karhunen-Loeve expansion cannot be used to represent it. Since the solution process is a function of the material properties, nodal concentrations, $u(\theta)$, can be formally expressed as some nonlinear functional of the set $\{\xi_i(\theta)\}$ used to represent the material stochasticity. It has been shown (Cameron and Martin, 1947) that this functional dependence can be expanded in terms of polynomials in gaussian random variables, referred to as Polynomial Chaoses. Namely,

$$
u(\theta) \;=\; u_0 \Gamma_0 + \sum_{i_1=1}^{\infty} u_{i_1} \Gamma_1\left(\xi_{i_1}(\theta)\right)
$$

$$
+ \sum_{i_1=1}^{\infty} \sum_{i_2=1}^{i_1} u_{i_1 i_2} \Gamma_2\left(\xi_{i_1}(\theta), \xi_{i_2}(\theta)\right) + \dots \qquad (3)
$$

In this equation, the symbol $\Gamma_n(\xi_{i_1}, \dots, \xi_{i_n})$ denotes the Polynomial Chaos (Wiener, 1938; Kallianpur, 1980) of order n in the variables $(\xi_{i_1}, \dots, \xi_{i_n})$. These are generalizations of the multidimensional Hermite polynomials to the case where the independent variables are functions measurable with respect to the Wiener measure. Introducing a one-to-one mapping to a set with ordered indices denoted by $\{\Psi_i(\theta)\}$ and truncating the Polynomial Chaos expansion after the Pth term, equation (3) can be rewritten as,

$$
u(\theta) = \sum_{j=0}^{P} u_j \Psi_j(\theta) \; . \qquad (4)
$$

The polynomials chaos are orthogonal in the sense that their inner product $\langle \Psi_j \Psi_k \rangle$, which is defined as the statistical average of their product, is equal to zero for $j \neq k$. Moreover, they can be shown to form a complete basis in the space of second order random variables. A complete probabilistic characterization of the process $u(\theta)$ is obtained once the deterministic coefficients u_j have been calculated. A given truncated series can be refined along the random dimension either by adding more random variables to the set $\{\xi_i\}$ or by increasing the maximum order of polynomials included in the Polynomial Chaos expansion. The first refinement takes into account higher frequency random fluctuations of the underlying stochastic process, while the second refinement captures strong non-linear dependence of the solution process on this underlying process.

It should be noted at this point that the Polynomial Chaos expansion can be used to represent, in addition to the solution process, stochastic processes that model non-gaussian material properties. The processes representing the material properties are thus expressed as the output of a nonlinear system to a gaussian input.

In the next section, the Karhunen-Loeve and the Polynomial Chaos expansions are integrated into the equations governing the behavior of typical problems in mechanics, and procedures are developed for evaluating an expansion of the solution process with respect to the Polynomial Chaos basis.

3 THE STOCHASTIC FINITE ELEMENT METHOD

3.1 Overview

In this section, the formulation is presented in terms of generic time-dependent equations that are representative of many problems in mechanics. The linear problem is treated first in order to keep the notation concise, with the nonlinear problem treated in a later section. Consider the equation,

$$\frac{\partial}{\partial t} u(\mathbf{x}, t, \theta) = \mathcal{Q}_a \left[u(\mathbf{x}, t, \theta) \right] + f(\mathbf{x}, t, \theta), \qquad x \in \mathcal{B}, \quad t \in [0, T], \quad \theta \in \Omega \quad (5)$$

where $\mathcal{Q}_a[.]$ is a linear differential operator whose coefficients $a(\mathbf{x}, t, \theta)$ can be modeled as random fields exhibiting random fluctuations both in space and time, and Ω is the set of elementary events. By virtue of the randomness of the differential operator, the solution to the differential equation will itself be a random process. Separating the random component of $\mathcal{Q}_a[.]$ from its deterministic component results in the following equation,

$$\frac{\partial}{\partial t} u(\mathbf{x}, t, \theta) = (\mathcal{L} + \mathcal{R}_\alpha) \left[u(\mathbf{x}, t, \theta) \right] + f(\mathbf{x}, t, \theta) \quad (6)$$

where now $\mathcal{L}[.]$ indicates a deterministic linear operator whose coefficients represent the average values of the random coefficients, and $\mathcal{R}_\alpha[.]$ denotes a random linear operator whose coefficients $\alpha(\mathbf{x}, t, \theta)$ are zero mean random fields. In the above, it is assumed that the random coefficients $a(\mathbf{x}, t, \theta)$ appear as multiplicative constants in the explicit expression of the operator $\mathcal{Q}_a[.]$. Extensions to more complicated situations have already been carried out (Ghanem and Dham, 1998; Ghanem, 1998a, 1999; Ghanem and Red-Horse, 1999). For clarity of presentation, however, the simpler, linear case is used in this section.

Conceptually, given the joint probabilistic information about $\mathcal{R}_\alpha[.]$ and $f(\mathbf{x}, t, \theta)$, the solution is completely determined when the joint probabilistic information about $\mathcal{R}_\alpha[.]$, $f(\mathbf{x}, t, \theta)$ and $u(\mathbf{x}, t, \theta)$ is obtained. Traditionally, a number of simplifying assumptions are introduced and a partial solution of the problem, restricted to the second order statistics of the solution, is sought. The process $\alpha(\mathbf{x}, t, \theta)$ can be thought of as representing the stochastic fluctuations about its mean of some property of the system described by the operator $\mathcal{R}_\alpha$. In general, the process $\alpha(\mathbf{x}, t, \theta)$ is continuous with respect to $\mathbf{x}$ although the form of the corresponding functional dependence is not explicitly known. This fact presents a major difficulty in relation to the numerical treatment of the problem, specially when these processes are integrated with respect to **x**, as is customary in finite element techniques. The difficulty may be overcome by using the Karhunen-Loeve expansion (Loeve, 1977). Equation (6) then becomes

$$\frac{\partial}{\partial t} u(\mathbf{x}, t, \theta) = (\mathcal{L} + \mathcal{R}_\xi) [u(\mathbf{x}, t, \theta)] + f(\mathbf{x}, t, \theta) , \tag{7}$$

where the subscript ξ on $\mathcal{R}$ indicates that the problem has now been recast in a space spanned by a denumerable set of random variables. Using the deterministic finite element method this partial differential equation is approximated by an ordinary differential equation in R^N of the form,

$$\frac{d}{dt}\mathbf{u}(t, \theta) = [\mathbf{L} + \mathbf{R}_\alpha] \, \mathbf{u}(t, \theta) + \mathbf{f}(\theta) , \tag{8}$$

where the argument θ for the $N \times N$ matrix $\mathbf{R}$ and the N-dimensional vectors $\mathbf{u}(t, \theta)$ and $\mathbf{f}$ shows their random dependence. Symbolically, the random solution vector may be expressed as

$$\mathbf{u}(t, \theta) = \mathbf{u}_t[\{\xi(\theta)\}] , \tag{9}$$

where $\mathbf{u}_t[.]$ is a nonlinear vector functional of its argument. Expressed in this form, the solution can be viewed as the output of a nonlinear filter to white noise. This nonlinear filter is physically represented by the transformation, through the physical medium, of the scales of heterogeneity of the material properties into the scales of heterogeneity of the solution fields. The problem then can be restated as one of finding a convergent expansion, preferably optimal, of the functional $\mathbf{u}_t[.]$.

Next, the solution is expressed through its expansion with respect to the Polynomial Chaos system in the form,

$$u(\theta) = \sum_{i=0}^{\infty} u_i \, \Psi_i(\theta), \tag{10}$$

where u_i is some constant independent of θ, and $\{\Psi_i(\theta)\}_{i=1}^{\infty}$ is a basis in Θ. Using the following notation,

$$\mathbf{R}_0 = \mathbf{L} \qquad\text{and}\qquad \xi_0(\theta) = 1 , \tag{11}$$

equation (7) can be rewritten as

$$\frac{d}{dt}\mathbf{u}(t, \theta) + \left[\sum_{m=0}^{M} \xi_m(\theta) \, \mathbf{R}_m \right] \mathbf{u}(t, \theta) = \mathbf{f} . \tag{12}$$

In this equation, $\mathbf{R}_m$ can be thought of as the stiffness or conductance matrix associated with material property given by the mth scale of fluctuation.

Truncating equation (10) at the Pth polynomial, and substituting for $\mathbf{u}(t, \theta)$, equation (12) yields an expression for the resulting error

$$\epsilon = \mathbf{f} - \sum_{i=0}^{P} \Psi_i(\theta) \frac{d}{dt} \mathbf{u}_i(t) + \left[\sum_{m=0}^{M} \sum_{i=0}^{P} \xi_m(\theta) \Psi_i(\theta) \mathbf{R}_m \mathbf{u}_i(t) \right] \quad (13)$$

This error results from truncating the series in equation (10) after a finite number of terms, as well as from using a finite number of elements from the infinite set $\{\xi(\theta)\}_{i=1}^{\infty}$. The error, as expressed by equation (13), is minimized by requiring it to be orthogonal to the solution space, with respect to the inner product defined by the statistical averaging operator. A procedure similar to the one used in the previous section, for the deterministic finite element method, can be used here. Mathematically, this is equivalent to

$$(\epsilon, \ \Psi_s(\theta)) = 0 \qquad s = 0, 1, \ldots, \quad (14)$$

The orthogonality constraint results in a set of algebraic equations that can be solved for the vector coefficients $\mathbf{u}_i(t)$. From these coefficient, the spectral expansion of the solution vector is determined with respect to the Polynomial Chaos basis. From equation (10) it can be seen that all the probabilistic information concerning the random vector $\mathbf{u}(t, \theta)$ is contained in the expansion coefficients. Namely, the mean of the solution is equal to $\mathbf{u}_0$, and the covariance of the solution is given by $\mathbf{u}_i \mathbf{u}_i^T \sum_i \langle \Psi_i^2 \rangle$. So that once these coefficients have been computed, the probability distribution of the solution vector $\mathbf{u}(t, \theta)$ can be determined.

Repetitive application of the orthogonality requirement given by equation (14) for successive Polynomial Chaoses results in the matrix equation

$$\frac{d\mathbf{u}}{dt} + \mathbf{A} \mathbf{u} = \mathbf{F}, \quad (15)$$

where $\mathbf{u}$ is the $(P+1)N$-dimensional vector of coefficients, $\mathbf{A}$ is a $(P+1)N \times (P+1)N$ matrix constituted of block submatrices, where the (i, j) block is an $N \times N$ matrix given by

$$\mathbf{A}_{ij} = \sum_{m=0}^{M} \mathbf{R}_m \langle \Psi_i(\theta) \Psi_j(\theta) \xi_m(\theta) \rangle, \qquad i, j = 0, \ldots, b-1 \quad (16)$$

and $\mathbf{F}$ is a $(P+1)N$ dimensional vector given by

$$\mathbf{F}_i = \langle \mathbf{f} \Psi_i(\theta) \rangle, \qquad i = 0, \ldots, b-1. \quad (17)$$

Adopting a functional theoretic approach to formulate the problem has the benefit of providing a path to its solution that transcends any particular algorithm. For instance, other basis functions, besides the Polynomial Chaos bases can be used in the characterization of the random variables in the problem. This has led to the development of the hybrid stochastic finite elements concept whereby a P-term Polynomial Chaos expansion is used for the solution process, with the residual being represented in terms of a basis set $\Phi(\theta)$ that is different from the Chaos set. This results in the following representation of the solution,

$$\mathbf{u} = \sum_{i=0}^{P} \mathbf{u}_i \Psi_i + \sum_{i=0}^{M} \mathbf{u}_i^* \Phi_i . \tag{18}$$

The Monte Carlo simulation procedure can be associated with the set Φ_i being identified with the sequence of delta function, $\delta(\theta - \theta_i)$. These generalized functions are defined through their action on any random variable $f(\theta)$,

$$\langle f(\theta)\Phi_i(\theta) \rangle = f(\theta_i) , \tag{19}$$

and resulting in the realization $f(\theta_i)$ of f. The residual can now be constrained to being orthogonal to all the basis functions used in representing the solution process. These consist in the polynomial chaos bases and the generalized functions (Ghanem, 1998b).

3.2   Material Nonlinearities

Generalizing equation (5) to nonlinear operators and following through with space and time discretization, leads to equations having the following generic form,

$$\mathbf{K}(\mathbf{u}, \alpha(\theta))\mathbf{u}(\theta) = \mathbf{f}(\theta) , \tag{20}$$

where the nonlinear dependence of matrix $\mathbf{K}$ on the solution vector $\mathbf{u}$, and the random nature of the various quantities involved are explicitly shown. Moreover, the coefficient matrix $\mathbf{K}$ is random through its dependence on a vector stochastic process, α. Representing this process in terms of its Karhunen-Loeve expansion, and using the Polynomial Chaos expansion (Ghanem and Spanos, 1991) for the solution process, results in the following equation,

$$\mathbf{u} = \sum_{j=0}^{P} \mathbf{u}_j \Psi_j \tag{21}$$

$$\sum_{j=0}^{P} \Psi_j \mathbf{K}(\mathbf{u}, \xi(\theta))\mathbf{u}_j = \mathbf{f}(\theta) , \tag{22}$$

where the vector of uncorrelated random variables, ξ, now replaces the stochastic process α. The error in the above equation is constrained to be orthogonal to the approximating space. This requirement results in the following set of equations

$$\sum_{j=0}^{P} \langle \Psi_k \Psi_i \mathbf{K}(\mathbf{u}, \xi(\theta)) \rangle \mathbf{u}_j \;=\; \langle \Psi_k \mathbf{f}(\theta) \rangle \quad k = 0, 1, \ldots \tag{23}$$

Representing $\mathbf{K}$ in its Polynomial Chaos expansion, equation (22) becomes,

$$\mathbf{K} = \sum_{i=0}^{L} \Psi_i \mathbf{K}_i, \qquad \mathbf{K}_i = \frac{\langle \Psi_i \mathbf{K} \rangle}{\langle \Psi_i^2 \rangle}, \tag{24}$$

results in the following,

$$\sum_{j=0}^{P} \sum_{i=0}^{L} \langle \Psi_k \Psi_i \Psi_m \rangle \mathbf{K}_i \mathbf{u}_j \;=\; \langle \Psi_k \mathbf{f}(\theta) \rangle \quad k = 0, 1, \ldots \tag{25}$$

It should be emphasized here that representing $\mathbf{K}$ in terms of its Polynomial Chaos expansion is neither necessary nor necessarily most efficient. In particular, using equation (23) in conjunction with such techniques as stratified sampling, or latin hypercube sampling may lead to more efficient algorithms. These sampling techniques can be viewed as an efficient way to evaluate the integral implied by the averaging process (which is nothing more than an inner product in the space of random variables).

4 DECISION SUPPORT

An important feature of the development presented herein is the integration of decision support into the numerical simulation. This can be easily accomplished in the formulation since a series representation of the solution as a function of all the random quantities in the problem would have been evaluated. In this section, two procedures for utilizing that expansion in a decision making context are presented.

4.1 Probabilistic Characterization of the Solution

The representation of the solution as the Polynomial Chaos expansion, contains a complete probabilistic characterization. Indeed, while the mean value of the solution field is given by the zeroth order term in the expansion, the covariance matrix of the solution is readily obtained as,

$$\mathbf{R}_{uu} = \sum_{i=1}^{P} \langle \Psi_i^2 \rangle \mathbf{u}_i \mathbf{u}_i^T . \tag{26}$$

Higher order statistics can be obtained in a similarly simple format. More important than these statistical moments, though, is the possibility to efficiently generate a large statistical population of the solution process. For every simulation of the stochastic field representing the system's parameters, a corresponding simulation can be obtained for the random variables Ψ_i appearing in the polynomial chaos expansion. This permits the synthesis of the corresponding simulation of the solution process by simply evaluating the expansion. Once a statistically significant population has been generated, probabilities of rare events can be accurately evaluated through standard statistical techniques.

4.2   Optimal Sampling Locations

According to the foregoing, the vector solution field $\mathbf{u}(\theta)$ is obtained in the form,

$$\mathbf{u}(\theta) \;=\; \sum_{i=0}^{P} \mathbf{u}_i \, \Psi_i(\theta) \; . \tag{27}$$

Taking the derivative of the above equation with respect to $\alpha(t, \theta)$ results in,

$$\frac{\partial \mathbf{u}}{\partial \alpha(t,\theta)} \;=\; \sum_{i=0}^{M} \frac{\partial \mathbf{u}}{\partial \xi_i} \frac{\partial \xi_i}{\partial \alpha(t,\theta)} \; . \tag{28}$$

It should be noted here that the proper interpretation should be given to the derivatives appearing in equation (28). Indeed, these are derivatives with respect to random variables, and the usual mean square calculus does not apply to this case. Recalling that random variables are treated in the present context as elements of a Hilbert space, these derivatives should be understood as directional derivatives (Ghanem, 1999). Moreover, it can be shown that based on equation (1), the following relation holds between ξ_i, $\alpha(\mathbf{x}, t, \theta)$ and $\psi(\mathbf{x}, t, \theta)$,

$$\xi_i(\theta) \;=\; \frac{1}{\sqrt{\lambda_i}} \int_B \alpha(\mathbf{x}, t, \theta) \psi_i(\mathbf{x}, t) d\mathbf{x} \; . \tag{29}$$

Upon discretization, this last equation becomes,

$$\xi_i(\theta) \;=\; \frac{1}{\sqrt{\lambda_i}} \alpha^T(t,\theta) \psi_i(t,\theta) \; , \tag{30}$$

so that,

$$\frac{\partial \xi_i(\theta)}{\partial \alpha(t,\theta)} \;=\; \frac{1}{\sqrt{\lambda_i}} \psi_i^T(t,\theta) \; . \tag{31}$$

Moreover, using equation (27) results in,

$$\frac{\partial \mathbf{u}}{\partial \xi_i} \;=\; \sum_{j=0}^{P} \frac{\partial \Psi_j}{\partial \xi_i} . \mathbf{u}_j \; , \tag{32}$$

where the terms $\frac{\partial \Psi_j}{\partial \xi_i}$ can be readily evaluated based on Finally, equation (28) is written as,

$$\frac{\partial \mathbf{u}(t, \theta)}{\partial \alpha(t, \theta)} = \sum_{j=0}^{P} \sum_{i=0}^{M} \frac{\partial \Psi_j}{\partial \xi_i} \mathbf{u}_j(t) \psi_i^T(t, \theta) . \tag{33}$$

In this equation, the quantity $\mathbf{u}_j(t) \psi_i^T(t, \theta$ is deterministic while the quantity $\frac{\partial \Psi_j}{\partial \xi_i}$ is a random variable. An element $\frac{\partial \mathbf{u}_i(t, \theta)}{\partial \alpha_j(t, \theta)}$ in this Jacobian matrix represents the sensitivity of the solution $\mathbf{u}_i(t, \theta)$ at node i with respect to the material property $\alpha_j(t, \theta)$ at node j. Given the spatial distribution of the uncertainty in the material property, $\alpha(t, \theta)$, equation (33) can be used to determine the location of the solution $\mathbf{u}_i(t, \theta)$ that is least sensitive to this uncertainty. Also, spatial locations for sampling the material property $\alpha(t, \theta)$ can be determined in such a manner that they contain the most information regarding the solution process at a given point. These correspond to nodal values associated with a large mean value, and a small variance, of the Jacobian in equation (33). From equation (33), statistics and confidence intervals are developed for the optimal sampling locations. This is accomplished by simulating the right hand side of the equation, a very easy task now that the deterministic coefficients have been calculated. From these simulations, large sample statistics of the sensitivity coefficients can be calculated. Moreover, equation (33) can be used in an adaptive sampling scheme. Specifically, as predictions of $\mathbf{u}(t)$ are made in time, the location of the optimal samples is likely to change. This change will always be predicted by equation (33).

The optimal sampling locations, both of the material properties as well as of the solution process, are useful for identifying a reduced model of the overall system.

4.3 Adaptive Data Refinement

In addition to the above decision support concepts enabled by the present methodology, the possibility of a rational data refinement scheme, driven by the mechanics of the problem, can be formulated (Ghanem and Pellissetti, 2001). Indeed, as additional data is collected, the probabilistic characterization of the uncertain material parameters is updated, and their polynomial chaos decomposition can be modified accordingly. Equation (25) provides the starting point for estimating the sensitivity of the model-based predictions on small perturbations in the value of $\mathbf{K}_i$ and hence on the accuracy of the polynomial chaos decomposition of the material properties. This equation is rewritten here in a slightly different form as,

$$\sum_{j=0}^{N} \sum_{i=0}^{L} c_{ijk} \mathbf{K}_i \mathbf{u}_j = \mathbf{f}_k \quad \forall k , \tag{34}$$

where c_{ijk} denotes $\langle \Psi_i \Psi_j \Psi_k \rangle$.

The stiffness matrix consists in the expansion $\sum_{i=0}^{L} \mathbf{K}_i$, where $\mathbf{K}_0$ denotes the stiffness matrix corresponds to the mean material property, and the other terms represent the random fluctuations about the mean.

After solving this equation for $\mathbf{u}_j$, the sensitivity of the solution $\mathbf{u}_j$ with respect to changes of the material properties may be of interest. Of course, any change of the material parameters reflects itself in a variation of the stiffness matrix. In general, material properties are subject to spatial variations. Hence, it makes sense to consider local changes of the material. In an FEM-based analysis these will result in variations of the element stiffness matrices.

Taking the derivative of equation (34) with respect to $k_l^{(m)}$, where $k_l^{(m)}$ is the l-th term in the Karhunen-Loeve expansion of the material property of element m, results in,

$$\sum_{j=0}^{N} \sum_{i=0}^{L} c_{ijk} \mathbf{K}_i \frac{\partial \mathbf{u}_j}{\partial k_l^{(m)}} = -\sum_{j=0}^{P} c_{ljk} \frac{\partial \mathbf{k}_l^{(m)}}{\partial k_l^{(m)}} \mathbf{u}_j \quad \forall k , \tag{35}$$

The left hand side matrix in this last equation is identical to the left hand side matrix in equation (34), a fact that can be used to great advantage in developing efficient algorithms for solving these equations.

The solution, $\mathbf{u}$ can thus be expanded in a first order Taylor series as,

$$\mathbf{u}_j = \mathbf{u}_j^{hdp} + \sum_{l=0}^{P} \sum_{m} \frac{\partial \mathbf{u}_j}{\partial k_l^{(m)}}\bigg|_{\text{nominal}} \delta k_l^{(m)} + \epsilon_h + \epsilon_P . \tag{36}$$

The term $\mathbf{u}_j^{hdp}$ represents the approximation to the true solution associated with the current level of mesh, data, and polynomial chaos approximation, and the terms ϵ_h and ϵ_P represent, respectively, the errors associated with mesh and Polynomial Chaos approximations.

Clearly, given a certain amount of data (i.e. a sample space of a fixed size), statistics of events over this sample space can be evaluated with at most a certain accuracy that is monotonically related to the size of the sample. Equation (35) permits the quantification of that error in the solution that can be attributed to errors in computing these statistics ($\mathbf{K}_i$). This error (referred to hence as s-type error) is very useful as it permits the identification of a level of mesh refinement, and an associated h- or p-type error (Babuska and Rheinboldt, 1978; Oden et.al, 1986), that is consistent with the amount of data. Thus a certain size of the sample space should only justify a well quantifiable level of accuracy in the deterministic analysis.

5   CONCLUSIONS

As science gets more sophisticated in its probing of nature, either through more efficient computing or more detailed sensing, two key issues come to the foreground. The first of these addresses model validation and certification:

Given the present state of certainty in the data, how much confidence is justified in the prediction of the model. The second question addresses the worth of information: given a certain level of resources, how should they be allocated in order to maximize their effect on the confidence in the predictability of the model. Having models the confidence in which predictions is quantifiable and controllable is a significant shift in paradigm from deterministic, or even probabilistic, models that could only carry through the forward process of acting on the present state of knowledge and information. This is akin, in some ways, to adaptive mesh refinement in deterministic FEM whereby the level of approximation is adaptively refined or coarsened as suggested by the analysis. In the present case, mesh refinement would be replaced by data refinement. The format of the solution provided by the Polynomial Chaos expansion is such that analytical manipulations leading, for instance, to optimal sampling strategies, is possible.

Although the methodology presented in this paper is very computer intensive, it is expected that with the rapid development of computational resources, it will soon become routine to carry through mechanistic-based analyses all the way to the decision-making level.

6 ACKNOWLEDGMENTS

The financial support of the National Science Foundation under Grant No. CMS-9870005, of the Sandia National Laboratories, NM, under Contract No. BC-4596, the Office of Naval Research under grants number N00014-99-1-1038 and N00014-99-1-0900, and of DARPA under agreement number F30602-00-2-0612 is gratefully acknowledged.

7 REFERENCES

[1] Cameron, R.H. and Martin, W.T., "The orthogonal development of nonlinear functionals in series of Fourier-Hermite functionals", *Ann. Math*, Vol. 48, pp. 385-392, 1947.

[2] Babuska, I., and Rheinboldt, W.C., "Error estimates for adaptive finite element computations," *SIAM Journal of Numerical Analysis*, Vol. 15, No. 4, 1978.

[3] Ghanem, R., and Spanos, P., *Stochastic Finite Elements: A Spectral Approach*, Springer Verlag, 1991.

[4] Ghanem, R., and Dham, S., "Stochastic finite element analysis for multiphase flow in heterogeneous porous media," *Transport in Porous Media*, Vol. 32, pp. 239-262, 1998.

[5] Ghanem, R., "Scales of fluctuation and the propagation of uncertainty in random porous media," *Water Resources Research*, Vol. 34, No. 9, pp. 2123-2136, September 1998a.

[6] Ghanem, R., "Hybrid stochastic finite elements: coupling of spectral expansions with monte carlo simulations," *ASME, Journal of Applied Mechanics,* Vol. 65, pp. 1004-1009, 1998b.

[7] Ghanem, R., and Red-Horse, J., "Propagation of uncertainty in complex physical systems using a stochastic finite element approach," *Physica D,* Vol. 133, No. 1-4, pp. 137-144, 1999.

[8] Ghanem, R., "Ingredients for a general purpose stochastic finite elements formulation," *Computer Methods in Applied Mechanics and Engineering,* Vol. 168 Nos. 1-4, pp. 19-34, 1999.

[9] Ghanem, R., and Pellissetti, M., "Stochastic adaptive refinement: An s-type error estimato r," *Sixth International Conference on Structural Safety and Reliability, ICOSSAR'01* Newport Beach, CA, June 17-22, 2001.

[10] Kallianpur, G., *Stochastic Filtering Theory,* Springer-Verlag, Berlin, 1980.

[11] Loeve, M. *Probability Theory,* 4th edition, Springer-Verlag, New York, 1977.

[12] Oden, J.T., Demkowicz, L., Strouboulis, T., and Devloo, P. "Adaptive methods for problems in solid and fluid mechanics," pp. 249-280, *Accuracy Estimates and Adaptive Refinements in Finite Element Computations,* Edited by I. Babuska, O.C. Zienckiewicz, J. Gago, and E.R. de A. Oliveira, John Wiley & Sons, Ltd., 1986.

[13] Wiener, N., "The homogeneous chaos", *Amer. J. Math,* Vol. 60, pp. 897-936, 1938.

Author Index

Subject Index